Porous Polymer Science and Applications

Porous Polymer Science and Applications

Edited by

Inamuddin
Mohd Imran Ahamed
Rajender Boddula

CRC Press
Taylor & Francis Group
Boca Raton London New York

CRC Press is an imprint of the
Taylor & Francis Group, an **informa** business

First edition published 2022
by CRC Press
6000 Broken Sound Parkway NW, Suite 300, Boca Raton, FL 33487-2742

and by CRC Press
2 Park Square, Milton Park, Abingdon, Oxon, OX14 4RN

CRC Press is an imprint of Taylor & Francis Group, LLC

Library of Congress Cataloging-in-Publication Data
Names: Inamuddin, 1980- editor. | Ahamed, Mohd Imran, editor. | Boddula, Rajender, editor.
Title: Porous polymer science and applications / edited by Inamuddin, Mohd Imran Ahamed, Rajender Boddula.
Description: First edition. | Boca Raton : CRC Press, 2022. | Includes bibliographical references and index.
Identifiers: LCCN 2021050623 | ISBN 9780367770587 (hardback) | ISBN 9780367770686 (paperback) | ISBN 9781003169604 (ebook)
Subjects: LCSH: Polymers. | Porous materials.
Classification: LCC QD381 .P627 2022 | DDC 547/.7--dc23/eng/20220128
LC record available at https://lccn.loc.gov/2021050623

ISBN: 978-0-367-77058-7 (hbk)
ISBN: 978-0-367-77068-6 (pbk)
ISBN: 978-1-003-16960-4 (ebk)

DOI: 10.1201/9781003169604

Typeset in Times
by SPi Technologies India Pvt Ltd (Straive)

Contents

Preface

Porous polymer research is finding the attention of researchers worldwide. The porous polymers possess ultrahigh surface area and tailored porosity. They have the potential to integrate the advantages of both porous materials and polymers involved in advanced applications such as catalysis, gas storage, energy, and environmental sectors, etc. A wide variety of porous polymers have been developed including porous aromatic frameworks, hyper-crosslinked polymers, covalent organic frameworks, porous ionic polymers, conjugated microporous polymers, covalent triazine framework and porous polymer networks, etc.

Porous Polymer Science and Applications aims to provide recent advances in synthesis, tuning parameters, and applications of porous polymers. This book brings together reviews written by highly accomplished panels of experts working in the area of porous polymers. It encompasses basic studies and addresses topics of novel issues concerning the applications of porous polymers. It is an invaluable guide to students, professors, scientists, and R&D industrial experts working in the field of porous polymers and their applications.

Chapter 1 deals with an introduction to porous materials and their types. Various porous materials systems like microporous polymers, covalent organic frameworks, polymers of intrinsic microporosity, and hyper-crosslinked polymers with high surface area and porosity are discussed. The focus is given to communicate the advantages, drawbacks, features, and parameters of different fabrication methods reported in the literature.

Chapter 2 presents a detailed overview of hyper-crosslinked polymers (HCPs). It covers the synthetic and functionalization strategies, characterization techniques, structural and morphological studies, properties, and potential applications of these novel materials. It also describes the design and development of HCPs with a special focus on morphological properties.

Chapter 3 considers dry ionic polymers, the mechanism of hydration of functional groups, and different models of the structure of swollen polymers containing both hydrophilic and hydrophobic voids. The methods for investigations of the porous structure of dry and swollen ionic polymers are also discussed.

Chapter 4 compares the porous polymers manufactured in different conditions concerning operating characteristics. It presents the assessment of the collection of parameters needed in the assessment of porous polymer materials, definitions of their comparative priorities based on expert opinions, and the provision of a straightforward quantitative investigation.

Chapter 5 presents the process to prepare polymers at high pressure. The parameters and mechanisms of the processes, as well as techniques that involve supercritical fluids in drying and obtaining foaming, scaffolds, food packaging, emulsion templating, supports for catalysts materials, and metal–organic frameworks are discussed.

Chapter 6 talks about the importance, stability, and functionalization of porous polymers as heterogeneous catalysts along with their preparation strategies, namely co-polymerization and self-polymerization. It further focuses on the various applications of porous polymers including CO_2 capture and utilization, energy storage, heterogeneous catalysis, photocatalysis, and electrocatalysis.

Chapter 7 details various synthetic approaches for the fabrication of triazine porous frameworks. In particular, polymerization strategies, structural designs, and surface properties are briefly summarized. Further, different applications of triazine porous frameworks, including CO_2 capture, catalysis, and energy storage are also discussed.

Chapter 8 presents various applications of porous polymer in the separation process. It mostly describes advanced separation processes. The advanced separation processes which are mentioned in this chapter are focused on gas separation and wastewater treatment. Comparison of similar processes is also discussed.

Chapter 9 discusses the basic introduction about the synthesis of porous polymers, preparation of porous polymeric membranes using various techniques. How the morphology of membrane is considered to decide the different parameters and applications of the porous polymeric membrane in the area of the biomedical, fuel cell, separation, and sensing.

Chapter 10 discusses the various forms of solar cell systems which have been obtained from porous polymer. The porous polymers are the substances that were used directly in various components of solar cell systems to improve the efficiencies of their components. The metal–organic framework of polymer solar cells is an efficient source of energy in the near future.

Chapter 11 addresses recent advancements and approaches used for the H_2 production from porous polymers along with their synthesis techniques. A major emphasis is given on several challenges, advantages, and disadvantages of utilization of porous polymers for H_2 production with the future perspectives based on available reports.

Chapter 12 summarizes the molecular structures and physical and chemical properties of porous polymers associated with photocatalytic characteristics. The mechanism of action of porous polymers as a photocatalyst in various organic reactions is also presented. Moreover, this perspective highlights the unique fundamental and practical importance of porous photocatalysts.

Chapter 13 discusses the role of porous polymers in CO_2 capture to avert climate change. Different CO_2 mitigation strategies and types of porous organic polymers are discussed in detail. In addition, various factors that affect the CO_2 absorption and selectively in the porous polymers are also highlighted.

Chapter 14 discusses the antibacterial application of porous polymers in the food industry and the development of products with medical and pharmaceutical applications. The structural role of porous polymers is studied. Finally, synthesis strategies and coping procedures for resistant microbial infections are highlighted.

Editors

Dr. Inamuddin is working as an Assistant Professor at the Department of Applied Chemistry, Aligarh Muslim University, Aligarh, India. He earned his Master of Science degree in Organic Chemistry from Chaudhary Charan Singh (CCS) University, Meerut, India, in 2002. He earned his Master of Philosophy and Doctor of Philosophy degrees in Applied Chemistry from Aligarh Muslim University (AMU), India, in 2004 and 2007, respectively. Dr. Inamuddin has extensive research experience in multidisciplinary fields of Analytical Chemistry, Materials Chemistry, and Electrochemistry and, more specifically, Renewable Energy and Environment. He has worked on different research projects as project fellow and senior research fellow funded by University Grants Commission (UGC), Government of India, and Council of Scientific and Industrial Research (CSIR), Government of India. He received the Fast Track Young Scientist Award from the Department of Science and Technology, India, to work in the areas of bending actuators and artificial muscles. He has completed four major research projects sanctioned by the University Grant Commission, Department of Science and Technology, Council of Scientific and Industrial Research, and Council of Science and Technology, India.

He has published 196 research articles in international journals of repute and 19 book chapters in knowledge-based book editions published by renowned international publishers. He has published 145 edited books with Springer (U.K.), Elsevier, Nova Science Publishers, Inc. (U.S.A.), CRC Press Taylor & Francis Asia Pacific, Trans Tech Publications Ltd. (Switzerland), IntechOpen Limited (U.K.), Wiley-Scrivener, (U.S.A.) and Materials Research Forum LLC (U.S.A). He is a member of various journals' editorial boards. He is also serving as Associate Editor for journals (Environmental Chemistry Letter, Applied Water Science and Euro-Mediterranean Journal for Environmental Integration, Springer-Nature), Frontiers Section Editor (Current Analytical Chemistry, Bentham Science Publishers), Editorial Board Member (Scientific Reports-Nature), Editor (Eurasian Journal of Analytical Chemistry), and Review Editor (Frontiers in Chemistry, Frontiers, U.K.) He is also guest-editing various special thematic special issues to the journals of Elsevier, Bentham Science Publishers, and John Wiley & Sons, Inc. He has attended as well as chaired sessions in various international and national conferences.

Dr. Inamuddin has worked as a Postdoctoral Fellow, leading a research team at the Creative Research Initiative Center for Bio-Artificial Muscle, Hanyang University, South Korea, in the field of renewable energy, especially biofuel cells. He has also worked as a Postdoctoral Fellow at the Center of Research Excellence in Renewable Energy, King Fahd University of Petroleum and Minerals, Saudi Arabia, in the field of polymer electrolyte membrane fuel cells and computational fluid dynamics of polymer electrolyte membrane fuel cells. He is a life member of the Journal of the Indian Chemical Society. His research interests include ion exchange materials, a sensor for heavy metal ions, biofuel cells, supercapacitors, and bending actuators.

Dr. Mohd Imran Ahamed earned his PhD degree on the topic "Synthesis and characterization of inorganic-organic composite heavy metals selective cation-exchangers and their analytical applications," from Aligarh Muslim University, Aligarh, India in 2019. He has published several research and review articles in the journals of international recognition. Springer (U.K.), Elsevier, CRC Press Taylor & Francis Asia Pacific and Materials Research Forum LLC (U.S.A). He has completed his B.Sc. (Hons) Chemistry from Aligarh Muslim University, Aligarh, India, and M.Sc. (Organic Chemistry) from Dr. Bhimrao Ambedkar University, Agra, India. He has co-edited more than 20 books with Springer (U.K.), Elsevier, CRC Press Taylor & Francis Asia Pacific, Materials Research Forum LLC (U.S.A), and Wiley-Scrivener, (U.S.A.). His research work includes ion-exchange chromatography, wastewater treatment and analysis, bending actuator, and electrospinning.

Dr. Rajender Boddula works with the Chinese Academy of Sciences-President's International Fellowship Initiative (CAS-PIFI) at the National Center for Nanoscience and Technology (NCNST, Beijing). He earned a Master of Science in Organic Chemistry from Kakatiya University, Warangal, India, in 2008. He earned his Doctor of Philosophy in Chemistry with the highest honors in 2014 for work entitled "Synthesis and Characterization of Polyanilines for Supercapacitor and Catalytic Applications" at the CSIR-Indian Institute of Chemical Technology (CSIR-IICT) and Kakatiya University (India). Before joining the National Center for Nanoscience and Technology (NCNST) as CAS-PIFI research fellow, China, he worked as senior research associate and Postdoc at National Tsing-Hua University (NTHU, Taiwan) respectively in the fields of bio-fuel and CO_2 reduction applications. His academic honors include University Grants Commission National Fellowship and many merit scholarships, study-abroad fellowships from Australian Endeavour Research Fellowship, and CAS-PIFI.

He has published many scientific articles in international peer-reviewed journals and has authored about 20 book chapters, and he is also serving as an editorial board member and a referee for reputed international peer-reviewed journals. He has published edited books with Springer (UK), Elsevier, Materials Science Forum LLC (U.S.A.), Wiley-Scrivener, (U.S.A.) and CRC Press Taylor & Francis group. His specialized areas of research are energy conversion and storage, which include sustainable nanomaterials, graphene, polymer composites, heterogeneous catalysis for organic transformations, environmental remediation technologies, photoelectrochemical water-splitting devices, biofuel cells, batteries and supercapacitors.

Contributors

Zeeshan Abid
Department of Chemistry
Government College University Lahore
Lahore, Pakistan

Simran Aggarwal
Amity Institute of Click-Chemistry Research and Studies (AICCRS)
Noida, India

Mohd. Imran Ahamed
Department of Chemistry
Aligarh Muslim University
Aligarh, India

Giselle Cristine Melo Aires
Extraction Laboratory/Graduate Program in Food Science and Technology (LABEX/PPGCTA)
Federal University of Pará
Belém, Brazil

Muhammad Altaf
Department of Chemistry
Government College University Lahore
Lahore, Pakistan

Naushad Anwar
Department of Chemistry
Aligarh Muslim University
Aligarh, India

Belén Arcentales-Vera
Chemistry Department
School of Chemical and Engineering
Yachay Tech University
Urcuqui city, Ecuador

Raja Shahid Ashraf
Department of Chemistry
Government College University Lahore
Lahore, Pakistan

Figen Balo
Engineering Faculty
Department of Industrial Engineering
Fırat University
Elazığ, Turkey

Syqa Banoo
Department of Chemistry
Mangalayatan University
Beswan, India

Jhonatas Rodrigues Barbosa
Extraction Laboratory/Graduate Program in Food Science and Technology (LABEX/PPGCTA)
Federal University of Pará
Belém, Brazil

Masrat Bashir
Department of Chemistry
Aligarh Muslim University
Aligarh, India

Lisandra Bastidas
Chemistry Department
School of Chemical and Engineering
Yachay Tech University
Urcuqui city, Ecuador

Fernanda Wariss Figueiredo Bezerra
Extraction Laboratory/Graduate Program in Food Science and Technology (LABEX/PPGCTA)
Federal University of Pará
Belém, Brazil

Emilio Bucio
Department of Radiation Chemistry and Radiochemistry
Institute of Nuclear Sciences
National Autonomous University of Mexico
Mexico City, Mexico

Moises Bustamante-Torres
Biomedical Engineering Department
School of Biological and Engineering
Yachay Tech University
Urcuqui city, Ecuador

and

Department of Radiation Chemistry and Radiochemistry
Institute of Nuclear Sciences
National Autonomous University of Mexico
Mexico City, Mexico

Jorddy Neves da Cruz
Adolpho Ducke Laboratory, Botany Coordination
Museu Paraense Emílio Goeldi
Belém, Brazil

Mahuya Das
Greater Kolkata College of Engineering and Management Baruipur
Kolkata, India

Yuliya Dzyazko
V.I. Vernadskii Institute of General and Inorganic Chemistry
Kyiv, Ukraine

Lucas Cantão Freitas
Extraction Laboratory/Graduate Program in Food Science and Technology (LABEX/PPGCTA)
Federal University of Pará
Belém, Brazil

Raul Nunes de Carvalho Junior
Extraction Laboratory/Graduate Program in Food Science and Technology (LABEX/PPGCTA)
Federal University of Pará
Belém, Brazil

Sughra Gulzar
Department of Chemistry
Government College University Lahore
Lahore, Pakistan

Syed Arfan Haider
Department of Chemistry
Government College University Lahore
Lahore, Pakistan

Humaira Idrees
Institute of Polymer & Textile Engineering
University of the Punjab
Lahore, Pakistan

Shah Imtiaz
Department of Chemistry
Aligarh Muslim University
Aligarh, India

Rafi Ullah Khan
Institute of Polymer & Textile Engineering
University of the Punjab
Lahore, Pakistan

Anil Kumar
Department of Chemistry
Faculty of Technology and Science
Lovely Professional University
Phagwara, India

Ashish Kumar
Department of Chemistry
Faculty of Technology and Science
Lovely Professional University
Phagwara, India

Subhajit Kundu
Department of Chemical Technology
University of Calcutta
Kolkata, India

Marielba de Los Angeles Rodriguez Salazar
Extraction Laboratory/Graduate Program in Food Science and Technology (LABEX/PPGCTA)
Federal University of Pará
Belém, Brazil

Vivek Mishra
Amity Institute of Click-Chemistry Research and Studies (AICCRS)
Noida, India

Debarati Mitra
Department of Chemical Technology
University of Calcutta
Kolkata, India

Mehdi Mogharabi-Manzari
Pharmaceutical Sciences Research Center
Mazandaran University of Medical Sciences
Sari, Iran

Ali Motaharian
Food and Drugs Control Laboratory, Food and Drugs Administration
Birjand University of Medical Sciences
Birjand, Iran

Ammara Mushtaq
Department of Chemistry
Government College University Lahore
Lahore, Pakistan

Johnatt Allan Rocha de Oliveira
Institute of Health Sciences
Faculty of Nutrition (FANUT)
Federal University of Para (UFPA)
Belém, Brazil

Anatolii Omel'chuk
V.I. Vernadskii Institute of General and Inorganic Chemistry
Kyiv, Ukraine

Shubham Pandey
Amity Institute of Click-Chemistry Research and Studies (AICCRS)
Noida, India

Zahra Pakdin-Parizi
Department of Nuclear Medicine and Molecular Imaging
Razavi Hospital
Mashhad, Iran

Rafael Henrique Holanda Pinto
Extraction Laboratory/Graduate Program in Food Science and Technology (LABEX/PPGCTA)
Federal University of Pará
Belém, Brazil

Paul Maldonado Pinos
Department of Chemistry
School of Chemical and Engineering
Yachay Tech University
Urcuqui city, Ecuador

Shahla Rezaei
Pharmaceutical Sciences Research Center
Mazandaran University of Medical Sciences
Sari, Iran

Aneela Sabir
Institute of Polymer & Textile Engineering
University of the Punjab
Lahore, Pakistan

Masoud Salehipour
Department of Biology
Faculty of Biological Sciences
Parand Branch of Islamic Azad University
Parand, Iran

Muhammad Shafiq
Institute of Polymer & Textile Engineering
University of the Punjab
Lahore, Pakistan

Munazza Shahid
University of Education
Lahore, Pakistan

Nimra Shakeel
Department of Chemistry
Aligarh Muslim University
Aligarh, India

Luiza Helena da Silva Martins
Institute of Health and Animal Production (ISPA)
Federal Rural University of the Amazon (UFRA)
Belém, Brazil

Yuvika Sood
Department of Chemistry
Lovely Professional University
Phagwara, India

Lutfu S. Sua
School of Entrepreneurship and Business Administration, AUCA
Turkey

Abhinay Thakur
Department of Chemistry
Faculty of Technology and Science
Lovely Professional University
Phagwara, India

Faiza Wahad
Department of Chemistry
Government College University Lahore
Lahore, Pakistan

1 Introduction to Porous Polymers

Anil Kumar, Yuvika Sood, and Ashish Kumar
Department of Chemistry, Lovely Professional University, Phagwara, India

CONTENTS

1.1 INTRODUCTION

Porous polymers have been developed continuously since the introduction of foam rubber with a "sponge-like or cellular structure" from aqueous emulsion produced by Dunlop Rubber.[1] The particles of porous polymer, particularly those which are having a spherical shape are used for several practical applications like gas storage,[2–4] separations,[5, 6] catalysis,[7–11] organic pollutants,[12] and photoconductors[13], such as the materials to execute a smaller amount energy-intensive separation. Such performance occurs as a result of composite structures from synthetic materials and polymers. The attraction of these polymers is because of various properties like very large surface area and phenomenon risen from the structural aspect of the materials.[11] Main features like cross-linking and porous nature make it different from gel type polymer particles which arises due to distinct features like higher surface area, capacity to take solvent with different polarity. Porous polymers are typically explained by shape and pore size distribution. The size and features of pores vary depending on the application of the porous material to be produced.[14]

1.2 TYPES OF POROUS POLYMERS

According to IUPAC, porous polymers are categorized into numerous groups. Based on pore size, they are grouped into macroporous materials, mesoporous materials, and microporous materials having pore size more than 50 nm, between 2 and 50 nm, and less than 2 nm, respectively.[15]

Macro-porous Polymers: These types of porous polymers are prepared by methods like free radical cross linking copolymerization. Through this method, the polymerization of vinyl-divinyl monomers takes place as inert diluents. The structure of pores or macroporosity is governed by the diluent (solvent, no-solvent and linear polymers) acting as a pore to form an agent.[16] In the development of these polymers, the diluent is not resistant to monomers but is also a very poor polymerization solution for styrene and divinylbenzene. Phase separation occurs due to poor solubility in the diluent leading to a high internal surface filled with large holes.

Meso-porous Polymers: The members of this category of polymers are synthesized by a soft templating process that uses the organic-based surfactant as a template (for example, block co-polymer) and through hard templating method using inorganic materials as templates (for example, silica nanoparticles). Mesoporous polymers are used in various applications like free metal alloys, catalyst support, membrane synthesis, purification, and separation.

DOI: 10.1201/9781003169604-1

These types of polymers can be produced by large molecules that block the corresponding phases. The self-assembly of surfactant molecules with hydrophilic tail, hydrophobic head and group of planks produces porous structures (hexagonal or cubic) which are then converted into porous polymers using polymerization.[1]

Micro-porous Polymers: The members of this category of polymers have flexible and rotating polymers that allow increasing intramolecular and intermolecular interactions and fill the space in solid shape. Therefore, the existence of free space inside the content is not inhabited by polymer molecules due to various factors like molecular structure and physical state, etc. Therefore, interconnectivity and microporosity take place if the material owns a large free volume.[17] They can be built with the design of solid macromolecular structures. The pores can be formed by the way monomers are formed to combine during hyper cross-linking and polymerization. This type of polymeric material is further divided into seven types namely polymers of intrinsic microporosity (PIMs),[18, 19] hyper cross-linked polymers (HCPs), crystalline covalent organic frameworks (COFs),[20] polyacetylene,[21] covalent triazine frameworks (CTFs),[6] porous aromatic frameworks (PAFs)[22] and conjugated micro-porous polymers (CMPs).[23]

These polymers can be classified as insoluble or soluble in organic solvents on the basis of their solubility in a solvent. The numerous examples of PIMs dissolve easily while COFs, HCPs, CTFs, CMPs and PAFs are insoluble. HCPs are made up of very different chemicals with various properties and are cheap which leads to a lack of structural diversity.[24] PIMs as amorphous polymers have internal microporosity caused by ineffective packing of extremely rigid along with distorted polymeric chains.[5] It is well known that conjugated polyacetylene complements the structure of PIMs and is assembled for 20 years. PIMs and HCPs are naturally amorphous. However, COFs are defined as crystal-like structures formed from the covalent assembly of molecular elements into a unique structure connected by a strong covalent bond.[25–28] Covalent organic frameworks are different from other perforated polymers in that they are mostly crystalline and are similar to metal-organic frameworks (MOFs). The functional design of the conjugated CMP matrix is intensification in composite giving rise to various possibilities for opto-electronic devices and related applications.[29] CTFs are prepared by reduction of aromatic nitriles which have some unique characteristics i.e. aromatic (C=N) linkage (triazine unit), the absence of weak bonds except for the aromatic group, which endow themselves with high chemical stability and rich nitrogen contents, which bring great value for many practical applications such as separation and storage of gases, energy storage, photocatalysis and heterogeneous catalysis.[30, 31] The PAF is considered to be exceptional as it exceeds stability hinderance for expanded structures (COFs and MOFs) and the surface boundary layer of porous polymer networks bonded covalently (CTFs, HCPs and CMPs).[32, 33] The main difference between CMPs and PAFs leads to an increase in conjugation by C-C bond formation. CMPs maintain the composite of the whole matrix and 3D monomers nodes fail to explain extended conjugation in PAFs. PAFs that are important in the field of porous networks produce a high surface area with higher stability. A summary of the characteristics of the porous polymers is given in Table 1.1.

TABLE 1.1
Outline of the Types of Porous Polymers

	HCPs	PIMs	Polyacetylenes	COFs	CMPs	CTFs	PAFs
	Amorphous	Amorphous	Amorphous	Crystalline	Amorphous	Amorphous,	Amorphous
BET-Surface Area	2090[34]	1760[35]	850[36]	5084[37]	3180[38]	3270[39]	6462
Porosity	Micro-porous	Micro-porous, Macro-porous	Micro-porous, Macro-porous	Micro-porous	Micro-porous	Micro-porous	Micro-porous, Macro-porous

A certain component such as the surface area of the polymeric materials derived from the adsorption isotherm that represents the relation within the amount of trace gas emitted via the pressure and unit mass at persistent temperature. Brunauer-Emmett-Teller (BET)[40] as well as Langmuir [41, 42] theories are conjectures of research on the physisorption of gas molecules on the surface of polymeric materials. Langmuir's theory deals with the release of monolayer adsorption emissions and BET theory captures multilayer adsorption emissions. Another characteristic of these polymers is their physical and chemical stability. PAF has the unique stability of having a rigid biphenyl structure.[32] Stability shows important analyses for COFs with assumptions like linkage conversion[43, 44] and irreversible reactions.

1.3 SYNTHETIC METHODS FOR POROUS POLYMER NETWORK

The numerous properties such as mechanical strength, permeability, adsorption capacity and size exclusion depend largely on the polymerization and connectivity of polymer fibers. Various techniques of porous polymeric material are explained in Table 1.2.

Starting fabrication using the gas foaming technique is different from other techniques as it requires the usage of solvents, high temperature and residues left over after the completion of the process of damaging cells and tissues. Gases such as CO_2 and nitrogen are pressurized at high pressure until the polymers are saturated. The process of phase separation depends on the change in temperature which separates the solution of polymers forming two-phase liquid. The main advantage of this process is that it could be easily combined with more synthesis methods for designing three structures with control morphology. Meanwhile, some manufacturing techniques are based on the

TABLE 1.2
Techniques of Porous Polymeric Material

Method	Porogene	Pores size	Morphologies	Examples of Polymers
Foaming[45–47]	Solvents Gases Supercritical CO_2	100 nm–1 mm	Open/Closed pores	Polystyrene, polylactide, polyeurathen, polycarbonate
HIPE[48, 49] *Polymerization*	Emulsions	10 μm–100 μm	Isotropic, open porous	Poly(vinylbenzyl chloride)
Phase Separation[50]	Solvents	1 μm–1 mm	Open, porous and isotropic	Polystyrene, polyamide, polyepoxide, polyamide
Thermally induced[51, 52]	–	–	–	–
Chemically induced[53, 54]	–	–	–	–
Immersion techniques[55]	–	–	–	–
Template synthesis[56]				
Molecular imprinting	Molecules	1 nm–10 nm	Isotropic, Open closed pores	Poly(methacrylic acid), polyurethane, polystyrene
Micelles imprinting	Micelles	10 nm–50 nm	–	–
Colloid crystal Templating	Colloids crystals	50 nm–1 μm	–	–
Biocontinous micro emulsion	Bic. Microemulsions	5 nm–10 μm	–	–
Polymerization of Liquid mesophases	Lyotropic mesophases	5 nm–10 μm	Anisotropic	Poly(acrylic acid), Polyacryl(amide)
Other Techniques				
Mechanical stretching	Mechanical deformation	10 nm–100 nm	Anisotropic, open pores	Poly(propylene)
Interparticular cross – linking	Solvent	1 μm–10 μm	Isotropic, open pores	Polyacrylic acid

TABLE 1.3
Comparison of Different Synthesis Method with Benefits and Drawbacks[57]

Method of Fabrication	Benefits	Drawbacks
Porogen leaching	Control of porosity and geometry	Insufficient pore size
Gas foaming	Unbound from strong organic solvents, control of porosity	Less mechanical property
Phase separation	No decrease in molecular activity	Unable to control morphology
Solvent casting	Control of pore size and crystallinity	Less mechanical property and residual solvents

concept of porogen that acts as the owner of the pores sites and pores linkage. Comparison of different synthesis methods with their benefits and drawbacks is explained in Table 1.3.

The synthetic arrangement of polymers is already well established. A major limitation in the widespread use of porous polymeric material is the lack of in-depth processing until large scale films are not produced for emulsion. Pore geometry, surface area and pore size are the most predominant features of polymers on which their performance depends. The surface area of this network is controlled by pore size and pore surface. The design of the framework relies upon the type of reactions, the monomers and the condition of reaction used. The reactions for the synthesis of networks range from different cross coupling reactions, like Yamamoto-type Ullmann reactions for PAFs, Yamamoto-Ullmann and Suzuki-Miyaura for CMPs and Buchwald reactions for HCP synthesis; self-condensation reactions for CMPs; homocoupling reactions for CMPs and non-reversible condensation and COFs reactions for PIMs. The copolymerization reactions of HCPs synthesis, electrochemical (CMP), solvothermal (COF) and ionothermal (COF and PAF) have also been used to improve porosity, free volume and surface area of these frames.[6] The characteristics and functions of polymers are determined by their structures. Features such as coordinating ligands, monomers, functional groups and their interactions are characterized to provide the total features of these networks. Other reaction parameters such as concentration, temperature and nature of solvents also play a role during the fabrication methods depending on the structure of the polymer and have shown that increasing temperature improves the surface area for many porous polymers. Porous polymers are used for separation applications such as chromatographic materials, selective membranes and gel electrophoresis.[58] Also known to be used in the extraction of dyes,[31] desalination,[59] nanofiltration,[60] separation of gases,[61–63] etc. Determination of soluble 2D COFs, SCMPs leads to organic electronics and photoelectronics applications. Khayum *et al.*[64] described flexible COF thin sheets for small, portable electric motors. Covalent Organic Nanosheets (CONs) are very important in biomedical utilization because of their high proportions. For example – the average number of CONs prepared for the Langmuir-Blodgett process exceeded 20000.[65] The maximum high aspect ratio reported a large surface area of target sites while the high flexibility gives rise to unique dynamics.[66, 67] Mitra *et al.*[68] used artificial reduction of COFs that facilitated the progress of CONs delivery for chemotherapy drugs. Zhang *et al.*[69] introduced a fluorescence method to detect DNA sequences using CONs in bioinformatics and genomic analysis. CONs are gaining scope in biomedicine by identifying ionic nano-bio interfaces. The new research represents the use of iCONs in obtaining double-stranded DNA[70] and protein transport.[71] Based on biomedical research, CONs provide opportunities for further considerations, e.g.-cellular imaging and artificial enzymes.

1.4 CONCLUSION

This summary provides details of the synthetic processes that are widely used in polymeric material with an extensive range of applications. The information provided represents a solid keystone for investigators obsessed in working with polymers. It recapitulates the benefits and drawbacks of the respective process and provides a better understanding of choosing the appropriate process for a particular polymer. A new synthetic method has been introduced to produce a standard formulation procedure used in all liquid polymers to make their outer surface powder. For example, PMHS in porous form was started by the use of porous silicon scaffolds. The proposed method is straight forward, flexible and simple and its future work involves further separation using different characterization and the phenomenon of developing various new types of polymers with specific application properties.

REFERENCES

1. Silverstein, M. S. *Isr. J. Chem.* 2020, *60*, 140.
2. Budd, P. M.; Butler, A.; Selbie, J.; Mahmood, K.; McKeown, N. B.; Ghanem, B.; Msayib, K.; Book, D.; Walton, A. *Phys. Chem. Chem. Phys.* 2007, *9*, 1802.
3. Germain, J.; Fréchet, J. M. J.; Svec, F. *Small* 2009, *5*, 1098.
4. Chen, Q.; Liu, D. P.; Luo, M.; Feng, L. J.; Zhao, Y. C.; Han, B. H. *Small* 2014, *10*, 308.
5. Mc Keown, N. B.; Budd, P. M. *Chem. Soc. Rev.* 2006, *35*, 675.
6. Dawson, R.; Cooper, A. I.; Adams, D. J. *Prog. Polym. Sci.* 2012, *37*, 530.
7. Kamiya, K.; Kamai, R.; Hashimoto, K.; Nakanishi, S. *Nat. Commun.* 2014, *5*, 1.
8. Iwase, K.; Yoshioka, T.; Nakanishi, S.; Hashimoto, K.; Kamiya, K. *Angew. Chemie - Int. Ed.* 2015, *54*, 11068.
9. Thomas, P.; Pei, C.; Roy, B.; Ghosh, S.; Das, S.; Banerjee, A.; Ben, T.; Qiu, S.; Roy, S. *J. Mater. Chem. A* 2015, *3*, 1431.
10. Du, X.; Sun, Y.; Tan, B.; Teng, Q.; Yao, X.; Su, C.; Wang, W. *Chem. Commun.* 2010, *46*, 970.
11. Schmidt, J.; Weber, J.; Epping, J. D.; Antonietti, M.; Thomas, A. *Adv. Mater.* 2009, *21*, 702.
12. Ren, H.; Ben, T.; Sun, F.; Guo, M.; Jing, X.; Ma, H.; Cai, K.; Qiu, S.; Zhu, G. *J. Mater. Chem.* 2011, *21*, 10348.
13. Ding, X.; Guo, J.; Feng, X.; Honsho, Y.; Guo, J.; Seki, S.; Maitarad, P.; Saeki, A.; Nagase, S.; Jiang, D. *Angew. Chemie - Int. Ed.* 2011, *50*, 1289.
14. Hartsell, E. L. *J. Undergrad. Mater. Res.* 2005, *1*, 20.
15. Wu, D.; Xu, F.; Sun, B.; Fu, R.; He, H.; Matyjaszewski, K. *Chem. Rev.* 2012, *112*, 3959.
16. Zheng, Y.; Gun'Ko, V. M.; Howell, C. A.; Sandeman, S. R.; Phillips, G. J.; Kozynchenko, O. P.; Tennison, S. R.; Ivanov, A. E.; Mikhalovsky, S. V. *ACS Appl. Mater. Interfaces* 2012, *4*, 5936.
17. Berro, S.; El Ahdab, R.; Hajj Hassan, H.; Khachfe, H. M.; Hajj-Hassan, M. *J. Nanomater.* 2015, *2015*.
18. McKeown, N. B.; Hanif, S.; Msayib, K.; Tattershall, C. E.; Budd, P. M. *Chem. Commun.* 2002, *2*, 2782.
19. MacKintosh, H. J.; Budd, P. M.; McKeown, N. B. *J. Mater. Chem.* 2008, *18*, 573.
20. Ockwig, N. W.; Co, A. P.; Keeffe, M. O.; Matzger, A. J.; Yaghi, O. M. 2005, *310*, 1166.
21. Masuda, T.; Isobe, E.; Higashimura, T.; Takada, K. *J. Am. Chem. Soc.* 1983, *105*, 7473.
22. Ben, T.; Ren, H.; Ma, S.; Cao, D.; Lan, J.; Jing, X.; Wang, W.; Xu, J.; Deng, F.; Simmons, J. M.; Qiu, S.; Zhu, G. *Angew. Chemie* 2009, *121*, 9621.
23. Jiang, J. X.; Su, F.; Trewin, A.; Wood, C. D.; Campbell, N. L.; Niu, H.; Dickinson, C.; Ganin, A. Y.; Rosseinsky, M. J.; Khimyak, Y. Z.; Cooper, A. I. *Angew. Chemie - Int. Ed.* 2007, *46*, 8574.
24. Tsyurupa, M. P.; Davankov, V. A. *React. Funct. Polym.* 2002, *53*, 193.
25. Waller, P. J.; Gándara, F.; Yaghi, O. M. *Acc. Chem. Res.* 2015, *48*, 3053.
26. Kandambeth, S.; Dey, K.; Banerjee, R. *J. Am. Chem. Soc.* 2019, *141*, 1807.
27. Colson, J. W.; Dichtel, W. R. *Nat. Chem.* 2013, *5*, 453.
28. Rager, S.; Jakowetz, A. C.; Gole, B.; Beuerle, F.; Medina, D. D.; Bein, T. *Chem. Mater.* 2019, *31*, 2707.
29. Lee, J. S. M.; Cooper, A. I. *Chem. Rev.* 2020, *120*, 2171.
30. Liu, M.; Guo, L.; Jin, S.; Tan, B. *J. Mater. Chem. A* 2019, *7*, 5153.
31. Krishnaraj, C.; Jena, H. S.; Leus, K.; Van Der Voort, P. *Green Chem.* 2020, *22*, 1038.
32. Chenot, C.; Robiette, R.; Collin, S. *J. Agric. Food Chem.* 2019, *67*, 4002.
33. Ben, T.; Qiu, S. *CrystEngComm* 2013, *15*, 17.
34. Ahn, J. H.; Jang, J. E.; Oh, C. G.; Ihm, S. K.; Cortez, J.; Sherrington, D. C. *Macromolecules* 2006, *39*, 627.
35. Ghanem, B. S.; Hashem, M.; Harris, K. D. M.; Msayib, K. J.; Xu, M.; Budd, P. M.; Chaukura, N.; Book, D.; Tedds, S.; Walton, A.; McKeown, N. B. *Macromolecules* 2010, *43*, 5287.

36. Satilmis, B.; Uyar, T. *ACS Appl. Polym. Mater.* 2019, *1*, 787.
37. Baldwin, L. A.; Crowe, J. W.; Pyles, D. A.; McGrier, P. L. *J. Am. Chem. Soc.* 2016, *138*, 15134.
38. Holst, J. R.; Stöckel, E.; Adams, D. J.; Cooper, A. I. *Macromolecules* 2010, *43*, 8531.
39. Kuhn, P.; Forget, A.; Su, D.; Thomas, A.; Antonietti, M. *J. Am. Chem. Soc.* 2008, *130*, 13333.
40. Sing, K. *Colloids Surfaces A Physicochem. Eng. Asp.* 2001, *187–188*, 3.
41. Armbruster, M. H.; Austin, J. B. *J. Am. Chem. Soc.* 1938, *60*, 467.
42. Langmuir, I. *J. Franklin Inst.* 1917, *184*, 721.
43. Zhang, B.; Wei, M.; Mao, H.; Pei, X.; Alshmimri, S. A.; Reimer, J. A.; Yaghi, O. M. *J. Am. Chem. Soc.* 2018, *140*, 12715.
44. Kandambeth, S.; Mallick, A.; Lukose, B.; Mane, M. V.; Heine, T.; Banerjee, R. *J. Am. Chem. Soc.* 2012, *134*, 19524.
45. Xing, Z.; Wang, M.; Du, G.; Xiao, T.; Liu, W.; Qiang, D.; Wu, G. *J. Supercrit. Fluids* 2013, *82*, 50.
46. Salerno, A.; Di Maio, E.; Iannace, S.; Netti, P. A. *J. Supercrit. Fluids* 2011, *58*, 158.
47. Tsivintzelis, I.; Angelopoulou, A. G.; Panayiotou, C. *Polymer (Guildf).* 2007, *48*, 5928.
48. Fortifica, D. E. E. D. E. E. D. E. 1996, *126*.
49. Pulko, I.; Krajnc, P. *Macromol. Rapid Commun.* 2012, *33*, 1731.
50. Kiefer, J.; Hilborn, J. G.; Hedrick, J. L. *Polymer (Guildf).* 1996, *37*, 5715.
51. Matsuyama, H.; Teramoto, M.; Kuwana, M.; Kitamura, Y. *Polymer (Guildf).* 2000, *41*, 8673.
52. Nam, Y. S.; Park, T. G. *Biomaterials* 1999, *20*, 1783.
53. Wang, L. M.; Gong, W. L.; Ewing, R. C.; Webert, W. J. *398*, 233.
54. Li, J.; Du, Z.; Li, H.; Zhang, C. *Polymer (Guildf).* 2009, *50*, 1526.
55. Di Luccio, M.; Nobrega, R.; Borges, C. P. *Polymer (Guildf).* 2000, *41*, 4309.
56. Hoa, M. L. K.; Lu, M.; Zhang, Y. *Adv. Colloid Interface Sci.* 2006, *121*, 9.
57. Subia, B.; Kundu, J.; Kundu, S. *Tissue Eng.* 2010.
58. Righetti, P. G. *J. Chromatogr. A* 1995, *698*, 3.
59. Yang, H.; Yang, L.; Wang, H.; Xu, Z.; Zhao, Y.; Luo, Y.; Nasir, N.; Song, Y.; Wu, H.; Pan, F.; Jiang, Z. *Nat. Commun.* 2019, *10*, 1.
60. Shinde, D. B.; Sheng, G.; Li, X.; Ostwal, M.; Emwas, A. H.; Huang, K. W.; Lai, Z. *J. Am. Chem. Soc.* 2018, *140*, 14342.
61. Bhavsar, R. S.; Mitra, T.; Adams, D. J.; Cooper, A. I.; Budd, P. M. *J. Memb. Sci.* 2018, *564*, 878.
62. Fu, J.; Das, S.; Xing, G.; Ben, T.; Valtchev, V.; Qiu, S. *J. Am. Chem. Soc.* 2016, *138*, 7673.
63. Das, S.; Ben, T.; Qiu, S.; Valtchev, V. *ACS Appl. Mater. Interfaces* 2020, *12*, 52899.
64. Khayum, M. A.; Vijayakumar, V.; Karak, S.; Kandambeth, S.; Bhadra, M.; Suresh, K.; Acharambath, N.; Kurungot, S.; Banerjee, R. *ACS Appl. Mater. Interfaces* 2018, *10*, 28139.
65. Cheng, Y.; Pu, Y.; Zhao, D. *Chem. - An Asian J.* 2020, *15*, 2241.
66. Zhang, S.; Sunami, Y.; Hashimoto, H. *Nanomaterials* 2017, *7*, 1.
67. Takeoka, S.; Okamura, Y.; Fujie, T.; Fukui, Y. *Pure Appl. Chem.* 2008, *80*, 2259.
68. Mitra, S.; Sasmal, H. S.; Kundu, T.; Kandambeth, S.; Illath, K.; Díaz Díaz, D.; Banerjee, R. *J. Am. Chem. Soc.* 2017, *139*, 4513.
69. Peng, Y.; Huang, Y.; Zhu, Y.; Chen, B.; Wang, L.; Lai, Z.; Zhang, Z.; Zhao, M.; Tan, C.; Yang, N.; Shao, F.; Han, Y.; Zhang, H. *J. Am. Chem. Soc.* 2017, *139*, 8698.
70. Mal, A.; Mishra, R. K.; Praveen, V. K.; Khayum, M. A.; Banerjee, R.; Ajayaghosh, A. *Angew. Chemie - Int. Ed.* 2018, *57*, 8443.
71. Kong, W.; Wan, J.; Namuangruk, S.; Guo, J.; Wang, C. *Sci. Rep.* 2018, *8*, 8.

2 Hyper-crosslinked Polymers

Belén Arcentales-Vera and Lisandra Bastidas
Yachay Tech University, Urcuqui city, Ecuador

Moises Bustamante-Torres
Yachay Tech University, Urcuqui city, Ecuador; National Autonomous University of Mexico, Mexico City, Mexico

Paul Maldonado Pinos
Yachay Tech University, Urcuqui city, Ecuador

Emilio Bucio
National Autonomous University of Mexico, Mexico City, Mexico

CONTENTS

DOI: 10.1201/9781003169604-2

2.1 INTRODUCTION

2.1.1 Overview

Polymer science is one of the fastest-growing and most impactful fields because of its scientific, technological, and industrial potential. In this context, porous materials have emerged as candidates for designing and developing novel materials, including adhesives, adsorbents, catalysts, nanocomposite materials, etc. This interest is stimulated because of their particular characteristics such as pore size, pore shape, and pore surface reactivity, as well as the different resulting morphologies.

Among the different types of porous polymers, porous organic polymers (POPs) stand out as promising materials due to their rigid structure, high thermal stability, inherent porosity, and moderate–high surface area that exceeds that of the well-known zeolites and porous silicates (Côté et al. 2005). POPs are generally classified into two categories depending on long-range order: crystalline polymers, and amorphous polymers such as hyper-crosslinked network materials (Zhang et al. 2020).

Hyper-crosslinked polymers (HCPs) were one of the first classes of amorphous organic porous polymers investigated (Davankov and Tsyurupa 2010). Since reported work in the 1970s, HCPs have been thoroughly investigated. The investigations have led to know different methods of development and functionalization of the HCPs. These methodologies have enabled the synthesis of porous polymer chains with specific features and functionalities.

The unique characteristics of HCPs make them versatile materials that have been employed for numerous and attractive fields such as renewable energy (i.e., gas storage and CO_2 capture), environmental science (e.g., capture and removal of pollutants), biomedical field (e.g., drug delivery), etc. The latest trends in synthesis, functionalization, characterization, and applications of these novel materials are presented in the following sections.

2.1.2 Porous Polymer

Porous polymers describe all the polymers that contain small empty volumes or spaces. Generating the empty volume and then fixing it in place results in the most challenging feature to achieve in porous polymers. However, a wide range of polymerization reactions promotes the generation and fixation of pores. The most common pore generation strategies include macromolecular design (Dawson, Cooper, and Adams 2012), self-assembly (Wang et al. 2014), phase separation (Bai et al. 2007), solid and liquid templating (Das et al. 2018), and sol–gel formation (Moon et al. 2010). These methods provide access to polymers with different pore size such as macropores (greater than 50 nm), mesopores (between 2 and 50 nm), and micropores (less than 2 nm) (Silverstein 2020).

POPs, a class of polymers possessing micropores, are composed of highly rigid polymer networks fixed with strong covalent bonds that prevent porous architecture's collapse and maintain inner cavities (Germain, Fréchet, and Svec 2007). POP networks can be synthesized with different structures, including covalent organic frameworks (COFs), HCPs, polymers of intrinsic microporosity (PIM), conjugated microporous polymers (CMPs), and so on (Wu et al. 2021). Among those POPs, HCPs stand out due to their simplicity and attractive synthesis (Woodward 2020). Something remarkable about the polymeric nature of POPs is their easy processability that allows controlling the micro morphologies such as nanoparticles (quasi-zero-dimensional), hollow capsules, membranes (two-dimensional, 2D), and monolithic blocks (three-dimensional, 3D) (Tan and Tan 2017).

2.1.3 Crosslinking

Depending on the degree of crosslinking, polymer chains play several roles and exhibit different structures. Figure 2.1 depicts the internal system of uncrosslinked, crosslinked, and highly crosslinked networks. Uncrosslinked polymers generally have significant viscous flow components, while crosslinked and highly crosslinked networks exhibit a rigid structure.

2.2 HYPER-CROSSLINKED POLYMERS

In the early 1970s, HCPs were described for the first time. Since then, HCPs have been used for applications in gas storage, catalytic supports, drug delivery, and chromatographic separation (Tsyurupa and Davankov 2002, 2006). Crosslinking is often based on Friedel–Crafts reactions which generate strong linkages, yielding networks with a high degree of crosslinking that prevent the nanoporous structure from collapsing. However, researchers also found the possibility to use direct condensation/polymerization reactions and knitting rigid aromatic monomers as alternatives to the traditional post-synthetic hyper-crosslinking methodology (Weber and Meng 2014; Li et al. 2011).

HCPs are materials with unique properties such as specific surface area, regulated pore topology, and the capacity to incorporate specific functionalities into the porous polymer networks (Tan and Tan 2017). The synthesis of HCPs is simple, versatile, and employs a wide range of aromatic monomers, leading to produce polymeric chains having several pore structures and functionalities that trigger improvements in surface area and unique features. Other advantages of synthesizing HCPs involve the low cost of reagents, easy control, and handling reaction conditions leading to high yields of polymer production.

2.3 SYNTHESIS METHODS OF HCPs

The enormous demand for new materials to provide technological breakthroughs pushes the scientific investigation forward to produce HCPs with specific features. These materials are commonly prepared in three main ways: (1) post-crosslinking of preformed polymeric precursors, (2) direct polycondensation of monomers, and (3) knitting by using external crosslinkers. The development and optimization of the different synthetic routes have greatly extended the variety of suitable starting blocks, enabling the exploration of new hyper-crosslinked materials with prominent features and applications.

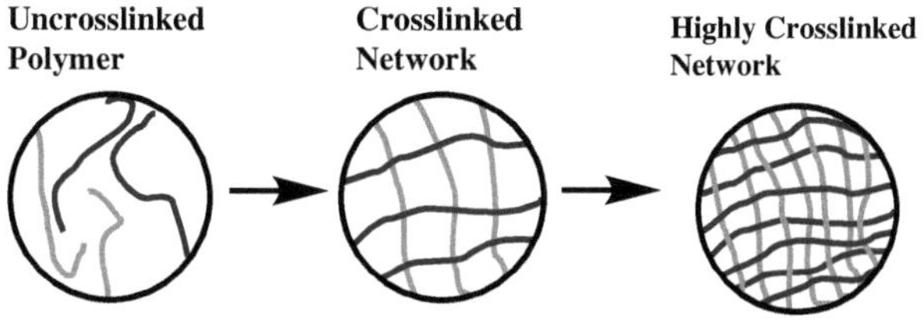

FIGURE 2.1 Polymer segments and crosslinking network.

2.3.1 Post-crosslinking Polymer Precursors

Hyper-crosslinked polystyrene, the first representative of HCP networks, was first reported during the 1970s by Davankov and Tsyurupa (Davankov, Rogoshin, and Tsyurupa 1973). The synthesis of this type of HCP, also known as Davankov-type resins, differs from the process of formation of conventional macroporous systems. The latter are obtained fundamentally by copolymerization of monomers, while the hyper-crosslinked networks are prepared by intensive post-crosslinking of solvated/swollen polymer precursors using external crosslinkers (Davankov and Tsyurupa 1990; Davankov, Rogoshin, and Tsyurupa 1974). Such a process is shown schematically in Figure 2.2.

Specifically, linear polystyrene (PS) or slightly crosslinked gel-type polystyrene-based precursors (generally styrene/divinylbenzene (St-DVB) copolymers) are dissolved or swollen in a thermodynamically favorable solvent followed by the addition of an external electrophile and a Friedel–Crafts catalyst. In this way, the solvated polystyrene chains are crosslinked by forming rigid bridges between phenyl rings. These rigid linkages fix the chains in their expanded conformations even after the solvent is removed (Germain, Fréchet, and Svec 2007; Tsyurupa and Davankov 2006; Davankov and Tsyurupa 2010). Networks based on linear polystyrene or crosslinked St-DVB copolymers (0.3–2% DVB) show intrinsic microporosity. In contrast, networks based on macroporous polystyrenes exhibit a bimodal pore size distribution (PSD). That is, in addition to the macropores initially present in the precursor, micropores formed during the hyper-crosslinking process are also present (Davankov and Tsyurupa 1990; Ahn et al. 2005). The resulting hyper-crosslinked networks are characterized by their low packing density and very high apparent surface area (up to 2000 m^2/g) as a result of a significant amount of highly interconnected micropores (Ahn et al. 2005).

Many external bifunctional reagents such as 4,4′-bis-(chloromethyl)biphenyl (BCMBP) (Davankov, Rogoshin, and Tsyurupa 1974), monochlorodimethyl ether (MCDE) (Pastukhov, Tsyurupa, and Davankov 1999), p-dichloroxylene (DCX) (Davankov, Rogoshin, and Tsyurupa 1974; Grassie and Gilks 1973), tetrachloromethane (CCl_4) (Hradil and Králová 1998; Gadwdzik and Osypiuk 2001), 1,4-bis-(p-chloromethylphenyl)-butane (DPB) (Tsyurupa, Lalaev, and Davankov 1984), and trifunctional compound 1,3,5-tris-(chloromethyl)-mesitylene (TCMM) (Tsyurupa, Lalaev, and Davankov 1984) have been commonly used as crosslinking agents. Under Friedel–Crafts reaction conditions, these compounds alkylate two aromatic rings (or three if TCMM is employed), generating a final network containing rigid bridges. The length of these rigid bridges is two phenyl rings greater than the length of the crosslinking substance (Davankov and Tsyurupa 2010). The

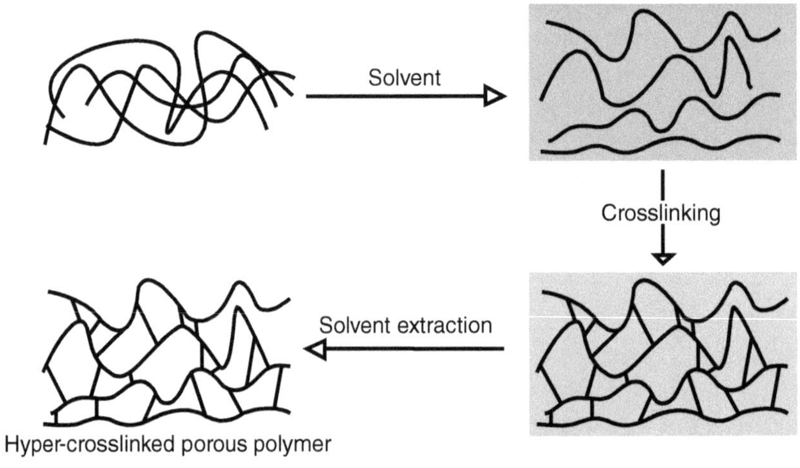

FIGURE 2.2 Schematic representation of post-crosslinking synthesis method.

(Adapted from Germain, Fréchet, and Svec 2007).

conformational mobility of the final hyper-crosslinked network and its outstanding features such as surface area, porosity, and swelling capacity are markedly affected by the length and rigidity of the cross-bridges formed between polymeric chains. Indeed, polystyrene networks crosslinked with MCDE show porosity at a crosslinking degree of 25% with an apparent inner surface area of 240 m^2/g. At a higher crosslinking degree (60–100%), the networks exhibit a greater surface area (1000 m^2/g). Importantly, the bridges of the diphenylmethane-type produced by MCDE are the most rigid ones. The networks crosslinked with trifunctional compound TCMM reveal higher porosity at lower amounts of the crosslinker agent (700 m^2/g apparent surface area at crosslinking degree of only 25%). This results from the enhanced rigidity of such networks since one molecule of TCMM links simultaneously three polystyrene chains. By contrast, DPB is not able to construct networks with limited conformational mobility due to its four flexible methylene units facilitating conformational rearrangements, thus, these networks lack porosity at any crosslinking degree (Tsyurupa and Davankov 2002; Davankov and Tsyurupa 1990).

In a different approach, the use of chloromethylated St-DVB copolymers (containing 5% DVB) as starting material was reported for the synthesis of hyper-crosslinked polystyrene. In this two-step process, the chloromethyl substituent yields an internal electrophile to form methylene bridges, creating additional crosslinking. This HCP material exhibited a surface area up to approximately 1000 m^2/g (Veverka and Jeřábek 1999). Subsequently, further exploration was carried out to synthesize hyper-crosslinked resin beads with optimized surface area. For this, poly(vinylbenzyl chloride-divinylbenzene) copolymers (VBC-DVB) with different DVB content (2 and 20%) and ratios of monomers were directly prepared by suspension polymerization. Then, these resins were hyper-crosslinked by varying the reaction conditions. In this study, the catalytic activities of $AlCl_3$, $SnCl_4$, and $FeCl_3$ were compared. It was found that $FeCl_3$ was the most effective catalyst for the hyper-crosslinking reactions yielding networks with the higher superficial areas in all cases. This previous result may be explained by a superior solubility–molecular size relationship. Besides, under suitable parameters, hyper-crosslinked materials produced from gel-type precursors showed higher surface areas (up to 2090 m^2/g) than those arising from porous precursors (up to 1055 m^2/g). This finding can be associated with the formation of a more regular microporous structure from gel-type species than that produced from porous precursors since the high swelling of gel-type polystyrene chains might allow the synthesis of doubly bridged structures, as shown in Figure 2.3. Moreover, the final products were compared based on their precursor's monomer ratio. In general, the surface area dramatically decreases when the amount of styrene in the co-monomer increased due to the reduced level of chloromethyl functionality. For instance, the precursor with 2% DVB and a St:VBC ratio of 3:1, gives rise to a hyper-crosslinked network with a low surface area of 2 m^2/g. Surprisingly,

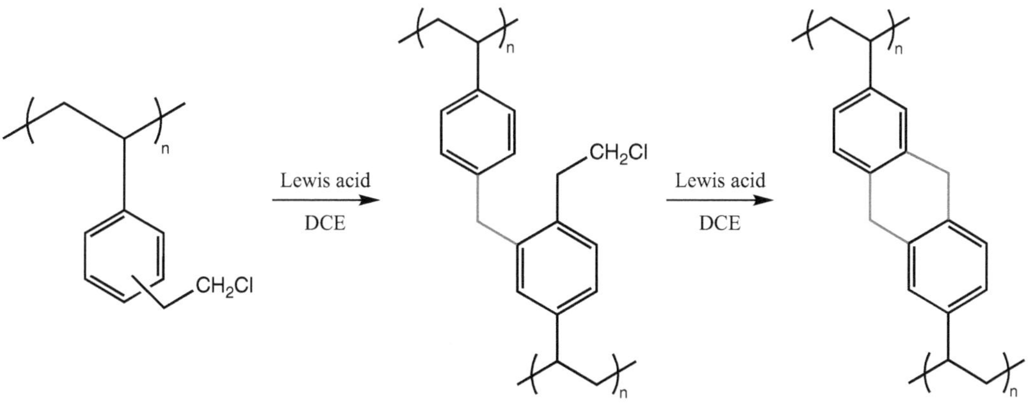

FIGURE 2.3 Hyper-crosslinking of DVB-VBC monomer via generation of internal electrophile.

(Adapted with permission from Ahn et al. 2005. Copyright (2005) American Chemical Society).

when using the precursor with 20% DVB content and identical St:VBC ratio, it exhibited a surface area of 474 m^2/g, which is considerably higher than the previous one, suggesting that the precursor's divinylbenzene content is a crucial factor affecting the PSD and surface area of these materials (Ahn et al. 2005).

Continuing to explore hierarchically porous polymers (HPPs), an effective synthetic strategy for HPPs with a bimodal pore structure was conducted through combining the self-assembly of blocks with hyper-crosslinking. The polymerization of VBC with DVB in the presence of polylactide macro-chain transfer agent (PLA-CTA) formed a crosslinked block polymer precursor via reversible addition–fragmentation chain transfer (RAFT). When polymerization reaction achieved a critical conversion point, the system was induced to microphase separation into bicontinuous microdomains of PLA and poly(VBC-DVB), which was arrested by in situ crosslinking due to the addition of DVB in the reaction mixture. Subsequently, the hyper-crosslinking process generated micropores in the poly(VBC-DVB) microdomain and, at the same time, selectively removed the PLA microdomain to yield a well-defined mesoporous framework containing micropores. Moreover, it was demonstrated that by varying the molar mass of the PLA, precise control over the mesopore size (6–15 nm) could be obtained (Seo et al. 2015).

Other types of hyper-crosslinked networks based on polysulfone, polyarylate, polyaniline, and polypyrrole have also been prepared by the post-crosslinking approach. The synthesis of hyper-crosslinked polysulfone illustrated in Figure 2.4a was carried out by self-crosslinking of bromomethylated polysulfone through Friedel–Crafts reaction. Nevertheless, the ether linkage of

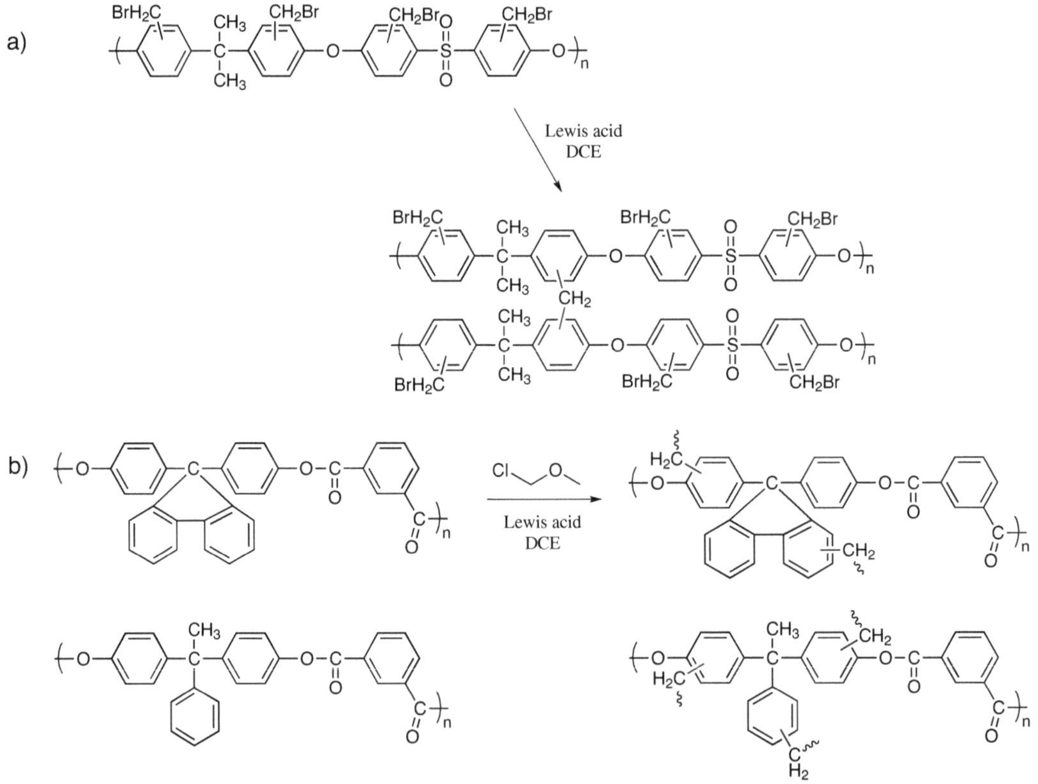

FIGURE 2.4 Crosslinking of (a) polysulfone and (b) polyaryalate.

(Adapted with permission from Tsyurupa and Davankov 2002. Copyright (2002) American Chemical Society).

polysulfone provides higher conformational flexibility to the structure, which in combination with the low degree of crosslinking (due to low activity of sulfone groups), leads to materials with low apparent inner surface areas (up to 72 m^2/g) (Tsyurupa and Davankov 2002). The hyper-crosslinking of polyarylate precursors was carried out with chloromethyl methyl ether (CMME). The resulting polyarylate networks are shown in Figure 2.4b. Polyarylates with more rigid chains yielded networks with high porosity; however, the highest surface area reported was only 380 m^2/g (Tsyurupa and Davankov 2002).

Polyaniline has been crosslinking through the N-alkylation route using diiodoalkanes as crosslinkers. Diiodoalkanes are preferred over dihaloalkanes containing dichloro or dibromo as these compounds are very reactive, compensating for the reduced reactivity of polyaniline precursors. Hyper-crosslinked polyanilines obtained with diiodomethane and paraformaldehyde crosslinkers displayed the highest Brunauer–Emmett–Teller (BET) surface areas reaching values of 632 m^2/g and 390 m^2/g, respectively. Besides, for these materials, it was shown that by increasing the concentration of polyaniline in the reaction mixture, the surface area of these polymers could be increased (Germain, Fréchet, and Svec 2007). Moreover, nanoporous hyper-crosslinked polyaniline was developed utilizing diiodomethane in a microwave-assisted synthesis, resulting in a surface area of 1059 m^2/g (Sharma et al. 2015). In addition to diiodoalkanes, polyhalogenobenzenes have also been used for hyper-crosslinking polyaniline through Buchwald or Ullmann coupling reactions. Both of these processes form networks of aromatic rings linked through nitrogen atoms. Hyper-crosslinked polyaniline with the highest surface area (316 m^2/g) was obtained by the Buchwald approach using diiodobenzene as the crosslinker (Germain, Svec, and Fréchet 2008). A similar strategy was used to prepare nanoporous hyper-crosslinked polypyrroles using diiodomethane, triiodomethane, and triiodoborane to form networks with crosslinking units of different sizes (CH_2, CH, and B links, respectively). Polypyrroles crosslinked through CH_2 links displayed a high BET surface area of 732 m^2/g (Germain, Fréchet, and Svec 2009).

2.3.2 Direct One-Step Polycondensation

HCPs can also be obtained by the self-condensation of small molecules having functional groups. This method started when Tsyurupa and Davankov (2002) described the polycondensation of p-dichloroxylene (p-DCX) in DCE with $SnCl_4$ as the catalyst; this material's surface area reached a value of 1000 m^2/g. This approach was further developed by Wood et al. (2007). They synthesized various HCPs via Friedel–Crafts-catalyzed self-condensation reaction using simple bischloromethyl aromatic monomers including DCX, BCMBP, and 9,10-bis(chloromethyl)anthracene (BCMA). Their chemical structures are represented in Figure 2.5. The obtained polymers were evaluated, showing essentially microporous with relatively high BET surface areas. Networks produced from 100% BCMBP revealed a larger surface area (1874 m^2/g) than p-DCX- and BCMA-based networks, which only reached 1391 and 921 m^2/g, respectively. The monomer and catalyst

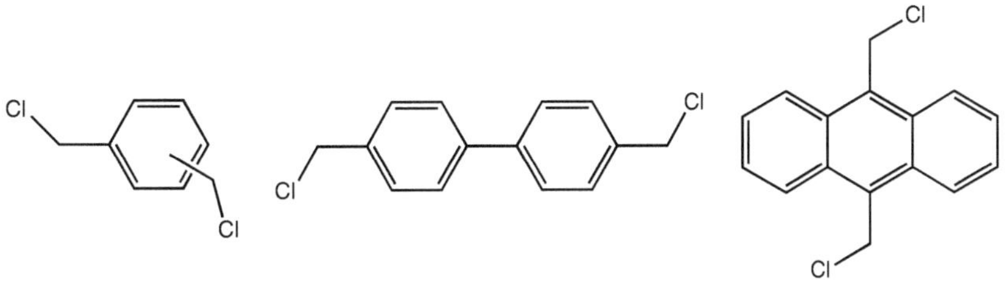

DCX BCMBP BCMA

FIGURE 2.5 Molecular structures of the monomers used to synthesize HCP networks.

(Adapted with permission from Wood et al. 2007. Copyright (2007) American Chemical Society).

ratios and the reaction conditions significantly affect the specific surface area and pore size of the hyper-crosslinked networks. As is the case of BCMBP and p-DCX copolymers, the incorporation of 25% mol/mol p-DCX resulted in an even higher surface area. Furthermore, Schwab et al. (2011) reported the copolymerization of BCMBP with a set of non-functionalized fluorene-based co-monomers including fluorene (FLUO), dibenzofuran (DBF), and others. The highest BET surface area was exhibited from the polymer coming from DBF that contains a 10% molar fraction. In general, the incorporation of 10 mol% co-monomer within the polymer networks resulted in slightly lower surface areas than the BCMBP homopolymer network reported in previous works (1874 m^2/g). Networks containing 25 mol% of co-monomer exhibited substantially lower BET surface areas (e.g., 1190 m^2/g for DBF) which could be attributed to the lower crosslinking density.

This strategy has been extended to the synthesis of hyper-crosslinked hybrid materials by self-condensation of benzyl chloride-containing precursors. Highly porous hyper-crosslinked siloxane-organic hybrids were successfully obtained. Solid-state ^{29}Si MAS NMR revealed that both the destruction of the D4R units and the organic groups' condensation occurred simultaneously during the synthesis (Chaikittisilp et al. 2011).

In addition to the chloromethyl groups, aromatic hydroxymethyl monomers can also be used directly as building blocks to produce porous frameworks through the self-condensation process. Luo et al. (2013) reported a series of simple hydroxyl-based HCPs, which were prepared by Friedel–Crafts self-condensation of 1,4-benzenedimethanol (BDM) and benzyl alcohol (BA). In these reactions, the hydroxymethyl groups were converted into methylene bridges resulting in an extensive cross-linking between molecules. The resulting self-condensation products afforded BET surface areas above 845 m^2/g using the bifunctional BDM monomer and 742 m^2/g using the monofunctional BA monomer. Similarly, polycondensed benzyl chloride (BC) was also synthesized, giving rise to a comparable surface area of 740 m^2/g. Surprisingly, these results demonstrated that monofunctional monomers can also yield HCPs with relatively high surface areas, which is contrasted with previous reports where multifunctional monomers were considered crucial for hyper-crosslinking.

Following this, a set of functional microporous organic polymers (MOPs) were prepared from different building aromatic monomers with diverse functionalities (such as 1,3,5-triphenylbenzene (TPB), naphthalene, pyrene, and meso-tetraphenylporphyrin) by a simple and cost-effective method. This method is based on the Scholl coupling reaction that implies the elimination of two aryl-bound hydrogen atoms and subsequent generation of aryl-aryl bonds with the aid of a Friedel–Crafts catalyst (certain reactions necessarily require the presence of hydrogen chloride to occur) (Grzybowski et al. 2013). The resulting materials exhibited ultra-micro and microporous structures. Among all of these, the copolymer of TPB with pyrrole yielded the highest BET surface area (1421 m^2/g). Also, it showed the best CO_2 adsorption properties reaching a CO_2 uptake capacity as high as 20.4 wt.% at 1 bar/273.15 K. Compared with MOPs synthesized by TPB and different co-monomers, homopolymer-based networks showed relatively lower surface areas. However, the homopolymer TPB network with a surface area of only 1254 m^2/g exhibited a more abundant ultra-microporous structure and thus, achieved the maximum H_2 uptake ability. These previous data indicated that while the structural properties are crucial for favorable H_2 adsorption, the functional groups play a more significant role in CO_2 uptake. Moreover, heterogeneous catalysts with high activity were designed by embedding small ligands (triphenylphosphine and 2,2-bipyridine) into the resulting microporous polymers. The proposed strategy may provide a general approach to produce a variety of microporous polymers with multifunctional properties for various applications in the future, including heterogeneous catalysis, gas storage, and energy applications (Li et al. 2014).

Highly porous polymers were successfully prepared using low-cost and readily available aromatic hydrocarbons as building blocks via $AlCl_3$-mediated coupling polymerization. The synthesis reactions were performed using $AlCl_3$ in refluxing DCM as solvent. The chosen reaction solvent had a significant impact on the surface area and porosity of the resulting networks since it was demonstrated that DCM not only acts as an optimal solvent but also provides methylene crosslinks to some extent. Consequently, polymer networks derived from TPB and biphenyl monomers exhibited a greatly enhanced microporosity with high BET surface areas. Considering this strategy's outstanding characteristics, this work opens a new path for constructing highly microporous polymers with exceptional surface areas (Msayib and McKeown 2016).

2.3.3 Knitting Rigid Aromatic Building Blocks by External Crosslinkers

In 2011, Li et al. (2011) proposed a versatile and unique method to synthesize HCPs consisting of "knitting" rigid aromatic building blocks. In this new method, formaldehyde dimethyl acetal (FDA) is used as the external crosslinker, which represents a less toxic option since the only by-product derived from the reaction is methanol instead of the common chlorinated compounds. This strategy depicted in Figure 2.6 is based on Friedel–Crafts reaction, which involves the use of simple aromatic compounds as building blocks. In this first work, hyper-crosslinked networks with high surface area and predominantly microporous structure were synthesized by knitting various monomers like benzene, biphenyl, TPB, methylbenzene, chlorobenzene, and phenol. Variations of the molar ratio of FDA with respect to the monomer could effectively adjust the porous properties and the surface area of the resulting polymers from 897 to 1391 m^2/g for benzene-based networks and 195 to 815 m^2/g for biphenyl-based networks. The highest value of surface area achieved (~1400 m^2/g) corresponded to a crosslinker to benzene molar ratio of 3:1.

Many other aromatic monomers have been demonstrated to be viable precursors to prepare HCPs by following the "knitting" method. Among these, commercial triptycene (Dawson et al. 2011) and tetraphenylmethane (C. Zhang et al. 2015) 3D monomers were employed to obtain HCPs displaying high surface areas (1426 and 1470 m^2/g, respectively). A series of TPM-based networks were prepared by progressively increasing the relative amount of FDA during the synthesis reaction. A molar ratio of 16:1 FDA:TPM was found to be optimal as it indicated the highest surface area (1314 m^2/g), which is attributed to the extensive crosslinking and substitution of the aromatic rings (Errahali et al. 2014).

Aromatic heterocyclic monomers can also be used as building blocks to form HCPs by the knitting strategy, as demonstrated by Luo et al. (2012). They synthesized microporous heterocyclic polymers from thiophene, pyrrole, and furan under the typical reaction conditions (FDA, $FeCl_3$, in DCE). Despite having relatively low surface areas, these polymers showed superior adsorption selectivity and capacity due to heteroatoms' presence in the networks. Moreover, the knitting strategy has also been successfully applied to fused-heterocyclic rings (e.g., carbazole, triazine (Zhu et al. 2014), indole, benzothiophene, dibenzothiophene, benzofuran, dibenzofuran (Saleh et al. 2014), and tricarbazolyltriptycene (Zhai et al. 2016)), and to inorganic group polyhedral oligomeric

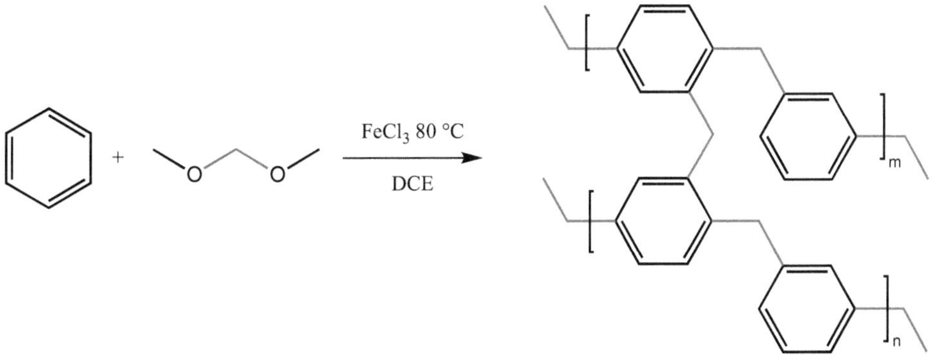

FIGURE 2.6 Scheme of the knitting strategy.

(Adapted with permission from B. Li, Gong, Wang, et al. 2011. Copyright (2011) American Chemical Society).

silsesquioxane (POSS) (S. Wang et al. 2015) to yield hyper-crosslinked networks with predominantly microporous structures.

Furthermore, there are other chemical crosslinking agents such as 1,4-dimethoxybenzene (DMB) (Tan et al. 2015), paraformaldehyde, glyoxylic acid (H. Wang et al. 2016), tetrachloromethane (D. Zhang et al. 2015), and dibromo-p-xylene (Bhunia, Banerjee, and Bhaumik 2015), which could be used in place of FDA in the knitting strategy. Several hyper-crosslinked resins were obtained by direct knitting of poly-p-phenylenes with diverse molecular weights in the presence of FDA and CCl_4 crosslinkers. The surface areas and pore volumes of both HCPs were detected to increase initially and then fall again as the molecular weight of the polymer precursors increased. However, employing FDA as an external crosslinker resulted in higher surface areas (up to 1019 m^2/g) (D. Zhang et al. 2015).

In the last years, with the introduction of mechanochemistry, it has been demonstrated that crosslinking can also be achieved through a solvent-free procedure (Grätz et al. 2019; Lee, Kurihara, and Horike 2020; Rubin Pedrazzo et al. 2020). This greener synthetic pathway afforded microporous HCPs with optimal surface areas up to 782 m^2/g and high yields for both Friedel–Crafts knitting and Scholl coupling reaction (Lee, Kurihara, and Horike 2020).

In general, the knitting strategy is considered to be a reliable and efficient polymerization method. Given the merits of using cost-effective reagents, mild reaction conditions, and the facility to integrate a wide variety of monomers with different functionalities, HCPs are provided with significant advantages for large-scale production and application.

2.4 STRUCTURE AND MORPHOLOGY OF HCPs

Crosslinked networks obtained from Friedel–Crafts alkylation reaction form a very rigid and firmly fixed amorphous skeleton topology. The complexity of the amorphous porous structure becomes challenging to create accurate simulations of the HCPs; consequently, it is pretty common to probe representative clusters that follow the experimental synthetic routes. The simulated polymerization starts with cages that contain monomers, followed by previously determined crosslinking commands until generating a network. Two factors are important to form microporous networks: to ensure the pores' formation and secure the network in the expanded state, which depends on the high degree of crosslinking (Huang and Turner 2018).

Understanding molecular polymeric construction at nanoscale formation based on the numerous synthetic methods for developing materials is quite tricky. However, investigation ligated to material morphology and how it affects the polymer properties, applications, and functionalities can be surpassed using computational approaches. Literature reports take simulations as valuable evidence to support the experimental data, showing that virtual synthesis leads to a deeper comprehension of the skeleton morphology and pore formation mechanism of HCPs (Kupgan, Liyana-Arachchi, and Colina 2016).

Generally, different types and amounts of monomers and crosslinkers and varying synthesis parameters lead to obtaining resins with notably distinct characteristics. The same principle could be applied for molecular optimization simulations of hyper-crosslinked polystyrene networks, showing that the steric hindrance and speed of crosslinking affect the final network arrangement. Abbott and Colina (2014) reported the formation of poly (styrene-co-vinyl benzyl chloride) HCPs prepared via virtual synthesis and using cross-linker content (VBC) ranging from 25 to 100%. At high crosslinking degrees, results show that resins obtain extraordinary porosity, a tendency that is in agreement with that obtained experimentally. Furthermore, simulations permit visualization of each stage during virtual synthesis due to the direct control in the degree of crosslinking adjusting the amount of VBC. VBC concentration tunes the network's structural properties, and by adjusting the density, it is possible to control the polymer swelling.

Further research by computational methodologies in polymers show in tetraphenylmethane (TPM)-based network, the possibility to predict isotherms that concord with low-pressure experimental results, showing a close description of the narrow microporous region but underestimated the mesoporous part (Errahali et al. 2014).

2.4.1 Nanoparticles

In recent years, the synthesis of uniform hyper-crosslinked nanoparticles has been achieved; their nanometric size and unique structure containing micropores provide extraordinary properties to be exploited in drug delivery (Liu et al. 2009) or gas sorption (Uemura et al. 2006). Li et al. (2010) used post-crosslinking emulsion to create microporous nanoparticles of poly(VBC-DVB) whose particle size can be tuned from 36 to 131 nm. Ouyang et al. (2013) obtained monodispersed microporous PS nanospheres with diameters below 190 nm. Using an unreactive coating of PS as a protective layer can be in-situ formed when hyper-crosslinking initiates to minimize the undesired inter-sphere crosslinking. Another approach suggests synthesizing polymers with controlled nano-morphologies through solvent-induced self-assembly strategy of poly (ethylene oxide)-b-polystyrene (PEO-b-PS) diblock copolymer (Gao et al. 2019).

PS was used as the main building block to synthesize amino-functionalized hyper-crosslinked polymer nanoparticles (AHCPNPs) for carbon dioxide uptake. These AHCPNPs were obtained via one-step hyper-crosslinking and subsequent ammonolysis reaction. AHCPNPs showed a core-shell structure with a specific surface area of 507.64 m^2/g, large surface-functionalized groups, and a size range between 63 and 153 nm (Pan et al. 2020). Branchi et al. (2018) used polymeric nanoparticles for vanadium redox flow batteries, synthesized via Friedel–Crafts alkylation and posterior post-functionalization. The resulting nanoparticles were in the order of 10–30 nm and exhibited a high surface area above 1000 m^2/g. Another interesting approach was reported by Guerritore et al. (2020). Mainly, they achieved styrene-based hyper-crosslinked nanocomposites (70–130 nm) by bulk polymerization using Friedel–Crafts reaction.

2.4.2 Hollow Capsules

Hollow capsules are attractive due to their hollow cavity inside that can be used as nanoreactors or as drug delivery materials because of their exceptional encapsulation efficiency to confine chemicals and prompt-controlled release (Sant, Poulin, and Hildgen 2008). Typically, the templating strategy is an extensively employed process to create the core shell, and then, the core is removed depending on the type of template used. Hard and soft templates usually employ silica nanoparticles, surfactants with small molecular sizes, and others (Tan and Tan 2017).

Morphine delivery was studied using functionalized hollow microporous organic capsules (HMOCs) with a co-monomer containing a carboxyl group that reinforces the nanoparticles' colloidal and chemical stability. The synthesis was carried out using an emulsion polymerization technique (Razzaque et al. 2017). Another research in the same field uses HMOCs for Doxorubicin delivery to cancer cells. This approach used silica as a core and was synthesized using the Stöber method; it also serves as a template covalent attachment for folic acid that further synthesizes conjugate with the surface of HMOCs. The spherical polymer particles (shell structure) were hyper-crosslinked utilizing Friedel–Crafts (Razzaque et al. 2020).

Razzaque et al. (2019) used hyper-crosslinked HMOCs as electrochemical sensors. Dithiocarbamate (DTC) functionalized HMOCs were prepared (HMOCs-DTC) by using silica nanoparticles as a template. SiO_2 nanoparticles also serve as an anchor for vinyl groups attached to the surface, acting as a seed in emulsion polymerization that produces the core-shell particles (SiO_2@PS-DVB-MMA). Then, the obtained core-shell particles were crosslinked in DCE by employing Friedel–Crafts reaction and removing silica core. The HMOCs-DTC were supported on a glassy carbon electrode, showing excellent stability, high surface area, appropriate microporous architecture, and sensitive sensing ability for Cu (II) ions.

2.4.3 2D Membranes

Polymeric membranes are an important research theme because of their excellent molecular level separation and high efficiency in the chemical industry (Cheng et al. 2018), gas separation membranes (Lau et al. 2015), and catalysis (Y. Wang et al. 2016). Membrane technology that does not involve a phase change has exhibited high energy efficiency benefits. Hou et al. (2020) demonstrated the possibility to control the size of HCP additives for better membrane separation. In this context, the molecular weight/particle size of HCPs was gradually increased while combining with PTSM and synthesized V125 liquid HCP polymer with growing amounts of crosslinker FDA and iron chloride as a catalyst in DCM. As a result, they obtained V125E. V125E, the most dispersive additive, produces membranes with higher selectivity and permeability.

Shi et al. (2020) created nanofilms with controllable hyper-crosslinked porous polymers with excellent gas storage ability (CO_2). They make an easy and innovative methodology to produce two-dimensional nanofilms with highly ordered porous arrays, combining covalent layer by layer self-assembly methodology followed by hyper-crosslinked treatment. As a result, nanofilms with different thicknesses exhibit BET surface area up to 605 m^2/g and better micro- and mesoporosity after hyper-crosslinking.

2.4.4 Monoliths

Porous monolithic materials are generally required in energy (Liu et al. 2021) and environmental applications (Chae et al. 2019). Monoliths contain as part of their precise structure macro-, meso-, and microporosity. The macropores inter-connected enhance mass transportation, whereas the micropores contribute as active sites for reactions that occur in the interface (Tan and Tan 2017).

Usually, hyper-crosslinked monoliths' synthesis starts with creating precursors that were originally synthesized. Then, the hyper-crosslinking procedure can be continued by the reported techniques such as post-crosslinking or crosslinking with external crosslinkers.

A model photothermal material should present the next properties: interconnected pores for water transport, light-absorbing structure, and insulating assets. Therefore, HCPs are promising candidates for photothermal materials because of their outstanding porous and physicochemical properties. For instance, Liu et al. (2021) firstly report the synthesis of monoliths for photothermal usage constructed by HCPs nanotube for solar steam generation. The monolith was fabricated using benzene and methylbenzene as building blocks by Friedel–Crafts reaction. Further softening by sodium alginate and poly-pyrrole results in the polymer named PPy-M-HCPs showing superhydrophobicity, high light absorption, and improved mechanical characteristics.

New technology such as 3D printing is used due to its versatility, high speed, and low cost. Fused deposition modeling is a printing operation that consists of melting thermoplastic polymeric filaments for further extrusion and discharge. Hock and Rose (2020) provided a new method to build structured monoliths comprising microporous materials to be used in catalytic reactors and adsorbers. The synthesis functionalizes the inner surface with HCPs; composite filaments are prepared for direct printing.

2.5 HCPs PROPERTIES

In recent years, HCPs have been considerably studied due to their remarkable characteristics. A significant number of studies have shown how varying certain aspects, such as the monomer nature, reagents, or reaction conditions during the preparation of HCPs, affect their resulting properties. HCPs have been demonstrated to be suitable for a great extent of fields owing to their 3D structures. Therefore, understanding the structure-property relationships plays an essential role in exploiting the capacity of these materials for important applications. In this section, more detailed considerations have been given to specific properties such as porosity, surface area, and swelling behavior of HPCs.

2.5.1 Polymer Surface

The surface allows the interaction with other compounds and provides enough points to form a bond. Surface properties are related to the nature of the material, porosity, chemistry, etc (Zidan, Greene, and Seyfoddin 2019). Besides, depending on the type of building block, the resulting HCPs can be hydrophilic, hydrophobic, or a combination of both, known as amphiphilic.

2.5.1.1 Hydrophilicity

Hydrophilic materials have a great affinity to interact with water, leading to solubility and swelling ability (Schmidt 2019). This type of polymer presents polar functional groups into their structure, such as hydroxyl (–OH) and carboxyl (–COOH). Nevertheless, the molecular weight (MW) and crosslinking degree play an essential role in the hydrophilicity property. For example, the hydrophilicity is reduced at a high MW and high-crosslinking degree (Deng et al. 2018).

HPCs can be obtained from various hydrophilic monomers. For instance, fluoride removal in water was treated with hydrophilic hyper-crosslinked porous polymers synthesized from monomers 2,2-biphenol and bisphenol-A. The whole resulting matrix was hydrophilic, and the hyper-crosslinking drastically changed the acidity of protonation sites in the material structure (Robshaw et al. 2020).

2.5.1.2 Hydrophobicity

Hydrophobic materials are those that, at the macroscopic surface, are not wetted by water. These polymers contain hydrocarbons chains or non-polar groups (Brash 1979). Some examples of hydrophobic monomer precursors for HCPs synthesis are styrene (St), VBC, and DVB, resulting in extremely hydrophobic HCPs (Wang, Ou, and Huang 2019). Such materials have a significant affinity to interact with hydrophobic aromatic compounds such as benzene and naphthalene. For instance, a hydrophobic HCP was synthesized from polystyrene-divinylbenzene. As a result of its hydrophobic nature, it interacts with analytes through π–π interaction and van der Walls' forces of the aromatic rings that make up the sorbent structure (Fontanals, Marcé, and Borrull 2010).

2.5.1.3 Amphiphilicity

Amphiphilic molecules are composed of both polar and non-polar regions. HPC materials can be synthesized using hydrophilic and hydrophobic precursors for adsorption capacity toward both hydrophobic and hydrophilic pollutants (Tu et al. 2020). Recently, a hyper-crosslinked porous cyclodextrin polymer (PBCD-B-D) with an amphiphilic character was prepared for organic micropollutants removal. The amphiphilic behavior was modulated by controlling the ratio of benzylation which changes the number of hydroxyl groups on cyclodextrin (Tu et al. 2020).

2.5.2 Porosity and Surface Area

Porosity is crucial during the characterization of surfaces with microstructures. The pore system is a communicating or non-communicating pores matrix that can be divided into three different porosity types (macroporosity, mesoporosity, and microporosity) (Issaadi et al. 2018). Figure 2.7 depicts the

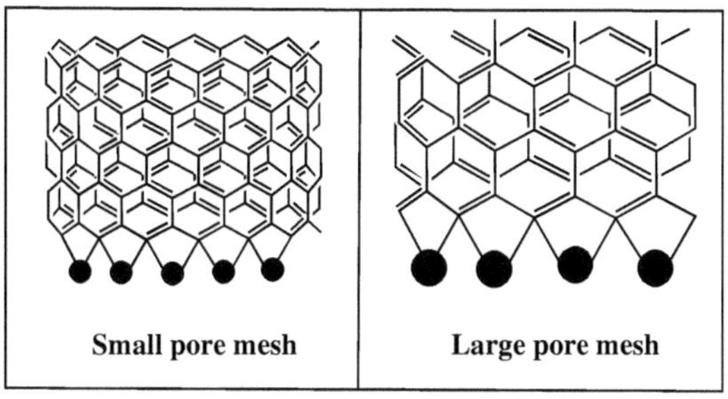

FIGURE 2.7 Representation of the pore size in two different meshes.

inner structure related to the size of the pores into a mesh. Generally, HCPs exhibit microporous and mesoporous networks. According to adsorption isotherms measurements, HCP resins show a considerable adsorption degree at low partial pressure ($p/p_0 < 0.1$), which is in agreement with the microporous structure. Depending on the form of the precursor polymer and the solvent employed for HCPs formation, the isotherms may also indicate specific features, at higher relative pressures, related to mesoporosity ($p/p_0 = 0.5$–0.9) and microporosity ($p/p_0 = 0.9$–1) (Silverstein, Cameron, and Hillmyer 2011).

In HCPs, the high degree of crosslinking leads to the formation of a microporous structure. As mentioned above, the synthesis of HCP resins is predominantly based on Friedel–Crafts alkylation, a reaction that allows rapid kinetics to create strong bonds, thereby forming a hyper-crosslinked network with predominantly porous structures (Tsyurupa and Davankov 2006).

Concerning surface area, HCPs belong to a novel generation of permanently porous polymers with high specific surface areas up to 2000 m^2/g (Bratkowska et al. 2010). Remarkable research even evidenced the synthesis of hyper-crosslinked siloxane-organic hybrid with ultrahigh surface area (~2500 m^2/g) and large pore volume (3.3 cm^3/g) (Chaikittisilp et al. 2011).

2.5.3 Swelling Behavior

Generally, crosslinked polymers in contact with solvents possess the ability to swell and do not dissolve (as shown in Figure 2.8). This property allows HCPs to be used in a vast number of applications, including ion exchange, removal of toxic pollutants, and solid-phase extraction (Wilson et al. 2017). Therefore, understanding the correlation of theory behavior with the real polymer is essential.

Several factors such as initial monomer concentration, the degree of monomer crosslinking, and the solvating state of the growing polymer chains affect the HCPs swelling. Polymer polarity can also alter the HCPs swelling, whereas it is independent of solvent properties because of weak interchain interactions in the network (Wilson et al. 2017).

For instance, hyper-crosslinked polystyrenes are capable of swelling in contact with both good (toluene) and poor solvents (water, methanol, and hexane). The volume of polystyrene resins tends to decrease in the order toluene > n-hexane > methanol > water. Figure 2.9 illustrates the PS structure and some solvents' structures such as 1-decanol, myristic acid, and certain alkanes, where the PS is soluble.

Binary solvents (acetone/hexane) vary the polymer swelling depending on solvents concentration (O'Kane and Sherrington 1990). In addition, the temperature influences the swelling of PS, which absorbs more solvent at lower temperatures than at higher temperatures (Bernardo and Vesely 2010).

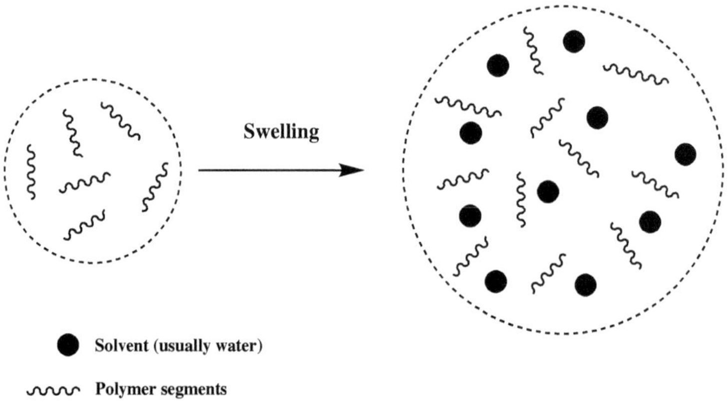

FIGURE 2.8 Swelling behavior of polymeric matrix.

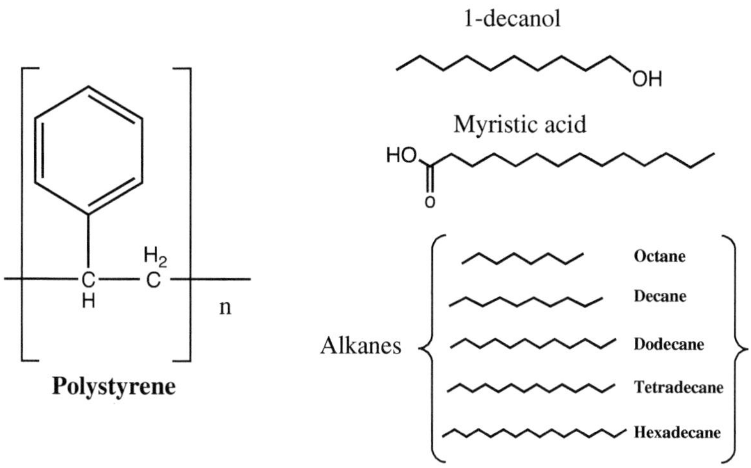

FIGURE 2.9 PS and some non-polar solvents.

This unique swelling behavior makes these materials promising candidates as adsorbents (Davankov and Tsyurupa 2010).

2.5.4 Thermomechanical Properties

Understanding and improving the thermomechanical properties of polymeric materials is essential for their future industrial applications. Regarding the polymerization method, as crosslinking increases, solvent resistance improves, cohesive strength and tensile properties improve (Everaerts and Clemens 2002).

Crosslinking classical theory fundamentally explains the loss of rubber elasticity during gradual crosslinking (Davankov and Tsyurupa 2010). It is because the chain length between two adjacent joints becomes shorter than the chain segment length, providing a certain degree of mobility. Therefore, a highly-crosslinked network with a short final distance between the joints should not swell in the solvent and remain glassy until its degradation temperature (Davankov and Tsyurupa 2010). Nevertheless, as mentioned in Section 2.5.3, HCPs have the ability to swell, which could be associated with unusually high network mobility. The improved mobility of HCP networks must also manifest itself under the action of external mechanical forces along with the temperature. Besides, the glass transition temperature (T_g) shifts quickly toward the thermal decomposition region as the degree of crosslinking increases.

The thermal stability of HCPs has been evaluated through TGA analysis. The results show no glass transition temperature below its temperature of decomposition (T_d) owing to the nature of their crosslinking structures. For example, the thermal analysis of hyper-crosslinked porous polyporphyrin (HCP-TCPP) and its corresponding monomer tetracarbazolylporphyrin (TCPP) were compared. HCP displayed excellent thermal stability with residual masses of 71.8% at 800°C, whereas the monomer suffered a 35% mass loss at around 635°C (Feng et al. 2017).

Regarding hyper-crosslinked polystyrene networks (HCP-PS), they are highlighted by their non-classical physical state since these networks belong neither to typical glassy materials nor to typical elastomers (Pastukhov, Tsyurupa, and Davankov 1999). Ranging the temperature from −70°C to 300°C, HCP-PS shows high deformation ability without glassy state characteristics. Moreover, these resins exhibit large values of deformations and reversibility of deformations, which are features associated with a rubberlike state. However, HCP-PS cannot be considered in this latter state because long heating or solvent treatments are required for the non-elastic part of deformation (Zhang, Shen, and Fan 2007).

2.6 FUNCTIONALIZATION OF HCPs

The performance and physicochemical properties of HCP materials can be enhanced by functionalizing their original network, which leads to broader applications such as biomedicine, catalysis, environmental remediation, and adsorption. The functionalization strategy is of particular interest because the resulting hydrophobic surface of HCPs is a disadvantage for the adsorption of polar compounds (Zhou, Man, and Huang 2018). Therefore, functionalized HCPs can be achieved through the incorporation of functional groups including hydroxyl (–OH), amine ($–NH_2$, –NH–), and carboxyl (–COOH) (Zeng et al. 2020), and heteroatoms including sulfur, nitrogen, boron, and phosphorous into the polymeric frameworks (Mohamed et al. 2020). Chemical strategies for modifying the HCP involve post-polymerization reactions or the employment of functional monomers, as schematized in Figure 2.10.

Regarding post-modification reactions, these can be performed before or after the hypercrosslinking process (Fontanals et al. 2015). Some of the most common functional groups introduced on the polymer surfaces are amine ($–NH_2$, –NH) and its derivatives which promote the acid–base interaction with acidic compounds (Shao and Huang 2017). Moreover, hydrogen bonding makes a significant contribution in improving the adsorption of polar aromatic compounds (Zhang et al. 2018). For example, DVB was employed as the precursor for the preparation of an HCP network (PDVBpc) by suspension polymerization, followed by Friedel–Crafts alkylation in the presence of $FeCl_3$ as a catalyst. Afterward, the addition of tetraethylenepentamine (TEPA) leads to the formation of a functional HCP where amines groups successfully contribute to the adsorption of CO_2 (Liu et al. 2018).

Macroporous low-crosslinked chloromethylated polystyrene (MP-CMPS) is one of the most widely used monomers for HCPs synthesis (Wang, Mao, and Huang 2018). As a result of the presence of chloromethylated residues, this type of polymer lacks active sites to some extent. Therefore, chloromethyl residues can be substituted via nucleophilic substitution by functional groups. For instance, HCPs were prepared via Friedel–Crafts reaction, using MP-CMPS as the starting material and $FeCl_3$ as the catalyst. The resulting polymers undergo a nucleophilic substitution where

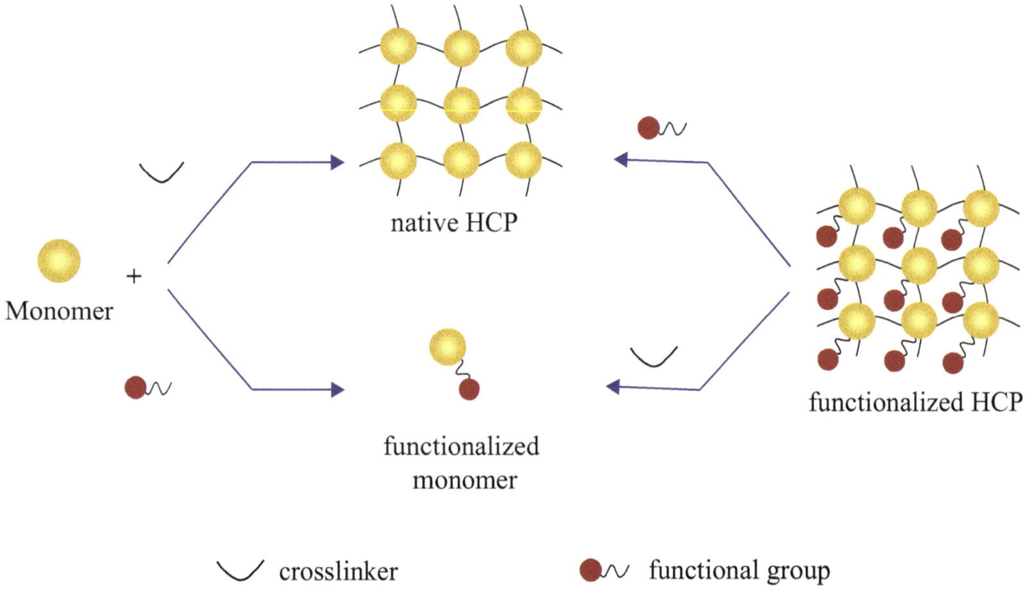

FIGURE 2.10 Schematic representation of functionalization methods of HCPs through post-polymerization reaction (top) and the employment of functional monomers (bottom).

the residual chlorine was replaced by methoxy and phenoxy groups. This functionalization makes possible the enhancement adsorption of salicylic acid from an aqueous medium (Zhou, Man, and Huang 2018).

However, since the HCPs network is composed of aromatic rings lacking reactive sites, it is often difficult to modify HCP materials through post-polymerization (Su et al. 2019). Therefore, another suitable method for HCPs functionalization involves the direct synthesis of precursor monomers that already contain polar functionalization. Polar compounds such as methylacrylate (MA), vinylimidazole (VIM), ethylene glycol dimethyl acrylate (EGDMA), and vinylpyrrolidone (VP) are widely used as functional monomers (Shao and Huang 2017; Zhang et al. 2018).

A recent study reported the synthesis of HCPs from low-crosslink chloromethylated polystyrene, the same monomer used by Zhou et al. However, in this investigation, the monomer was previously functionalized for efficient aniline adsorption capacity. The functionalized monomer employed was synthesized by the nucleophilic substitution under an alkaline medium, where the chlorine atoms of the MP-CMPS were substituted for the hydroxyl groups of hydroquinone. After, Friedel–Crafts alkylation was performed employing FDA as the crosslinker. The adsorption reached is explained by the interaction of the oxygen content (13.04 wt.%) of the resulting HCPs with aniline by means of hydrogen bonding, and acid–base interaction (Zeng et al. 2020).

The functionalization of the HCP resins with nanostructures has also attracted increasing attention. Indeed, hyper-crosslinked nanocomposites with carbon nanotubes (Castaldo et al. 2017), graphene sheets and functionalized graphene oxide (Zhao et al. 2015), and fullerenes (Volkova, Summanen, and Smirnov 2003) have been reported. This surface functionalization allows obtaining composites that combine HCP resins and inorganic supports characteristics, suggesting promising applications (Stepacheva et al. 2020).

2.7 CHARACTERIZATION OF HCPs

The characteristics of HCPs can be investigated by using various techniques. For composition and structural characterization, Fourier Transform Infrared (FTIR) Spectroscopy, solid-state Nuclear Magnetic Resonance (ssNMR), Energy Dispersive X-rays Spectroscopy (EDX), and elemental analysis are powerful techniques employed; morphological analysis can be explored by both Scanning Electron Microscopy (SEM) and high-resolution Transmission Electron Microscopy (TEM) techniques. Besides, several methods can be used to acquire porosity information, such as gas adsorption, mercury porosimetry, size-exclusion chromatography (SEC), and inverse size-exclusion chromatography (ISEC). Swelling and thermogravimetric analysis (TGA) are also generally used for the characterization of HCP. As follows, we detailed a brief description for each one.

2.7.1 Compositional and Structural Characterization

FTIR Spectroscopy is utilized to elucidate the functional groups present in the molecular structure in HCPs. The basis of FTIR is that each molecule of the functional group tends to vibrate at specific wavelengths of light. One of the advantages of this analysis is that it can prove and monitor the monomers incorporated in the polymerization or in the post-synthetic treatments applied to HCP resins (Fontanals et al. 2015).

FTIR was used to compare the styrene-based hyper-crosslinked resin structure before and after functionalization with multi-walled carbon nanotubes. The intensity of the peak at 1270 cm^{-1} attributed to the chloromethyl groups dramatically decreased after the hyper-crosslinking which proved the changes in structure because of the decreasing of C–Cl group as a result of its consumption that gives rise to the functionalized network (Subri et al. 2018).

Another of the most widely used techniques for materials' structure characterization is solution NMR spectroscopy. However, in highly porous polymers, it is impossible to use this technique due to a large number of carbon and hydrogen atoms (Law and Sherrington 1998). Therefore, ssNMR is a suitable technique that plays a fundamental role to provide structural and dynamic information of polymers such as structure, molecular mobility, bonding interaction, and crosslinking degree.

The signals depicted by ssNMR are broad due to the anisotropic interactions in solid samples. In order to eliminate this issue, it is essential to modulate the mechanical spinning of the sample at the angle of 54.741 with respect to the magnetic field, a phenomenon known as magic angle spinning (MAS). Solid-state cross-polarization (CP) improves the signal-to-noise (S/N) ratio of ssNMR by transferring polarization from abundant spin (^{1}H) to dilute spin (^{13}C) nuclei (Jarrells et al. 2020). It provides essential information such as molecular interactions, polymorphism, and chemical compositions of the molecules (Qiu and Ben 2015). Therefore, ^{1}H–^{13}C cross-polarization magical angle spinning (CP/MAS) NMR techniques are usually applied for a reliable study of HCPs.

Most HCPs are composed of aromatic carbons which are usually seen at around 120–140 ppm in the spectrum. ssNMR is a very practical technique to determine alterations in the polymer structure when functional groups are incorporated. For example, G. Liu et al. (2015) prepared different types of hyper-crosslinked networks by the crosslinking of benzene/biphenyl/1,3,5-triphenylbenzene via Friedel–Crafts reaction. The ^{13}C CP-MAS NMR gave signals at 139 and 130 ppm corresponding to aromatic carbons. The signal near 35 ppm was assigned to the methylene indicating the presence of the monomer moiety within the HCP.

Characterization can be complemented by EDX, and elemental analysis to determine the elemental composition of the materials. The principle behind EDX is the employment of a beam of electrons in which the atoms within the sample are excited, producing wavelengths of X-rays which are characteristics of each element. On the other hand, elemental analysis involves the combustion of the samples, followed by the analysis of the resultant products in order to determine the CHNS/O composition. Gao et al. (2016) performed both analyses to compare the composition of the obtained HCP and its raw materials. Results confirmed the incorporation of sulfur and chlorine, heteroatoms that enhanced gas adsorption capability.

2.7.2 Morphological Characterization

SEM and TEM analysis are versatile techniques used to observe the morphological features of the obtained polymers. Both techniques are based on the use of a higher-energy electron beam to construct nanoscale images. SEM principle is based on the radiation of an electron beam to the sample to generate detailed surface information. Generally, the SEM approach has a resolution between 1.5 and 20 nm. Otherwise, TEM images are obtained when the electron beam penetrates and interact with the sample structure. This analysis offers an enhanced resolution of up to 0.1 nm.

Despite the advantages of the SEM and TEM techniques, the results obtained are only a 2D projection of the surface. Then, in order to obtain a better analysis of the morphological features of materials, SEM is combined with novel tools such as Focused Ion-Beam (FIB) and Serial Block-Face (SBF) to generate 3D imaging (Alvarez et al. 2019). In porous materials, FIB-SEM and SBF-SEM are two potential approaches that provide more detailed information on porosity and its distribution at the micro and nanoscale.

Morphological analysis is very crucial for HCPs because of their porosity and different reported morphologies such as nanoparticles, hollow capsules, 2D membranes, and monoliths (Section 2.4). Indeed, Müllner et al.(2012) reported the morphological structure of a poly(styrene–divinylbenzene) (PS–DVB) monolith based on SBF-SEM analysis, which revealed a macropore space with two pores domains with lengths of 7.32 and 0.73 μm.

2.7.3 Porosity and Surface Area Analysis

HCPs have attracted great interest in research because of their porous structure that can be exploited in potential applications. Therefore, the study of the pore structure and surface area is a very important task. Despite the variety of techniques for analyzing porous structures, only a few can be applied to HCPs since their networks can be flexible, and thus, the porous structure can change (Tsyurupa and Davankov 2006). The method most used for the measure of these parameters is the gas adsorption method which involves measuring the amount of gas adsorbed on the sample surface. Generally, the gas most utilized for the analysis is N_2 at cryogenic temperature (77 K), although adsorptives such as CO_2, Ar, He, and CH_4, are also employed (Taylor et al. 2020).

For HCPs, several studies have been done with the aim of understanding the porous structure of the synthesized porous materials. Indeed, N_2 adsorption/desorption analysis for porous hyper-crosslinked polystyrene sorbents has revealed isotherms type IV, which is characteristic of conventional rigid mesoporous materials (Davankov and Tsyurupa 2010; Gatti et al. 2019). Lee et al. (2006) and Zhang et al.(2014) synthesized hyper-crosslinked resins, where the obtained isotherms were classified as type I, which is in agreement with the obtained microporous networks. Other researchers have also reported isotherms nearly close to type II evidencing resin where micropores are predominant, and the macropores are also present (Peng et al. 2020; Castaldo et al. 2017).

Nitrogen adsorption isotherm is the most widely used method for determining the total pore volume (V_p) of meso- and microporous adsorbents (Sellergren and Hall 2001). Specifically, V_p is derived from the assumption that adsorbent pores are filled with the liquified adsorbate (Tsyurupa and Davankov 2006), that is, at a relative pressure close to unity ($0.95 < p/p_0 < 1$).

Information related to the surface area is also obtained from the adsorption isotherms using mathematical models such as BET and Langmuir theories. For the same sample of porous material, the values provided by these theories may differ by as much as 30% (Tsyurupa and Davankov 2006).

BET is the most common method used to determine the surface area of porous solids from nitrogen adsorption isotherms (Davankov and Tsyurupa 2010). It is based on the correlation of the amount of adsorbed gas by the adsorbent with its surface area (Ambroz et al. 2018). This model extended the monolayer physical adsorption theory (Langmuir model) to a multilayer adsorption theory. Despite the extensive use of the BET method, some researchers have discussed the limitations for its application to microporous materials unless certain conditions are considered (Zou, Fan, and Liu 2020; Ambroz et al. 2018). Porous surface area can also be reported based on the Langmuir model, which, as was mentioned, assumes the formation of monolayer adsorption.

The gas adsorption method also provides information about PSD. To assess the PSD of porous materials, density functional theory (DFT), and non-local density functional theory (NLDFT) can be employed (Taylor et al. 2020). These methods conduct a more detailed and accurate study of the PSD of mesoporous and microporous materials because of their rigorous theoretical basis. Through these DFT methods, it is possible to customize different adsorbates, materials, and pore morphologies (Landers, Gor, and Neimark 2013).

Moreover, there are other approaches for porosity characterization, such as mercury porosimetry, SEC, and ISEC. Mercury porosimetry is the standard technique used to analyze mesopores and macropores. This method is based on the penetration of mercury into the pores as a function of the applied pressure. It covers pores at the usual high pressure of 200 MPa (mesopores) to at atmospheric pressure (macropores) (Rodríguez-Reinoso and Sepúlveda-Escribano 2001).

SEC is a technique used for separating substances according to their molecular size or hydrodynamic volume (Moldoveanu and David 2013). It can be used to measure both the size and the polydispersity of a synthesized polymer; that is, it can find the distribution of the sizes of polymer molecules (Giridhar, Manepalli, and Apparao 2017). Moreover, ISEC determines the PSD of a column packing material by monitoring the residence times of solutes of varying molecular diameters (Goto and McCoy 2000).

2.7.4 Other Analysis

Swelling tests are performed to measure the swelling ratio (SR) of resins. The theory of swelling of polymer gels assumes that the network immersed in a liquid attains the thermodynamic equilibrium with the surrounding solution after a finite time (Sakar et al. 2007). The swelling ratio in percentage is calculated through the following equation (Eq. 2.1):

$$\text{Swelling}(\%) = \left(\left(A_1 - A_2\right)/A_1\right) \times 100 \quad (2.1)$$

where A_1 is the initial weight (dry polymer), meanwhile A_2 corresponds to the swollen polymer.

Moreover, several solvents are used for swelling tests, including (i) polar solvents and (ii) non-polar solvents. Polar solvents have a relevant influence on the covalent structure of polar molecules. Such can be divided into protic and aprotic. Polar protic solvents are molecules that have opposite poles, meaning negative and positive charges. These solvents present covalent bonds such as O–H, N–H, which readily form hydrogen bonds (H-bonds). Therefore, polar protic solvents have either oxygen or nitrogen directly attached to a hydrogen atom. Some common examples are water (H_2O), methanol (MeOH), ethanol (EtOH), and isopropanol (Kappert et al. 2019). Moreover, polar aprotic solvents are still polar solvents, with a charge separation in the molecule. The main feature is that they will not be able to donate hydrogen for an H-bond. Examples of aprotic solvents are acetone, methyl ethyl ketone, and ethyl acetate (EA) (Kappert et al. 2019).

Otherwise, non-polar solvents lack partial charges due to similar electronegativities in their bonds because there are no exciting functional groups. Alkanes or perfluoroalkanes are probably the most typical representatives of such solvents (Klamt 2005). These solvents are ideal for hydrophobic molecules with any polarity.

TGA analysis is also another commonly used technique for the characterization of HCP materials. This technique provides important information such as thermal stability, molecular structure, and crosslinking degree of structures (Rodríguez-Baeza, Neira, and Aguilera 2003). It is based on the change in mass (loss) of samples as temperature increases. It is well-known that thermal stability increases with the crosslinking degree. Therefore, HCPs are characterized by remarkable thermal stability (Sadak 2021).

2.8 APPLICATIONS

In recent years, the development of effective synthesis and functionalization strategies of HCPs have been intensively studied because of their promising potential in numerous applications, including gas storage capacity, environmental remediation, catalysis, drug delivery, sensing, and others.

2.8.1 Storage Capacity

Design and development of materials with appropriate adsorption properties to storage gases is of great interest to promote renewable energy technologies. Among the porous organic systems, HCPs have received significant attention for gas storage such as hydrogen and methane, and CO_2 capture.

2.8.1.1 Storage of Hydrogen

In an effort to seek low/non-polluting fuels to replace fossil fuels, hydrogen emerges as an efficient alternative. The use of hydrogen as a fuel has several benefits like its availability and a large amount of produced energy. Another advantage of hydrogen is that it is a clean source releasing only water as a by-product. Thus, hydrogen is an environmental-friendly alternative that offers the potential to reduce or eradicate carbon emissions from fossil fuels. Despite the numerous studies to develop adsorbent materials for hydrogen storage, safe storage is still a huge challenge (Tan and Tan 2017). One of the approaches to store hydrogen consists of using physisorption with porous polymers that possess large storage capacity and reversible sorption (Tan and Tan 2017). Surface area is a property that plays a crucial role in the adsorption behavior because a greater specific surface area indicates a high hydrogen adsorption capacity.

Properties relevant to hydrogen storage were evaluated from a microporous hyper-crosslinked polystyrene resin. The polymer displayed the capacity for adsorbing hydrogen up to 3 wt.% at 77 K and a pressure of 15 bar, representing a hydrogen uptake value of approximately 60% higher compared to other reported high-performance organic polymers at that moment (Lee et al. 2006). Several functionalities have been introduced into hyper-crosslinked materials to enhance the hydrogen adsorption properties (Fu et al. 2018).

2.8.1.2 Storage of Methane

One of the fossil fuels that emit less-polluting gases than coal or oil is natural gas. It is primarily composed of methane, which has many qualities that make it an alternative future fuel because of its availability, low cost, and relatively clean combustion (Tan and Tan 2017). Methane requires the design and development of a highly efficient storage strategy. Thus, the thermal stability of HCP materials, combined with their large surface area, cost-effectiveness, and easy-to-scale synthesis, offers promising potential for fuel gas storage (Errahali et al. 2014).

In order to test the methane adsorption capacity, a variety of HCPs were synthesized by Friedel–Crafts. Methane uptakes were evaluated at different pressure conditions (15, 20, and 36 bar), achieving the maximum CH_4 uptake of 5.2 mmol/g at 298 K and a pressure of 20 bar. The adsorption ability of these types of HCPs is similar to other materials with microporosity (Wood et al. 2008). To date, the design of porous materials suitable for methane storage, meeting all US Department of Energy (DOE) requirements, remains a big challenge (Rodríguez-Reinoso and Silvestre-Albero 2019).

2.8.1.3 CO_2 Capture

Increasing atmospheric CO_2 levels have attracted growing attention to find new clean energy sources as well as CO_2 capture technologies. POPs are suitable materials for CO_2 capture because of their characteristics such as low density, large surface area, thermal stability, tunable pore size and structure, and easy functionalization (Wang, Zhou, and Yuan 2017). In this context, HCPs have been widely studied as adsorbents to reduce CO_2 concentration.

For example, CO_2 capture was tested in different types of MOP networks. Regarding the obtained HCP, it was synthesized by crosslinking tetraphenylmethane using FDA in the presence of $FeCl_3$. It evidenced CO_2 adsorption of 2.95 mmol/g (273 K and 1 bar), which represents a measured uptake lower than a "clicked" tetrahedral network (3.86 mmol/g) but higher than a non-functional network (2.65 mmol/g) even though the surface area of the latter was greater (4077 m^2/g) (Dawson et al. 2011).

Interestingly, other researchers have evaluated CO_2 adsorption and its consequent conversion into organic compounds. An example is the synthesis of imidazolium salt-modified porous HCPs, where the maximum CO_2 uptake exhibited is 14.5 wt.% at 273 K and 1 bar. These functionalized materials showed high activities to convert CO_2 into cyclic carbonates, perhaps due to the synergistic effect of the salt used and the microporosity (J. Wang et al. 2015).

2.8.2 Environmental Remediation

In recent years, the development of new materials to remove biological contaminants and organic and inorganic pollutants that cause significant environmental damage and human impact has attracted a great deal of attention. Porous materials offer the potential for efficient removal because of their high surface areas, and incorporation of functional sites (Samanta et al. 2019). HCPs adsorbents containing a hydrophobic surface have a strong sorption affinity for substances with low polarity, such as volatile organic compounds in the gas phase or organic pollutants from aqueous solutions (Schute and Rose 2015).

Davankov-type hyper-crosslinked nanoparticles with selective adsorption capability were prepared. The nanoparticles coated solid-phase microextraction (SPME) showed potential adsorption toward organic pollutants including highly hydrophobic alkanes, benzene molecules, among others.

Due to the efficient adsorption performance, the HCP-coated fibers were applied to determine polycyclic aromatic hydrocarbons in water samples giving satisfactory results (S. Liu et al. 2015).

In recent research, Samanta et al. (2017) explored the behavior of an HCP for the entrapment of cationic dyes (i.e., methylene blue (MB), crystal violet (CV), and rhodamine B (RB)) from an aqueous medium. Alkoxide-based HCP (HCP-91) synthesized through Friedel–Crafts reaction was modified with the incorporation of sodium cations (HCP-91@Na) through NaOH treatment for inducing ion exchange. Results showed efficient selective adsorption of cationic dyes because of the ionic selectivity. Furthermore, adsorption times are dependent on the size of the cationic dyes, that is, MB took the least time, and RhB takes the most time to complete removal from water.

Removal of organic pollutants and heavy metal ions have been also investigated using HCPs as adsorbents. For the synthesis, an HCP from 2-hydroxyterephthalic acid (2-HTA) was produced via Friedel–Crafts reactions. 2-HTA was employed to improve the polymer's adsorption selectivity due to –OH and –COOH groups. The resulting polymer exhibited excellent adsorption capacity of organic contaminants such as amines, dyes (RB and MO), and phenols. Moreover, the tests for uptaking of metal ions including Ni^{2+}, Cu^{2+}, Cd^{2+}, and Pb^{2+} showed a higher potential (even 178.27 mg/g for Pb^{2+}) compared with other frequently used adsorbents, such as chitosan and activated carbon (Yang et al. 2021).

2.8.3 Heterogeneous Catalysis

Heterogeneous catalysts have unique features such as reactivity, stability, easy separation, robustness, and good recyclability that convert them into materials of particular interest in research laboratories and industrial applications. It is estimated that 90% of all chemical processes occur on the surface of heterogeneous catalysts (Lu, Murray, and Zhu 2018). Advances in heterogeneous catalysis have demonstrated the potential of HCPs as superior supports for heterogeneous catalysis because of their permanent porosity (S. Liu et al. 2015).

The design and synthesis of HCPs as catalysts have undergone rapid growth. For instance, a practical approach to introduce active sites into the hollow sphere structures is to confine the metal core to the inner space to form a yolk-shell structure. This structure provides a nanoreactor system that can be used in shape-selective catalysis. Therefore, the catalytic decomposition of cyclohexyl hydroperoxide (CHHP) was tested through the synthesis of a gold yolk-shell type polymer (Au@PON). The results showed the effectiveness of the catalytic behavior because in the absence of the catalyst, the conversion of CHHP was only 15%, whereas, with the use of Au@PON, 90% conversion was reached (Shi et al. 2014).

2.8.4 Drug Delivery

Hollow microporous organic capsules (HMOCs), containing a porous and hollow morphology, are considered potential materials for drug delivery since the introduction of microporosity in the shells enhances their ability to encapsulate guest materials. Li et al. (2013) introduced a method to prepare HMOCs with HCPs shells. The applicability of these materials was tested through *in vitro* drug loading and release studies employing ibuprofen (IBU). The HMOCs showed high efficiency for drug loading in IBU hexane solution (1.68–2.04 g of ibuprofen/g) and presented a zero-order kinetic model indicating a constant drug release rate. In another study, hollow microporous organic nanorods (HMORs) were also synthesized to assess their efficacy in drug release since cells could preferentially absorb particles with these morphologies. Experimental results showed that, for this material, drug release is mostly determined by simple diffusion. Also, HMOCs can uptake 1.54 g/g (ibuprofen/HMOR) (Li, Jin, and Tan 2016). The introduction of desired functional ligands into the microporous organic capsules to improve the loaded drugs' bioavailability has also been studied. Recently, a folate-functionalized biocompatible HMOC has been synthesized with moderate potential for doxorubicin targeted delivery to cancer cells (Razzaque et al. 2020).

2.8.5 Sensing

HCPs have also been investigated as sensor substrates. Jiang, Fei et al. (2014) successfully prepared humidity-sensitive composites based on LiCl-loaded polymers. The porous polymers were synthesized via Friedel–Crafts alkylation since the cross-linking could improve the stability of the sensors. The impedances of humidity sensors based on pure and LiCl-loaded polymers were obtained for comparison. The pure polymer sensor did not show response until 50% of relative humidity. By contrast, the optimal sensor based on 4 wt.% LiCl-loaded polymers displayed excellent sensing properties. The impedance of this sensor decreased almost four orders of magnitude in the whole humidity range (11–95%). In another work, the porous polymer was treated with lithium hydroxide instead of LiCl salt. This new modification significantly increased the humidity sensitivity of the resulting lithium-modified polymers (Jiang, Kuang et al. 2014).

2.8.6 Other Applications

HCP materials are also being investigated for energy storage applications because of their high surface areas, their highly porous structures, and their framework with extended π-conjugated units. Different kinds of HCPs have been synthesized from versatile building blocks such as tetraphenylanthraquinone (Abuosoaud et al. 2020), tetraphenylethene, and triphenylamine derivatives (Mohamed et al. 2020) for high-performance capacitors. These materials exhibited excellent supercapacitor performance and cycling stability attributed to their high surface areas and the heteroatoms incorporation in their frameworks. For instance, Mohamed et al. (2020) synthesized a multifunctional HCP based on tetraphenylethene (TPE). According to electrochemical measurements, the resulting material showed a high specific capacitance with exceptional cycling stability. Therefore, the resulting TPE-HCP could be considered as a good material for energy storage.

Furthermore, chromatography is another field in which HCPs have been used as the stationary phase. Lu et al. (2016) fabricated a porous layer open tubular (PLOT) column for gas chromatography using a benzene knitting polymer (KAPs-1) for the first time. The resulting KAPs-1 coated capillary column showed a nonpolar nature with a high column efficiency for *n*-dodecane of 7769 plates/m. Moreover, the fabricated column exhibited high separation performance for a wide range of volatile organic compounds (VOCs), that cannot be resolved even by other stationary phases, since their close matching boiling points and molecular sizes represent a challenge. Despite the progress achieved in these materials, further research is mandatory to explore new areas of applications.

2.9 CONCLUSION

Unique properties of HCPs have drawn significant attention since the pioneering report in 1974 by Davankov. HCPs can be fabricated by three main synthetic approaches, that is, post-crosslinking of polymer precursors, self-condensation of monomers, and knitting with external crosslinkers. Furthermore, novel approaches have been developed for the proper surface functionalization of HCPs by employing a wide range of specific functional groups and monomers. The different strategies used to produce HCPs provide control over the polymer morphology and pore structure, including nanoparticles, hollow capsules, 2D membranes, and monoliths. These permanent microporous polymers possess fascinating features, including thermal stability, high adsorption capacity, swell ability in both good solvents and non-solvents, and their high-crosslinking degree generates specific surface areas as high as 2000 m^2/g. The combination of these fascinating properties leads to multifunctional applications. Gas storage of hydrogen and methane, CO_2 capture, environmental remediation, heterogeneous catalysis, drug delivery, and sensing are some of the most remarkable research areas where HCPs have been applied. It can be envisioned that in the near future, progress in green synthesis and functionalization of HCPs is the key component in the development of new materials with enhanced properties.

REFERENCES

Abbott, Lauren J., and Coray M. Colina. 2014. Formation of Microporosity in Hyper-Cross-Linked Polymers. *Macromolecules* 47 (15): 5409–15.

Abuosoaud, Mohamed Gamal, Xian Zhang, Tharwat Mansoure, Ahmed El-Mahdy, Chih-Feng Huang, Martin Danko, Zhong Xin, and Shiao-Wei Kuo. 2020. Hypercrosslinked Porous Organic Polymers Based on Tetraphenylanthraquinone for CO_2 Uptake and High-Performance Supercapacitor. *Polymer* 205: 122857.

Ahn, Jou-Hyeon, Jin-Eon Jang, Chang-Gun Oh, Son-Ki Ihm, Jamie Cortez, and David Sherrington. 2005. Rapid Generation and Control of Microporosity, Bimodal Pore Size Distribution, and Surface Area in Davankov-Type Hyper-Cross-Linked Resins. *Macromolecules* 39: 627–32.

Alvarez, Juan, Giovanni Saudino, Valentina Musteata, Poornima Madhavan, Alessandro Genovese, Ali Reza Behzad, Rachid Sougrat, Cristiana Boi, Klaus Viktor Peinemann, and Suzana P. Nunes. 2019. 3D Analysis of Ordered Porous Polymeric Particles Using Complementary Electron Microscopy Methods. *Scientific Reports* 9 (1): 13987.

Ambroz, Filip, Thomas J. Macdonald, Vladimir Martis, and Ivan P. Parkin. 2018. Evaluation of the BET Theory for the Characterization of Meso and Microporous MOFs. *Small Methods* 2 (11): 1800173.

Bai, Feng, Bo Huang, Xinlin Yang, and Wenqiang Huang. 2007. Synthesis of Monodisperse Porous Poly(Divinylbenzene) Microspheres by Distillation-Precipitation Polymerization. *Polymer* 48 (13): 3641–49.

Bernardo, Gabriel, and Drahosh Vesely. 2010. Anomalous Swelling of a Polystyrene Matrix in Organic Solvents. *Journal of Applied Polymer Science* 115 (4): 2402–8.

Bhunia, Subhajit, Biplab Banerjee, and Asim Bhaumik. 2015. A New Hypercrosslinked Supermicroporous Polymer, with Scope for Sulfonation, and Its Catalytic Potential for the Efficient Synthesis of Biodiesel at Room Temperature. *Chemical Communications* 51 (24): 5020–23.

Branchi, Mario, Matteo Gigli, Barbara Mecheri, Francesca Zurlo, Silvia Licoccia, and Alessandra D'Epifanio. 2018. Highly Ion Selective Hydrocarbon-Based Membranes Containing Sulfonated Hypercrosslinked Polystyrene Nanoparticles for Vanadium Redox Flow Batteries. *Journal of Membrane Science* 563 (October): 552–60.

Brash, J. L. 1979. Hydrophobic Polymers as Materials for Interfacing with Blood. In Xubiao Luo and Fang Deng (Eds.), *Assisted Circulation*, 506–19. Springer, Berlin Heidelberg.

Bratkowska, D., N. Fontanals, F. Borrull, P. A.G. Cormack, D. C. Sherrington, and R. M. Marcé. 2010. Hydrophilic Hypercrosslinked Polymeric Sorbents for the Solid-Phase Extraction of Polar Contaminants from Water. *Journal of Chromatography A* 1217 (19): 3238–43.

Castaldo, R., R. Avolio, M. Cocca, G. Gentile, M. E. Errico, M. Avella, C. Carfagna, and V. Ambrogi. 2017. Synthesis and Adsorption Study of Hyper-Crosslinked Styrene-Based Nanocomposites Containing Multi-Walled Carbon Nanotubes. *RSC Advances* 7 (12): 6865–74.

Chae, Ji Ae, Yuree Oh, Hea Ji Kim, Go Bong Choi, Kyoung Min Lee, Doyoung Jung, Yoong Ahm Kim, and Hyungwoo Kim. 2019. Preparation of Compressible Polymer Monoliths That Contain Mesopores Capable of Rapid Oil–Water Separation. *Polymer Chemistry* 10 (37): 5142–50.

Chaikittisilp, Watcharop, Masaru Kubo, Takahiko Moteki, Ayae Sugawara-Narutaki, Atsushi Shimojima, and Tatsuya Okubo. 2011. Porous Siloxane–Organic Hybrid with Ultrahigh Surface Area through Simultaneous Polymerization–Destruction of Functionalized Cubic Siloxane Cages. *Journal of the American Chemical Society* 133 (35): 13832–35.

Cheng, Xi Quan, Zhen Xing Wang, Xu Jiang, Tingxi Li, Cher Hon Lau, Zhanhu Guo, Jun Ma, and Lu Shao. 2018. Towards Sustainable Ultrafast Molecular-Separation Membranes: From Conventional Polymers to Emerging Materials. *Progress in Materials Science* 92: 258–83.

Côté, Adrien P., Annabelle I. Benin, Nathan W. Ockwig, Michael O'Keeffe, Adam J. Matzger, and Omar M. Yaghi. 2005. Chemistry: Porous, Crystalline, Covalent Organic Frameworks. *Science* 310 (5751): 1166–70.

Das, Annesha, Priyanka Ghosh, Saibal Ganguly, Dipali Banerjee, and Kajari Kargupta. 2018. Salt-Leaching Technique for the Synthesis of Porous Poly(2,5-Benzimidazole) (ABPBI) Membranes for Fuel Cell Application. *Journal of Applied Polymer Science* 135 (5): 45773.

Davankov, V. A., S. V. Rogoshin, and M. P. Tsyurupa. 1973. Macronet Polysterene Strucrures for Ionites and Method of Producing Same. *United States Patent Office*, issued 1973.

Davankov, V. A., S. V. Rogoshin, and M. P. Tsyurupa. 1974. Macronet Isoporous Gels through Crosslinking of Dissolved Polystyrene. *Journal of Polymer Science: Polymer Symposia* 47 (1): 95–101.

Davankov, V. A., and M. P. Tsyurupa. 1990. Structure and Properties of Hypercrosslinked Polystyrene-the First Representative of a New Class of Polymer Networks. *Reactive Polymers* 13 (1–2): 27–42.

Davankov, Vadim, and Maria P. Tsyurupa. 2010. *Hypercrosslinked Polymeric Networks and Adsorbing Materials: Synthesis, Properties, Structure, and Applications*. Elsevier.

Dawson, Robert, Andrew I. Cooper, and Dave J. Adams. 2012. Nanoporous Organic Polymer Networks. *Progress in Polymer Science* 37 (4): 530–63.

Dawson, Robert, Ev Stöckel, James R Holst, Dave J. Adams, and Andrew I Cooper. 2011. Microporous Organic Polymers for Carbon Dioxide Capture. *Energy & Environmental Science* 4 (10): 4239–45.

Deng, Fang, Xu Biao Luo, Lin Ding, and Sheng Lian Luo. 2018. Application of Nanomaterials and Nanotechnology in the Reutilization of Metal Ion From Wastewater. In Xubiao Luo and Fang Deng (Eds.), *Nanomaterials for the Removal of Pollutants and Resource Reutilization*, 149–78. Elsevier.

Errahali, M., G. Gatti, L. Tei, G. Paul, G. A. Rolla, L. Canti, A. Fraccarollo, et al. 2014. Microporous Hyper-Cross-Linked Aromatic Polymers Designed for Methane and Carbon Dioxide Adsorption. *Journal of Physical Chemistry C* 118 (49): 28699–710.

Everaerts, Albert I., and L.M. Clemens. 2002. Pressure Sensitive Adhesives. In Xubiao Luo and Fang Deng (Eds.), *Adhesion Science and Engineering*, 465–534. Elsevier.

Feng, Li Juan, Min Wang, Zhi Yong Sun, Yun Hu, and Zhen Tao Deng. 2017. Hypercrosslinked Porous Polyporphyrin by Metal-Free Protocol: Characterization, Uptake Performance, and Heterogeneous Catalysis. *Designed Monomers and Polymers* 20 (1): 344–50.

Fontanals, N., R. M. Marcé, F. Borrull, and P. A.G. Cormack. 2015. Hypercrosslinked Materials: Preparation, Characterisation and Applications. *Polymer Chemistry* 6 (41): 7231–44.

Fontanals, Núria, Rosa Marcé, and F. Borrull. 2010. Overview of the Novel Sorbents Available in Solid-Phase Extraction to Improve the Capacity and Selectivity of Analytical Determinations. *Contribution of Science* 6: 199–213.

Fu, Shuqing, Jinshui Yao, Zhizhou Yang, Hongqiang Sun, and Weiliang Liu. 2018. Silane-Based Hyper-Cross-Linked Porous Polymers and Their Applications in Gas Storage and Water Treatment. *Journal of Materials Science* 53 (14): 10469–78.

Gadwdzik, B., and J. Osypiuk. 2001. Modification of Porous Poly(Styrene-Divinylbenzene) Beads by Friedel-Crafts Reaction. *Chromatographia* 54 (5): 323–28.

Gao, Hui, Lei Ding, Hua Bai, Anhua Liu, Sizhong Li, and Lei Li. 2016. Pitch-Based Hyper-Cross-Linked Polymers with High Performance for Gas Adsorption. *Journal of Materials Chemistry A* 4 (42): 16490–98.

Gao, Tu Nan, Tao Wang, Wei Wu, Yali Liu, Qisheng Huo, Zhen An Qiao, and Sheng Dai. 2019. Solvent-Induced Self-Assembly Strategy to Synthesize Well-Defined Hierarchically Porous Polymers. *Advanced Materials* 31 (11): 1806254.

Gatti, Giorgio, Mina Errahali, Lorenzo Tei, Enzo Mangano, Stefano Brandani, Maurizio Cossi, and Leonardo Marchese. 2019. A Porous Carbon with Excellent Gas Storage Properties from Waste Polystyrene. *Nanomaterials* 9 (5): 726.

Germain, Jonathan, Jean M. J. Fréchet, and Frantisek Svec. 2007. Hypercrosslinked Polyanilines with Nanoporous Structure and High Surface Area: Potential Adsorbents for Hydrogen Storage. *Journal of Materials Chemistry* 17 (47): 4989–97.

Germain, Jonathan, Jean M. J. Fréchet, and Frantisek Svec. 2009. Nanoporous, Hypercrosslinked Polypyrroles: Effect of Crosslinking Moiety on Pore Size and Selective Gas Adsorption. *Chemical Communications* no. 12: 1526–28.

Germain, Jonathan, Frantisek Svec, and Jean M. J. Fréchet. 2008. Preparation of Size-Selective Nanoporous Polymer Networks of Aromatic Rings: Potential Adsorbents for Hydrogen Storage. *Chemistry of Materials* 20 (22): 7069–76.

Giridhar, Gurram, R. K.N.R. Manepalli, and Gudimamilla Apparao. 2017. Size-Exclusion Chromatography. In Xubiao Luo and Fang Deng (Eds.), *Thermal and Rheological Measurement Techniques for Nanomaterials Characterization*, edited by Sabu Thomas, Raju Thomas, Ajesh Zachariah, and Raghvendra Mishra, vol. 3, 51–65. Elsevier.

Goto, Motonobu, and Benjamin J. McCoy. 2000. Inverse Size-Exclusion Chromatography for Distributed Pore and Solute Sizes. *Chemical Engineering Science* 55 (4): 723–32.

Grassie, N., and J. Gilks. 1973. Thermal Analysis of Polystyrenes Crosslinked by P-Di(Chloromethyl)Benzene. *Journal of Polymer Science: Polymer Chemistry Edition* 11 (8): 1985–94.

Grätz, Sven, Sebastian Zink, Hanna Kraffczyk, Marcus Rose, and Lars Borchardt. 2019. Mechanochemical Synthesis of Hyper-Crosslinked Polymers: Influences on Their Pore Structure and Adsorption Behaviour for Organic Vapors. *Beilstein Journal of Organic Chemistry* 15: 1154–61.

Grzybowski, Marek, Kamil Skonieczny, Holger Butenschön, and Daniel T. Gryko. 2013. Comparison of Oxidative Aromatic Coupling and the Scholl Reaction. *Angewandte Chemie International Edition* 52 (38): 9900–9930.

Guerritore, Marco, Rachele Castaldo, Brigida Silvestri, Roberto Avolio, Mariacristina Cocca, Maria Emanuela Errico, Maurizio Avella, Gennaro Gentile, and Veronica Ambrogi. 2020. Hyper-Crosslinked Polymer Nanocomposites Containing Mesoporous Silica Nanoparticles with Enhanced Adsorption towards Polar Dyes. *Polymers* 12 (6): 1388.

Hock, Sebastian, and Marcus Rose. 2020. 3D-Structured Monoliths of Nanoporous Polymers by Additive Manufacturing. *Chemie Ingenieur Technik* 92 (5): 525–31.

Hou, Rujing, Rosemary O'Loughlin, James Ackroyd, Qin Liu, Cara M. Doherty, Huanting Wang, Matthew R. Hill, and Stefan J.D. Smith. 2020. Greatly Enhanced Gas Selectivity in Mixed-Matrix Membranes through Size-Controlled Hyper-Cross-Linked Polymer Additives. *Industrial and Engineering Chemistry Research* 59 (30): 13773–82.

Hradil, J., and E. Králová. 1998. Styrene-Divinylbenzene Copolymers Post-Crosslinked with Tetrachloromethane. *Polymer* 39 (24): 6041–48.

Huang, Jing, and S. Richard Turner. 2018. Hypercrosslinked Polymers: A Review. *Polymer Reviews* 58 (1): 1–41.

Issaadi, Nabil, Abdelkarim Aït-Mokhtar, Rafik Belarbi, and Ameur Hamami. 2018. Effect of Variability of Porous Media Properties on Drying Kinetics: Application to Cement-Based Materials. In Xubiao Luo and Fang Deng (Eds.), *Advances in Multi-Physics and Multi-Scale Couplings in Geo-Environmental Mechanics*, edited by Francois Nicot and Olivier Millet, 243–89. Elsevier.

Jarrells, Travis W., Deyi Zhang, Shaohua Li, and Eric J. Munson. 2020. Quantification of Monomer Units in Insoluble Polymeric Active Pharmaceutical Ingredients Using Solid-State NMR Spectroscopy I: Patiromer. *AAPS PharmSciTech* 21 (3): 116.

Jiang, Kai, Teng Fei, and Tong Zhang. 2014. Humidity Sensing Properties of LiCl-Loaded Porous Polymers with Good Stability and Rapid Response and Recovery. *Sensors and Actuators B: Chemical* 199: 1–6.

Jiang, Kai, Da Kuang, Teng Fei, and Tong Zhang. 2014. Preparation of Lithium-Modified Porous Polymer for Enhanced Humidity Sensitive Properties. *Sensors and Actuators B: Chemical* 203: 752–58.

Kappert, Emiel J., Michiel J.T. Raaijmakers, Kristianne Tempelman, F. Petrus Cuperus, Wojciech Ogieglo, and Nieck E. Benes. 2019. Swelling of 9 Polymers Commonly Employed for Solvent-Resistant Nanofiltration Membranes: A Comprehensive Dataset. *Journal of Membrane Science* 569: 177–99.

Klamt, Andreas. 2005. Fundamental Criticism of the Dielectric Continuum Approach. In Xubiao Luo and Fang Deng (Eds.), *COSMO-RS From Quantum Chemistry to Fluid Phase Thermodynamics and Drug Design*, 43–47. Elsevier.

Kupgan, Grit, Thilanga P. Liyana-Arachchi, and Coray M. Colina. 2016. Pore Size Tuning of Poly(Styrene-Co-Vinylbenzyl Chloride-Co-Divinylbenzene) Hypercrosslinked Polymers: Insights from Molecular Simulations. *Polymer* 99: 173–84.

Landers, John, Gennady Yu Gor, and Alexander V. Neimark. 2013. Density Functional Theory Methods for Characterization of Porous Materials. *Colloids and Surfaces A: Physicochemical and Engineering Aspects* 437: 3–32.

Lau, Cher Hon, Kristina Konstas, Aaron W. Thornton, Amelia C.Y. Liu, Stephen Mudie, Danielle F. Kennedy, Shaun C. Howard, Anita J. Hill, and Matthew R. Hill. 2015. Gas-Separation Membranes Loaded with Porous Aromatic Frameworks That Improve with Age. *Angewandte Chemie – International Edition* 54 (9): 2669–73.

Law, R.V., and D.C. Sherrington. 1998. Crosslinked Polymers. In Xubiao Luo and Fang Deng (Eds.), *Solid State NMR of Polymers*, edited by Ando Isao and Asakura Tetsuo, 1st ed., vol. 84, 509–88. Elsevier.

Lee, Jet-Sing M., Takuya Kurihara, and Satoshi Horike. 2020. Five-Minute Mechanosynthesis of Hypercrosslinked Microporous Polymers. *Chemistry of Materials* 32 (18): 7694–7702.

Lee, Jun Young, Colin D. Wood, Darren Bradshaw, Matthew J. Rosseinsky, and Andrew I. Cooper. 2006. Hydrogen Adsorption in Microporous Hypercrosslinked Polymers. *Chemical Communications* no. 25: 2670–72.

Li, Buyi, Ruini Gong, Wei Wang, Xin Huang, Wang Zhang, Huanmin Li, Chunxiao Hu, and Bien Tan. 2011. A New Strategy to Microporous Polymers: Knitting Rigid Aromatic Building Blocks by External Cross-Linker. *Macromolecules* 44 (8): 2410–14.

Li, Buyi, Zhenhong Guan, Xinjia Yang, Wei David Wang, Wei Wang, Irshad Hussain, Kunpeng Song, Bien Tan, and Tao Li. 2014. Multifunctional Microporous Organic Polymers. *Journal of Materials Chemistry A* 2 (30): 11930–39.

Li, Buyi, Xin Huang, Liyun Liang, and Bien Tan. 2010. Synthesis of Uniform Microporous Polymer Nanoparticles and Their Applications for Hydrogen Storage. *Journal of Materials Chemistry* 20 (35): 7444–50.

Li, Buyi, Xinjia Yang, Lingling Xia, Muhammad Irfan Majeed, and Bien Tan. 2013. Hollow Microporous Organic Capsules. *Scientific Reports* 3 (1): 2128.

Li, Qingyin, Shangbin Jin, and Bien Tan. 2016. Template-Mediated Synthesis of Hollow Microporous Organic Nanorods with Tunable Aspect Ratio. *Scientific Reports* 6 (1): 31359.

Liu, Fang, Weidong Liang, Chengjun Wang, Jingxian He, Chaohu Xiao, Zhaoqi Zhu, Hanxue Sun, and An Li. 2021. Superwetting Monolithic Hypercrosslinked Polymers Nanotubes with High Salt-Resistance for Efficient Solar Steam Generation. *Solar Energy Materials and Solar Cells* 221: 110913.

Liu, Fenglei, Shuoyu Wang, Guorong Lin, and Shuixia Chen. 2018. Development and Characterization of Amine-Functionalized Hyper-Cross-Linked Resin for CO_2 Capture. *New Journal of Chemistry* 42 (1): 420–28.

Liu, Guoliang, Yangxin Wang, Chaojun Shen, Zhanfeng Ju, and Daqiang Yuan. 2015. A Facile Synthesis of Microporous Organic Polymers for Efficient Gas Storage and Separation. *Journal of Materials Chemistry A* 3 (6): 3051–58.

Liu, Juewen, Alison Stace-Naughton, Xingmao Jiang, and C. Jeffrey Brinker. 2009. Porous Nanoparticle Supported Lipid Bilayers (Protocells) as Delivery Vehicles. *Journal of the American Chemical Society* 131 (4): 1354–55.

Liu, Shuqin, Darui Chen, Juan Zheng, Lewei Zeng, Jijun Jiang, Ruifeng Jiang, Fang Zhu, Yong Shen, Dingcai Wu, and Gangfeng Ouyang. 2015. The Sensitive and Selective Adsorption of Aromatic Compounds with Highly Crosslinked Polymer Nanoparticles. *Nanoscale* 7 (40): 16943–51.

Lu, Cuiming, Shuqin Liu, Jianqiao Xu, Yajuan Ding, and Gangfeng Ouyang. 2016. Exploitation of a Microporous Organic Polymer as a Stationary Phase for Capillary Gas Chromatography. *Analytica Chimica Acta* 902: 205–11.

Lu, Ping, Simone Murray, and Min Zhu. 2018. Electrospun Nanofibers for Catalysts. In Xubiao Luo and Fang Deng (Eds.), *Electrospinning: Nanofabrication and Applications*, 695–717. Elsevier.

Luo, Yali, Buyi Li, Wei Wang, Kangbing Wu, and Bien Tan. 2012. Hypercrosslinked Aromatic Heterocyclic Microporous Polymers: A New Class of Highly Selective CO_2 Capturing Materials. *Advanced Materials* 24 (42): 5703–7.

Luo, Yali, Shoucun Zhang, Yunxiang Ma, Wei Wang, and Bien Tan. 2013. Microporous Organic Polymers Synthesized by Self-Condensation of Aromatic Hydroxymethyl Monomers. *Polymer Chemistry* 4 (4): 1126–31.

Mohamed, Mohamed G., Ahmed, F. M. El-Mahdy, Tso-Shiuan Meng, Maha M. Samy, and Shiao-Wei Kuo. 2020. Multifunctional Hypercrosslinked Porous Organic Polymers Based on Tetraphenylethene and Triphenylamine Derivatives for High-Performance Dye Adsorption and Supercapacitor. *Polymers* 12 (10): 2426.

Moldoveanu, Serban C., and Victor David. 2013. Equilibrium Types in HPLC. In Xubiao Luo and Fang Deng (Eds.), *Essentials in Modern HPLC Separations*, 85–114. Elsevier.

Moon, Su-Young, Jae-Sung Bae, Eunkyung Jeon, and Ji-Woong Park. 2010. Organic Sol-Gel Synthesis: Solution-Processable Microporous Organic Networks. *Angewandte Chemie* 122 (49): 9694–98.

Msayib, Kadhum J., and Neil B. McKeown. 2016. Inexpensive Polyphenylene Network Polymers with Enhanced Microporosity. *Journal of Materials Chemistry A* 4 (26): 10110–13.

Müllner, Tibor, Armin Zankel, Claudia Mayrhofer, Herbert Reingruber, Alexandra Höltzel, Yongqin Lv, Frantisek Svec, and Ulrich Tallarek. 2012. Reconstruction and Characterization of a Polymer-Based Monolithic Stationary Phase Using Serial Block-Face Scanning Electron Microscopy. *Langmuir* 28 (49): 16733–37.

O'Kane, J. M., and D. C. Sherrington. 1990. Hysteresis-Like Behavior in the Swelling/Deswelling of Polystyrene Cross-Linked Resins Using Binary Solvent Mixtures. *Macromolecules* 23 (25): 5286–91.

Ouyang, Yi, Huimin Shi, Ruowen Fu, and Dingcai Wu. 2013. Highly Monodisperse Microporous Polymeric and Carbonaceous Nanospheres with Multifunctional Properties. *Scientific Reports* 3 (1): 1430.

Pan, Yaoyu, Ziqiang Xu, Wenze Tan, Yalin Zhu, Yun Wang, Peihang Li, Xueqin Chen, Zhengguang Sun, Cao Li, and Bingbing Jiang. 2020. Novel Amino-Functionalized Hypercrosslinked Polymer Nanoparticles Constructed from Commercial Macromolecule Polystyrene via a Two-Step Strategy for CO_2 adsorption. *New Journal of Chemistry* 44 (48): 21125–33.

Pastukhov, Alexander, M. P. Tsyurupa, and V. A. Davankov. 1999. Hypercrosslinked Polystyrene: A Polymer in a Non-Classical Physical State. *Journal of Polymer Science Part B: Polymer Physics* 37 (17): 2324–33.

Peng, Xiaoqiang, Pengpeng Yang, Kun Dai, Yong Chen, Xiaochun Chen, Wei Zhuang, Hanjie Ying, and Jinglan Wu. 2020. Synthesis, Adsorption and Molecular Simulation Study of Methylamine-Modified Hyper-Cross-Linked Resins for Efficient Removal of Citric Acid from Aqueous Solution. *Scientific Reports* 10 (1): 9623.

Qiu, Shilun, and Teng Ben. 2015. *Porous Polymers: Design, Synthesis and Applications*. Vol. 12. Royal Society of Chemistry.

Razzaque, Shumaila, Chen Cai, Qun Wei Lu, Feng Zhen Huang, Yu Sang Li, He Bin Tang, Irshad Hussain, and Bien Tan. 2017. Development of Functionalized Hollow Microporous Organic Capsules Encapsulating Morphine – An In Vitro and In Vivo Study. *Journal of Materials Chemistry B* 5 (4): 742–49.

Razzaque, Shumaila, Ying Cheng, Irshad Hussain, and Bien Tan. 2020. Synthesis of Surface Functionalized Hollow Microporous Organic Capsules for Doxorubicin Delivery to Cancer Cells. *Polymer Chemistry* 11 (12): 2110–18.
Razzaque, Shumaila, Kewei Wang, Irshad Hussain, and Bien Tan. 2019. Facile Synthesis of Hypercrosslinked Hollow Microporous Organic Capsules for Electrochemical Sensing of CuII Ions. *Chemistry – A European Journal* 25 (2): 548–55.
Robshaw, Thomas J., Alex M. James, Deborah B. Hammond, Jake Reynolds, Robert Dawson, and Mark D. Ogden. 2020. Calcium-Loaded Hydrophilic Hypercrosslinked Polymers for Extremely High Defluoridation Capacity: Via Multiple Uptake Mechanisms. *Journal of Materials Chemistry A* 8 (15): 7130–44.
Rodríguez-Baeza, Mario, C. Andrónico Neira, and J. Carlos Aguilera. 2003. Thermogravimetric Study of the Formation of Cross-Linked Structures in the Synthesis of Poly(Methylsiloxane). *Journal of the Chilean Chemical Society* 48 (2): 73–77.
Rodríguez-Reinoso, Francisco, and Antonio Sepúlveda-Escribano. 2001. Porous Carbons in Adsorption and Catalysis. In Xubiao Luo and Fang Deng (Eds.), *Handbook of Surfaces and Interfaces of Materials*, 309–55. Elsevier.
Rodríguez-Reinoso, Francisco, and Joaquín Silvestre-Albero. 2019. Nanoporous Materials for Gas Storage. In Xubiao Luo and Fang Deng (Eds.), *Green Energy and Technology*, edited by Kaneko Katsumi and Francisco Rodríguez-Reinoso, 227–330. Springer.
Rubin Pedrazzo, Alberto, Fabrizio Caldera, Marco Zanetti, Silvia Lucia Appleton, Nilesh Kumar Dhakar, and Francesco Trotta. 2020. Mechanochemical Green Synthesis of Hyper-Crosslinked Cyclodextrin Polymers. *Beilstein Journal of Organic Chemistry* 16: 1554–63.
Sadak, Ali Enis. 2021. A Comparative Gas Sorption Study of Dicarbazole-Derived Microporous Hyper-Crosslinked Polymers. *Microporous and Mesoporous Materials* 311: 110727.
Sakar, Dolunay, Ayfer Saraç, Ozlem Cankurtaran, and Ferdane Karaman. 2007. Swelling Behavior of Polystyrene Networks in Some Aliphatic Esters and Linear Polystyrene Solutions. *Polymer Bulletin* 58: 305–12.
Saleh, Muhammad, Han Myoung Lee, K. Christian Kemp, and Kwang S. Kim. 2014. Highly Stable CO_2/N_2 and CO_2/CH_4 Selectivity in Hyper-Cross-Linked Heterocyclic Porous Polymers. *ACS Applied Materials & Interfaces* 6 (10): 7325–33.
Samanta, Partha, Priyanshu Chandra, Aamod V. Desai, and Sujit K. Ghosh. 2017. Chemically Stable Microporous Hyper-Cross-Linked Polymer (HCP): An Efficient Selective Cationic Dye Scavenger from an Aqueous Medium. *Materials Chemistry Frontiers* 1 (7): 1384–88.
Samanta, Partha, Aamod V. Desai, Sumanta Let, and Sujit K. Ghosh. 2019. Advanced Porous Materials for Sensing, Capture and Detoxification of Organic Pollutants toward Water Remediation. *ACS Sustainable Chemistry and Engineering* 7 (8): 7456–78.
Sant, Shilpa, Suzie Poulin, and Patrice Hildgen. 2008. Effect of Polymer Architecture on Surface Properties, Plasma Protein Adsorption, and Cellular Interactions of Pegylated Nanoparticles. *Journal of Biomedical Materials Research – Part A* 87 (4): 885–95.
Schmidt, Bernhard V.K.J. 2019. Hydrophilic Polymers. *Polymers* 11 (4) article no. 693.
Schute, Kai, and Marcus Rose. 2015. Metal-Free and Scalable Synthesis of Porous Hyper-Cross-Linked Polymers: Towards Applications in Liquid-Phase Adsorption. *ChemSusChem* 8 (20): 3419–23.
Schwab, Matthias Georg, Angela Lennert, Jörg Pahnke, Gerhard Jonschker, Matthias Koch, Irena Senkovska, Matthias Rehahn, and Stefan Kaskel. 2011. Nanoporous Copolymer Networks through Multiple Friedel–Crafts-Alkylation—Studies on Hydrogen and Methane Storage. *Journal of Materials Chemistry* 21 (7): 2131–35.
Sellergren, Börje, and Andrew J. Hall. 2001. Chapter 2 Fundamental Aspects on the Synthesis and Characterisation of Imprinted Network Polymers. In Xubiao Luo and Fang Deng (Eds.), *Techniques and Instrumentation in Analytical Chemistry*, vol. 23, 21–57. Elsevier.
Seo, Myungeun, Soobin Kim, Jaehoon Oh, Sun-Jung Kim, and Marc A Hillmyer. 2015. Hierarchically Porous Polymers from Hyper-Cross-Linked Block Polymer Precursors. *Journal of the American Chemical Society* 137 (2): 600–603.
Shao, Lishu, and Jianhan Huang. 2017. Controllable Synthesis of N-Vinylimidazole-Modified Hyper-Cross-Linked Resins and Their Efficient Adsorption of p-Nitrophenol and o-Nitrophenol. *Journal of Colloid and Interface Science* 507: 42–50.
Sharma, Vivek, Asit Sahoo, Yogesh Sharma, and Paritosh Mohanty. 2015. Synthesis of Nanoporous Hypercrosslinked Polyaniline (HCPANI) for Gas Sorption and Electrochemical Supercapacitor Applications. *RSC Advances* 5 (57): 45749–54.
Shi, Pengju, Xueqin Chen, Zhengguang Sun, Cao Li, Ziqiang Xu, Xueliang Jiang, and Bingbing Jiang. 2020. Thickness Controllable Hypercrosslinked Porous Polymer Nanofilm with High CO_2 Capture Capacity. *Journal of Colloid and Interface Science* 563: 272–80.

Shi, Song, Chen Chen, Min Wang, Jiping Ma, Hong Ma, and Jie Xu. 2014. Designing a Yolk–Shell Type Porous Organic Network Using a Phenyl Modified Template. *Chemical Communications* 50 (65): 9079–82.
Silverstein, Michael S. 2020. The Chemistry of Porous Polymers: The Holey Grail. *Israel Journal of Chemistry* 60.
Silverstein, Michael S., Neil R. Cameron, and Marc Hillmyer. 2011. *Porous Polymers*. Hoboken, NJ: John Wiley & Sons.
Stepacheva, A. A., M. E. Markova, O. V. Manaenkov, A. V. Gavrilenko, A. I. Sidorov, M. G. Sulman, Yu Yu Kosivtsov, V. G. Matveeva, and E. M. Sulman. 2020. Modification of the Hypercrosslinked Polystyrene Surface. New Approaches to the Synthesis of Polymer-Stabilized Catalysts. *Russian Chemical Bulletin* 69 (4): 721–30.
Su, Pengcheng, Xu Zhang, Zehai Xu, Guoliang Zhang, Chong Shen, and Qin Meng. 2019. Amino-Functionalized Hypercrosslinked Polymers for Highly Selective Anionic Dye Removal and CO2/N2 Separation. *New Journal of Chemistry* 43 (44): 17267–74.
Subri, Nur Nida Syamimi, Peter A.G. Cormack, Siti Nurul Ain Siti Nurul, Luqman Chuah Abdullah, and Rusli Daik. 2018. Synthesis of Poly(Acrylonitrile-Co-Divinylbenzene-Co-Vinylbenzyl Chloride)-Derived Hypercrosslinked Polymer Microspheres and a Preliminary Evaluation of Their Potential for the Solid-Phase Capture of Pharmaceuticals. *Journal of Applied Polymer Science* 135 (2): 45677.
Tan, Liangxiao, Buyi Li, Xinjia Yang, Weixing Wang, and Bien Tan. 2015. Knitting Hypercrosslinked Conjugated Microporous Polymers with External Crosslinker. *Polymer* 70: 336–42.
Tan, Liangxiao, and Bien Tan. 2017. Hypercrosslinked Porous Polymer Materials: Design, Synthesis, and Applications. *Chemical Society Reviews* 46: 3322–3356.
Taylor, Dominic, Scott J. Dalgarno, Zhengtao Xu, and Filipe Vilela. 2020. Conjugated Porous Polymers: Incredibly Versatile Materials with Far-Reaching Applications. *Chemical Society Reviews* 49 (12): 3981–4042.
Tsyurupa, M. P., and V. A. Davankov. 2002. Hypercrosslinked Polymers: Basic Principle of Preparing the New Class of Polymeric Materials. *Reactive and Functional Polymers* 53 (2–3): 193–203.
Tsyurupa, M. P., and V. A. Davankov. 2006. Porous Structure of Hypercrosslinked Polystyrene: State-of-the-Art Mini-Review. *Reactive and Functional Polymers* 66 (7): 768–79.
Tsyurupa, M. P., V. V. Lalaev, and V. A. Davankov. 1984. On Reasons Determining Unusual Properties of Hypercrosslinked Styrene Polymers. *Doklady Akademii Nauk SSSR* 279: 156–60.
Tu, Yizhou, Guizhou Xu, Lu Jiang, Xuejiao Hu, Jun Xu, Xianchuan Xie, and Aimin Li. 2020. Amphiphilic Hyper-Crosslinked Porous Cyclodextrin Polymer with High Specific Surface Area for Rapid Removal of Organic Micropollutants. *Chemical Engineering Journal* 382: 123015.
Uemura, Takashi, Yousuke Hoshino, Susumu Kitagawa, Kaname Yoshida, and Seiji Isoda. 2006. Effect of Organic Polymer Additive on Crystallization of Porous Coordination Polymer. *Chemistry of Materials* 18 (4): 992–95.
Veverka, Pavel, and Karel Jeřábek. 1999. Mechanism of Hypercrosslinking of Chloromethylated Styrene-Divinylbenzene Copolymers. *Reactive and Functional Polymers* 41 (1): 21–25.
Volkova, N. N., E. V. Summanen, and L. P. Smirnov. 2003. Thermal Degradation of Hypercrosslinked Polystyrenes. *Polymer Science – Series A* 45 (10): 986–92.
Wang, Dali, Gangsheng Tong, Ruijiao Dong, Yongfeng Zhou, Jian Shen, and Xinyuan Zhu. 2014. Self-Assembly of Supramolecularly Engineered Polymers and Their Biomedical Applications. *Chemical Communications* 50 (81): 11994–2017.
Wang, Haijiang, Lili Pan, Wenxiu Deng, Guangjie Yang, and Xikui Liu. 2016. One-Pot Synthesis of Triptycene-Based Porous Organic Frameworks with Tailored Micropore Environments for Highly Efficient and Selective Amine Adsorption. *Polymer Journal* 48 (7): 787–92.
Wang, Jinquan, Waihong Sng, Guangshun Yi, and Yugen Zhang. 2015. Imidazolium Salt-Modified Porous Hypercrosslinked Polymers for Synergistic CO_2 Capture and Conversion. *Chemical Communications* 51 (60): 12076–79.
Wang, Shaolei, Liangxiao Tan, Chengxin Zhang, Irshad Hussain, and Bien Tan. 2015. Novel POSS-Based Organic–Inorganic Hybrid Porous Materials by Low Cost Strategies. *Journal of Materials Chemistry A* 3 (12): 6542–48.
Wang, Wenjing, Mi Zhou, and Daqiang Yuan. 2017. Carbon Dioxide Capture in Amorphous Porous Organic Polymers. *Journal of Materials Chemistry A*. 5 (4), 1334–1347.
Wang, Xiaomei, Xiang Mao, and Jianhan Huang. 2018. Hierarchical Porous Hyper-Cross-Linked Polymers Modified with Phenolic Hydroxyl Groups and Their Efficient Adsorption of Aniline from Aqueous Solution. *Colloids and Surfaces A: Physicochemical and Engineering Aspects* 558: 80–87.
Wang, Xiaomei, Han Ou, and Jianhan Huang. 2019. One-Pot Synthesis of Hyper-Cross-Linked Polymers Chemically Modified with Pyrrole, Furan, and Thiophene for Phenol Adsorption from Aqueous Solution. *Journal of Colloid and Interface Science* 538: 499–506.

Wang, Yuchao, Da Shi, Shengyang Tao, Wentong Song, Hongmin Wang, Xinkui Wang, Guangtao Li, Jieshan Qiu, and Min Ji. 2016. A General, Green Chemistry Approach for Immobilization of Inorganic Catalysts in Monolithic Porous Flow-Reactors. *ACS Sustainable Chemistry and Engineering* 4 (3): 1602–10.

Weber, Jens, and Qing Bo Meng. 2014. Microporous Polymers: Synthesis, Characterization, and Applications. In Xubiao Luo and Fang Deng (Eds.), *Encyclopedia of Polymer Science and Technology*, 1–49. Hoboken, NJ: John Wiley & Sons.

Wilson, Craig, Marcus J. Main, Nicholas J. Cooper, Michael E. Briggs, Andrew I. Cooper, and Dave J. Adams. 2017. Swellable Functional Hypercrosslinked Polymer Networks for the Uptake of Chemical Warfare Agents. *Polymer Chemistry* 8 (12): 1914–22.

Wood, Colin D., Tan Bien, Abbie Trewin, Niu Hongjun, Darren Bradshaw, Matthew J. Rosseinsky, Yaroslav Z. Khimyak, et al. 2007. Hydrogen Storage in Microporous Hypercrosslinked Organic Polymer Networks. *Chemistry of Materials* 19 (8): 2034–48.

Wood, Colin D., Bien Tan, Abbie Trewin, Fabing Su, Matthew J. Rosseinsky, Darren Bradshaw, Yan Sun, Li Zhou, and Andrew I. Cooper. 2008. Microporous Organic Polymers for Methane Storage. *Advanced Materials* 20 (10): 1916–21.

Woodward, Robert T. 2020. The Design of Hypercrosslinked Polymers from Benzyl Ether Self-Condensing Compounds and External Crosslinkers. *Chemical Communications* 56 (36): 4938–41.

Wu, Yingzhen, Na Xing, Sen Li, Leixin Yang, Yanxiong Ren, Yutao Liu, Xu Liang, et al. 2021. In Situknitted Microporous Polymer Membranes for Efficient CO_2 capture. *Journal of Materials Chemistry A* 9 (4): 2126–34.

Yang, Zujin, Guifang Wu, Qiuru Li, Hongxia Ai, Xingdong Yao, and Hongbing Ji. 2021. Removal of Various Pollutants from Wastewaters Using an Efficient and Degradable Hypercrosslinked Polymer. *Separation Science and Technology (Philadelphia)* 56 (5): 860–69.

Zeng, Xu, You Wang, Zhe Shu, and Jianhan Huang. 2020. Hydroquinone-Modified Hyper-Crosslinked Polymer and Its Adsorption of Aniline. *Chemical Engineering Journal Advances* 1: 100004.

Zhai, Tian-Long, Liangxiao Tan, Yi Luo, Jun-Min Liu, Bien Tan, Xiang-Liang Yang, Hui-Bi Xu, and Chun Zhang. 2016. Microporous Polymers from a Carbazole-Based Triptycene Monomer: Synthesis and Their Applications for Gas Uptake. *Chemistry – An Asian Journal* 11 (2): 294–98.

Zhang, Chun, Peng-Cheng Zhu, Liangxiao Tan, Jun-Min Liu, Bien Tan, Xiang-Liang Yang, and Hui-Bi Xu. 2015. Triptycene-Based Hyper-Cross-Linked Polymer Sponge for Gas Storage and Water Treatment. *Macromolecules* 48 (23): 8509–14.

Zhang, Di, Toms Tao, Junping Ju, Yanming Wang, Qihua Wang, and Tingmei Wang. 2015. Postmodification of Linear Poly-p-Phenylenes to Prepare Hyper-Crosslinked Polymers: Tuning the Surface Areas by the Molecular Weight. *Polymer* 60: 234–40.

Zhang, Ting, Guolong Xing, Weiben Chen, and Long Chen. 2020. Porous Organic Polymers: A Promising Platform for Efficient Photocatalysis. *Materials Chemistry Frontiers* 4: 332–353.

Zhang, Ting, Fa Zhou, Jianhan Huang, and Ruilin Man. 2018. Ethylene Glycol Dimethacrylate Modified Hyper-Cross-Linked Resins: Porogen Effect on Pore Structure and Adsorption Performance. *Chemical Engineering Journal* 339: 278–87.

Zhang, Xin, Shaohua Shen, and Liyuan Fan. 2007. Studies Progress of Preparation, Properties and Applications of Hyper-Cross-Linked Polystyrene Networks. *Journal of Materials Science* 42 (18): 7621–29.

Zhang, Yinghua, Yuda Li, Feng Wang, Yang Zhao, Chong Zhang, Xiaoyan Wang, and Jia Xing Jiang. 2014. Hypercrosslinked Microporous Organic Polymer Networks Derived from Silole-Containing Building Blocks. *Polymer* 55 (22): 5746–50.

Zhao, Wuxue, Zongsheng Hou, Zhaoquan Yao, Xiaodong Zhuang, Fan Zhang, and Xinliang Feng. 2015. Hypercrosslinked Porous Polymer Nanosheets: 2D RAFT Agent Directed Emulsion Polymerization for Multifunctional Applications. *Polymer Chemistry* 6 (40): 7171–78.

Zhou, Fa, Ruilin Man, and Jianhan Huang. 2018. Alkoxy-Modified Hyper-Cross-Linked Polymers with Hierarchical Porosity and Their Adsorption of Salicylic Acid from Aqueous Solution. *Industrial and Engineering Chemistry Research* 57 (37): 12420–28.

Zhu, Xiang, Shannon M. Mahurin, Shu-Hao An, Chi-Linh Do-Thanh, Chengcheng Tian, Yankai Li, Lance W. Gill, et al. 2014. Efficient CO_2 Capture by a Task-Specific Porous Organic Polymer Bifunctionalized with Carbazole and Triazine Groups. *Chemical Communications* 50 (59): 7933–36.

Zidan, Ghada, Carol Ann Greene, and Ali Seyfoddin. 2019. Formulation Design in Drug Delivery. In Xubiao Luo and Fang Deng (Eds.), *Engineering Drug Delivery Systems*, 17–41. Elsevier.

Zou, Jie, Chunyan Fan, and Xiu Liu. 2020. Effects of Molecular Cross-Sectional Areas of Adsorbed Nitrogen on the Brunauer–Emmett–Teller Analysis for Carbon-Based Slit Pores. *Langmuir* 36 (48): 14656–65.

3 Porous Ionic Polymers

Yuliya Dzyazko and Anatolii Omel'chuk
V.I. Vernadskii Institute of General and Inorganic Chemistry, Kyiv, Ukraine

CONTENTS

3.1 INTRODUCTION: A DISTINCTIVE FEATURE OF THE POROUS STRUCTURE OF IONIC POLYMERS

Porous ionic polymers are widely used for the solution of a wide spectrum of practical tasks. The main fields of their application are liquid purification from ionic species (ion exchange and electrodialysis), low-temperature fuel cells, and catalysis. As opposed to many inorganic ionic conductors, which demonstrate this property under high temperatures (Pogorenko et al. 2016, 2019), ion exchange polymers show high electrical conductivity under ambient conditions. The necessary requirement is a considerable content of water or other polar solvents. Properties of ionic polymers depend on their ion exchange capacity and porosity, the organization of which occurs mainly as a result of swelling. This feature complicates the study of their porous structure, since wide-spread methods, such as nitrogen adsorption-desorption and mercury intrusion, allow us to recognize pores only for dry polymers. Since solvated functional groups are responsible for porosity formation, the size and volume of pores strongly depend on the solvent. It means the unreliability of experimental results obtained with the abovementioned techniques since they are not related to the real operation conditions of ionic polymers. Untraditional porosometric methods are suitable for this purpose. Due to difficulties in determining the porous structure of ionic polymers, it was a subject of discussion for a long time. Nowadays, the generally accepted point of view is more or less developed. It is based on Hsu–Gierke model, which suggests porous structure as a continuous network of hydrophilic pores, namely nanosized clusters and narrower channels formed with functional groups (Hsu and Gierke 1983). Further, this model was slightly transformed and improved by other authors, but its main principles remained without changes.

In this chapter, we consider dry ionic polymers, the mechanism of hydration of functional groups, and different models of the structure of swollen polymers. A special method is needed in order to investigate all types of pores in solvated ionic polymers – this technique is also considered in this chapter.

DOI: 10.1201/9781003169604-3

3.2 IONIC POLYMERS IN DRY STATE

The structure of the dry ionic polymers is characterized by important peculiarities, which determine their structure in a swollen state. As shown in the example of Nafion 115 and Nafion 117 membranes (Figure 3.1), the polymer contains isolated heterogeneities. Their shape is spherical, diameter of the spheres is ≈1.5 nm, the distance between them is close to 2.7 nm (Allahyarov and Taylor 2007). According to (Eisenberg 1970; Dreifus 1986), each heterogeneity contains 2–8 pairs of fixed ion–counter ion.

These heterogeneities are formed during the polymer synthesis due to the dipole-dipole interaction that is caused by fixed and counter-ions (Figure 3.2). Based on the Allahyarov–Taylor model, no partial charges are attributed to O, C, and F atoms (Allahyarov and Taylor 2007). The charge is caused by sulfonate ($-SO_3$) groups and H^+ ions associated with them. The dipole moment is modeled according to the approach developed by Lennard-Jones. According to this approach, the distance between two unconnected species determines the potential energy of interaction between them. Compact, linear, and branched multiplets are formed as a result of dipole-dipole bonds.

Larger heterogeneities with a diameter of several nanometers are seen in the original TEM image of gel-like cation-exchange resin. Before the measurements, the resin was soaked with a very diluted $CuCl_2$ solution. The heterogeneities that are free from metal counter-ions are recognized as white round nanosized spots (Figure 3.3).

The heterogeneities provide pores even in dry ion exchange polymers, as shown below. However, their porous structure is poorly developed. According to Gierke, the diameter of these pores for Nafion reaches ≈1.8 nm, each pore contains 26 sulfonate groups (Gierke et al. 1981).

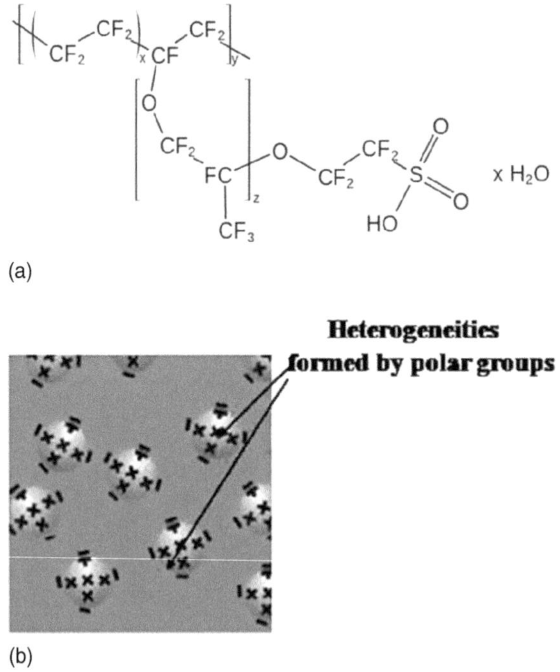

FIGURE 3.1 Structural formula of Nafion polymer (a) and heterogeneities in dry ion exchange polymer (b).

Part (b) is adapted from the review (Kononenko et al. 2017) according to the permission of Elsevier publisher, original scheme is in (Gebel 2000).

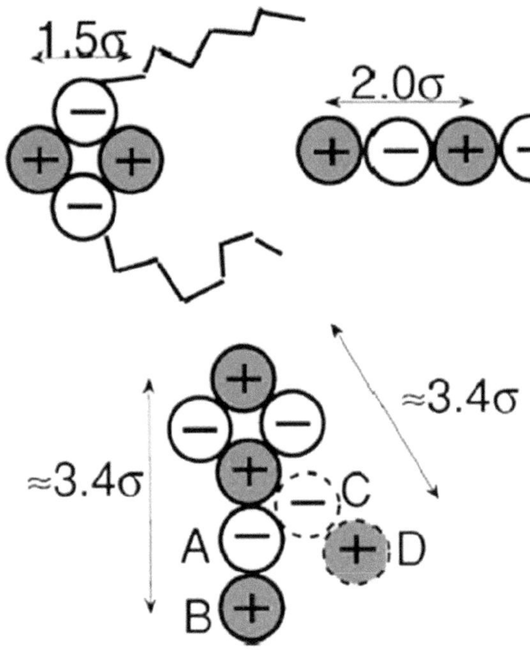

FIGURE 3.2 Multiplet structures of polymer containing polar groups (Allahyarov and Taylor 2007). Compact (on the left side, hydrophobic fragments are shown) and linear (on the right side) multiplets are given in the upper row, branched multiplet is in the bottom row. Other possible configuration of the A and D objects are dashed C and D objects respectively. Here σ is the monomer diameter.

Reproduced according to the permission of AIP Publisher.

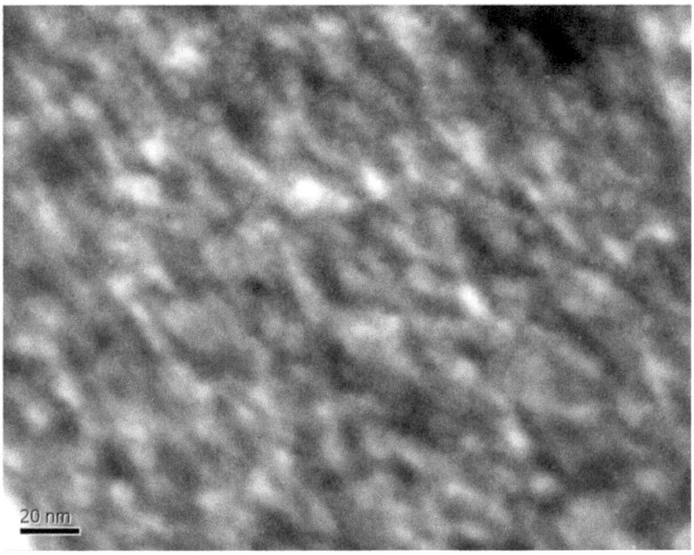

FIGURE 3.3 Original TEM image of dry strongly acidic cation exchange resin partially loaded with Cu^{2+} ions. Heterogeneities formed by functional groups are seen as white nanosized spots.

3.3 IONIC POLYMERS IN SWOLLEN STATE: HSU–GIERKE MODEL

Traditionally swollen ionic polymers were considered as a solution of water in solid (Samsonov and Pasechnik 1969; Hodgdon and Boyack, 1965; Borodin et al. 1965). It means one gel phase of water and polymer. Porous structure (macropores) was suggested only for macroporous resins (Kunin et al. 1962; Kun and Kunin 1967) and heterogeneous ion exchange membranes (Arnold and Koch 1966; Millar et al. 1964). Regarding macroporous ion exchange resins, purposeful formation of large pores occurs, when the polymer is synthesized. The macropores of heterogeneous membranes are cavities between ionic polymer and binder. Micro- and mesoporosity of ion exchange polymers were not considered. Even in the classical monograph by Helfferich, a number of isotherms of water adsorption are given, however, no calculation of pore size distribution is made (Helfferich 1995).

However, back in 1962, Glueckauf assumed a continuous network of aqueous fissures in polymer ion-exchangers. As supposed, their width is at least 7 nm, they expanded up to 200 nm and cross-connected by smaller fissures, the dimension of which is comparable with that of inter-chain distances (Glueckauf 1962). The concentration of counter-ions in the wider voids decreases rapidly with the distance from their walls. The distribution of counter-ions obeys a Poisson relation. Depending on the location, their concentration may differ from the average value by 3–4 times.

Nowadays, the generally accepted concept of the structure of swollen ion exchange polymers is the model developed by Hsu and Gierke (1983). They carried out the measurements for Nafion membrane using small-angle X-ray scattering (SAXS). The cluster-channel approach has been proposed: H^+ ions and water move from one globular nanosized cluster to other through channels, which are narrower than clusters and have a cylindrical shape (Figure 3.4). Clusters and channels are formed during swelling as a result of the reorganization of heterogeneities, which exist in dry ion exchange polymers (see above). A very important conclusion of this work is the different phases

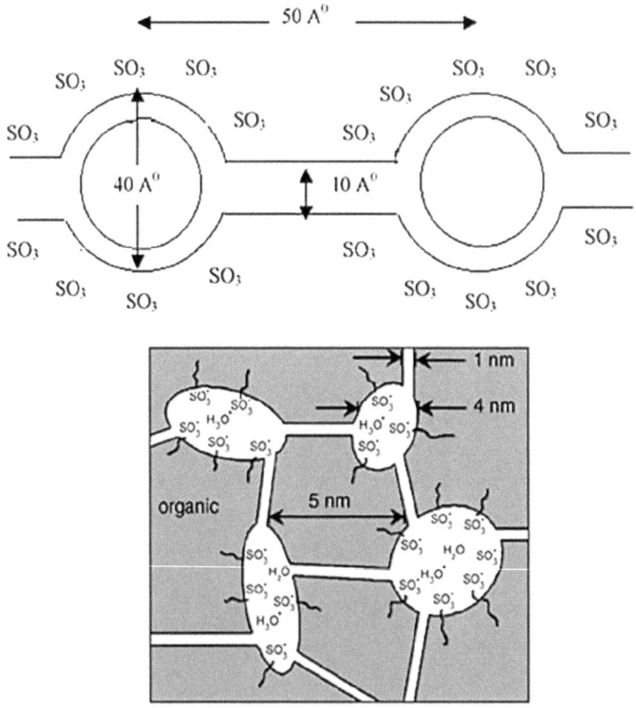

FIGURE 3.4 Cluster-channel structure of swollen Nafion membrane (Smitha et al. 2005).

Reproduced according to the permission of Elsevier publisher.

of polymer and water. In other words, swollen ion exchange materials are characterized by a porous structure. It means special porosimetric techniques can be applied to swollen polymers (see below). Since functional groups are located namely inside clusters and channels, these voids are considered as hydrophilic pores. There are no functional groups in some fragments of polymer chains, these regions are responsible for the formation of hydrophobic pores. Their size is larger than 10 nm. Additionally, ionic polymers contain also structure defects, their size is several microns. The largest pores of heterogeneous ionic membranes are the cavities between polymer and binder.

Now the SAXS technique is widely used for the investigation of different ion exchange polymers, cluster size is determined in this manner. Different materials are considered: polymers with a hydrophobic backbone containing short hydrophilic and hydrophobic chains (Dorenbos 2020), sulfonated poly(1,4-phenylene ether-ether-sulfone) intercalated with poly (2-acrylamido-2-methyl-1-propanes ulfonic acid) (Manohar et al. 2021), fluorene-based poly (arylene ether sulfone) containing grafted imidazolium cations (Guo et al. 2017). For instance, the cluster diameter has been estimated as 2.05–3.34 nm in the case of poly(arylene ether ketone) polymer (Lee et al. 2018).

It should be additionally stressed that the Hsu–Gierke approach suggests different phases of polymer and water, i.e. porous structure of ion exchange polymers.

3.4 MODIFICATIONS OF HSU–GIERKE MODEL: HYDRATION OF ION EXCHANGE POLYMERS

Since 1983 the cluster-channel model was developed and modified. Moreover, this approach was a reason for heated disputes. Problems such as size, shape, and continuality of hydrophilic pores were discussed. These discussions were caused by the impossibility of the visualization of clusters and channels. It is also necessary to take into consideration the timescale of the polymer hydration. Another difficulty is the possibility to apply a large variety of models to spatially averaged SAXS data. Thus, a number of approaches, which give a different interpretation of the measurements even for the same object (Nafion polymer), have been reported. Critical views of Hsu–Gierke model is analyzed in (Mauritz and Moore 2004).

The formation of cluster agglomerates (5–30 nm) was suggested (James et al. 2000). The hierarchical structure of clusters was also proved (Elliott et al. 2000). Similar cluster size (5.3) nm has been found for different perfluorosulfonic acid membranes; however, they are characterized by various cluster densities (Halim et al. 1994). Parallelism of elongated channels to Nafion crystallites is suggested in (Schmidt-Rohr and Chen 2008). The channels are elongated, their average diameter is 2.4 nm at 20 vol.% of water. The cross-section of the polymer crystallites was estimated as 25 nm^2. The mentioned approaches were based on SAXS measurements.

The method of small-angle neutron scattering (SANS) is also used for the research of ionic polymers. This technique allowed one to determine the profile of water content in the ionic polymer during the operation of PEMFC (proton exchange membrane fuel cell) (Mosdale et al. 1996). The profile has been found to depend on the humidity of hydrogen sent to the anode. Based on the data of SANS, conclusions about cluster shape have been made (Wescott et al. 2006). Deformation of clusters under high water content was suggested –they are elliptical and barbell-like under these conditions.

The approach explaining equilibrium hydration states of swollen ion exchange polymer has been reported by Mauritz (1995). During the polymer hydration, the clusters are expanded due to osmotic pressure, which is caused by the osmotic pressure of the “microsolutions” inside the clusters. The components of these “microsolutions” are fixed and counter-ions, namely they provide osmotic pressure. The contractile pressure of ionic polymer resists osmosis. This pressure is determined by Young’s modulus on the one hand and the possible cluster size on the other hand. When the osmotic and contractile pressures are equal, this state corresponds to equilibrium. Earlier the osmotic theory for polymer ion-exchangers was described in detail by Helfferich (1995). Nevertheless, clusters were not mentioned in this monograph. The kinetics of hydration of functional groups was studied using a SANS technique (Fumagalli et al. 2015). As suggested, the equilibrium state of hydration can never be achieved.

A four-stage model of the equilibrium between unbound counter-ions and counter-ions was considered in (Lowry and Mauritz 1980) just before Hsu and Gierke publication in 1983. These conclusions were made based on the investigations of the Nafion membrane, which was loaded with different ions, by means of IR spectroscopy. The shift of the peaks attributed to the stretching vibrations of sulfonate groups has been found (Table 3.1). As shown in Figure 3.5, the stages of hydration involve: (1) complete dissociation of ion pairs, (2)contact of ions with each other by primary hydration shells, (3) formation of outer-sphere complexes, and (4)formation of inner-sphere complexes (undissociated ion pairs).

The composition of hydrate complexes of alkaline metal ions sulfostyrene resin containing 10% of divinylbenzene was calculated by Soldatov et al. (2016a, 2018). As found for Li^+ ions, the Li^+-O distance between the molecular layers of water rapidly increases (Figure 3.6a). However, no second layer has been found (Vishnyakov and Neimark 2001). The formation of $Li(H_2O)_4^+$ hydrate complex provides low selectivity of resins toward these ions (Soldatov et al. 2016a). There are four water molecules in the first molecular layer of Li^+ and Na^+ ions, which are not in contact with sulfonate groups. Regarding K^+, Rb^+, and Cs^+ ions, their first layers of water molecules are not packed densely (figs. 6b–d, Soldatov et al. 2018). Besides water molecules (4–5), the layers contain also

TABLE 3.1
Position of Maxima (cm^{-1}) Corresponding to $-SO_3^-$Symmetric Stretch Vibration of Nafion Membrane in Different Ionic Forms

Ion	Hydrated Polymer	Dry Polymer
Li^+	1058	1073
Na^+	1058	1064
K^+	1058	1059
Rb^+	1057	1057

Source: According to the data of (Lowry and Mauritz 1980).

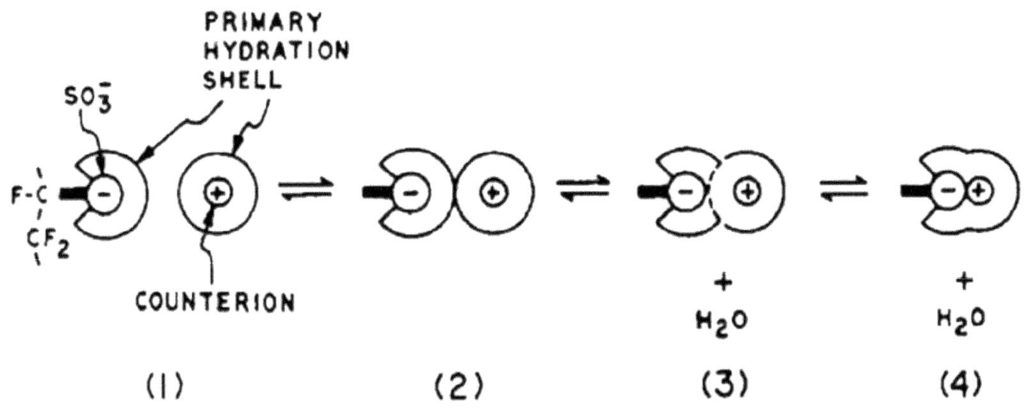

FIGURE 3.5 Formation of ion pairs of fixed ion – counter ion (Lowry and Mauritz 1980).

Reproduced according to the permission of ACS publisher.

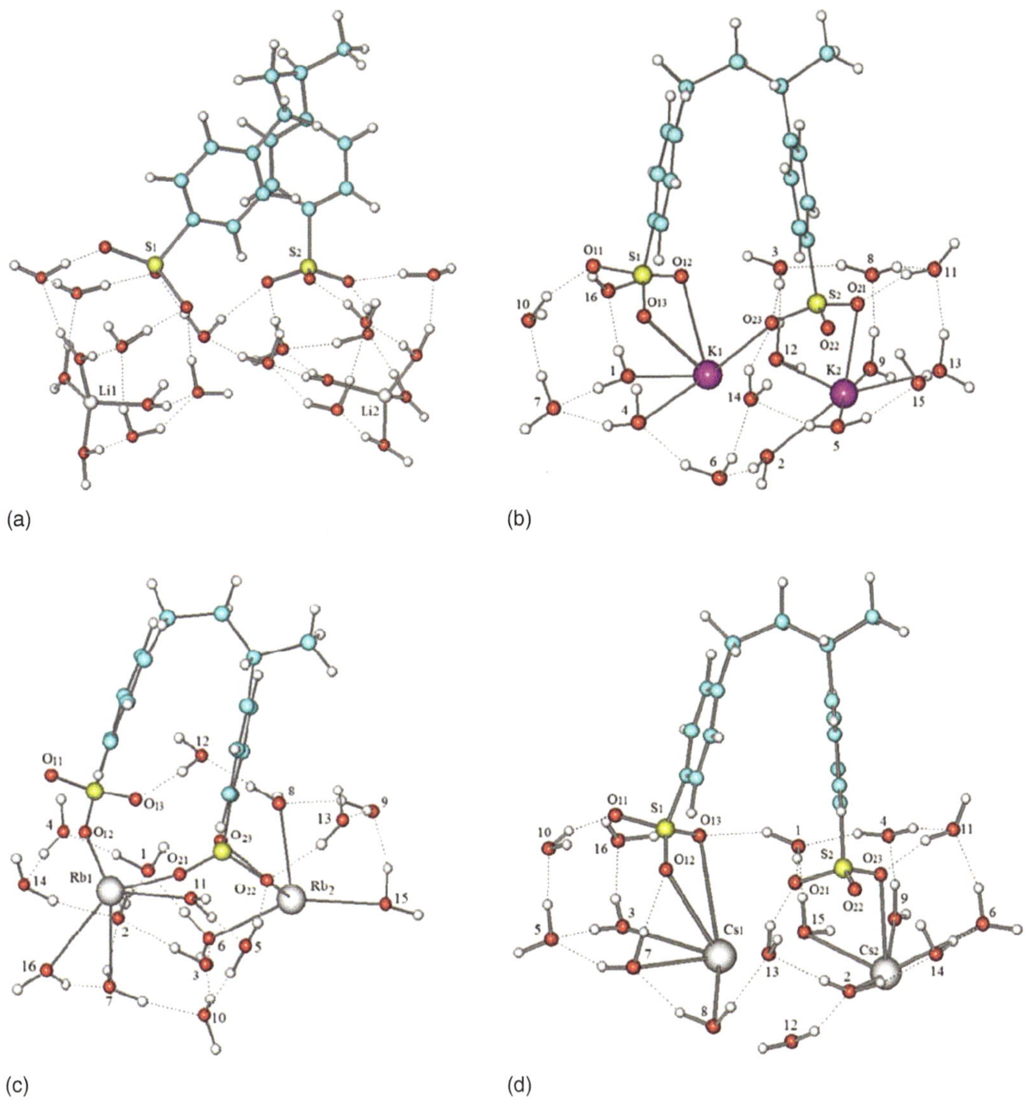

FIGURE 3.6 Hydrate complexes of Li^+ (a), K^+ (b), Rb^+ (c) and Cs^+ (d) cations with $–SO_3^-$ group of gel-like strongly acidic resin (a –Soldatov et al. 2016; b–d –Soldatov et al. 2018).

Reproduced according to the permission of Elsevier publisher.

1 or 2 oxygen atoms of sulfonate groups, which form electrostatic bonds with the cation. Due to this interaction, strongly acidic resins show higher selectivity toward large alkaline cations. Regarding H-form, four types of hydrates have been found (Soldatov et al. 2016b): (i) H_3O^+; (ii) $H_3O^+{\cdot}H_2O$; (iii) $H_3O^+{\cdot}2H_2O$; (iv) $H_3O^+{\cdot}6H_2O$. The data for weakly acidic resin and H^+, Li^+, Na^+, K^+, Mg^{2+}, Ca^{2+}, and Ba^{2+} ions are given in (Soldatov et al. 2013).

A lattice model has been reported in (Ioselevich et al. 2004). According to this approach, the ion exchange polymer consists of cages and voids between them (Figure 3.7). The cage is a tetrahedron (edges 3*a*) containing four smaller tetrahedral cavities (edges *a*). In the simplest structure, the cage includes four triangles and four hexagons. The neighboring cages have hexagonal planes, which form wide channels. These channels exist under low humidity. Regarding the more complex cages, the channels are smaller: they are formed under high water content.

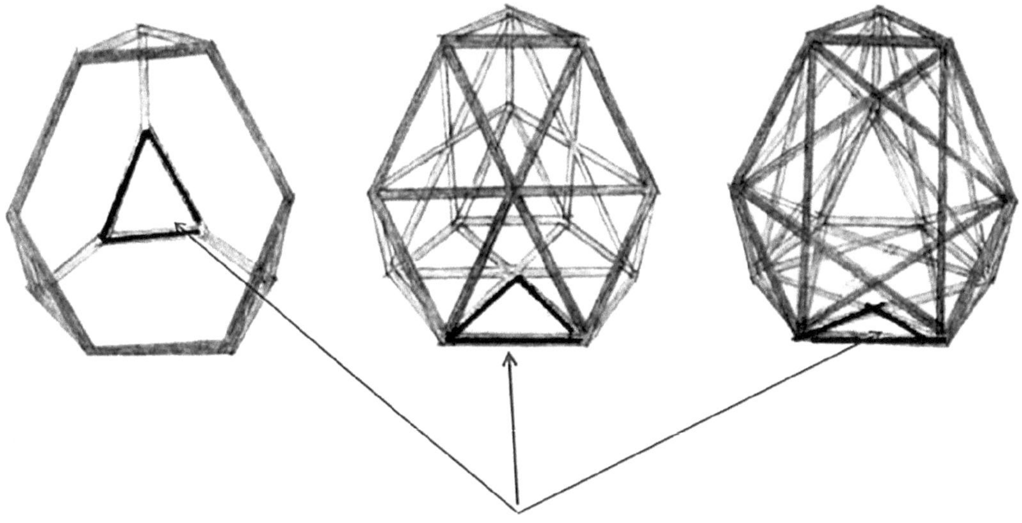

FIGURE 3.7 Cage model of ion exchange polymers (Ioselevich et al. 2004).

Redrawn according to the permission of ACS publisher.

The model of swelling has been developed in (Eikerling et al. 1997). First of all, water wets the inner surface of clusters and channels, functional groups are solvated. It is necessary that there should be 5–6 water molecules for the hydration of $-SO_3H$ groups of the Nafion membrane. In this case, the water structure is ice-like. It is assumed that the film of water is formed before swelling. During swelling, two types of pores are formed: (i) "red" pores containing water at the surface (this water provides hydration of functional groups) and (ii) "blue" pores containing surface water and also water possessing bulk properties. This conception is reviewed in the framework of bonded and free water, boundary water is also taken into consideration (Berezina et al. 2008). The criterion, which allows one to relate water to one or other type, is the bonding energy. It exceeds 1.7 kJ mol^{-1} (bonded water in pores, which are less than 3 nm). This value is less by two orders of magnitude compared with the hydration energy of ions. In fact, it is close to the energy of hydrogen bonds. Water of other type (boundary water) is located in pores with a diameter in the range of 3–10 nm. Larger pores contain also free water. These types of water were recognized with MSCP (method of standard contact porosimetry), which is described further.

Within the framework of the method of differential scanning calorimetry (DSC, see below), three types of water in ion exchange polymers are considered as non-freezing, freezable, and free water (Karlsson et al. 2002). Non-freezing water exists in dry polymers as residual water content. It is recognized with the method of nuclear magnetic resonance, and cannot be removed without polymer destruction. The non-freezing is strongly bonded with functional groups. The freezable water weakly interacted with the polymer chains; a broad melting range is characteristic of this type of water. Relatively to the MSCP, the freezable water involves free and probably boundary water. The melting point of free water in pores of ionic polymers is below 0°C.

Two types of non-freezing water have been found for poly(ether ether ketone) containing sulfonate groups (Al Lafi and Hay 2015). The first type is water associated with functional groups; the content of this water depends on ion exchange capacity of the polymer. Other type of water depends on the polymer cross-linkage.

The model of structure evolution of ion exchange polymers during swelling is considered by Gebel on the example of Nafion 117 and Nafion 115 membranes (Gebel 2000). The approach

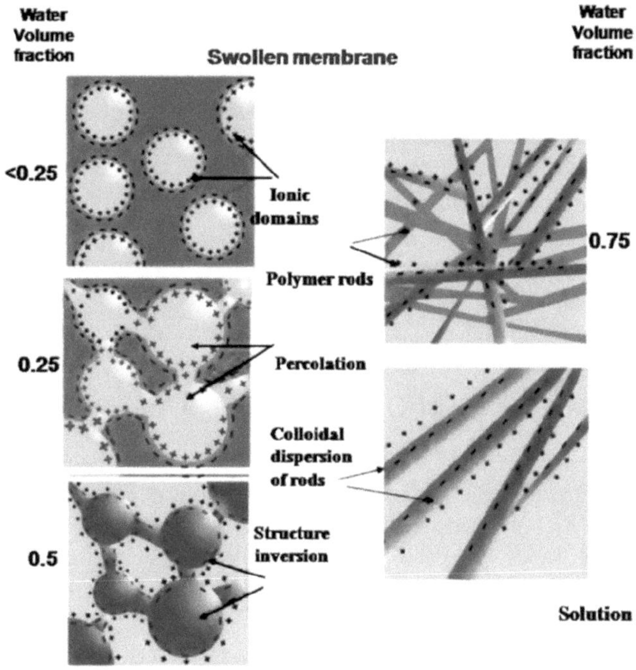

FIGURE 3.8 The model of swelling ion exchange polymer proposed in (Gebel 2000): light regions correspond to water, dark regions are attributed to polymer.

Redrawn from (Kononenko et al. 2017) according to the permission of Elsevier.

has been developed based on the SAXS and conductivity measurements. During swelling of the dry polymer, a modification of the structure of heterogeneities occurs: they become spherical water pools with a diameter of about 2 nm, which are isolated from each other (Figure 3.8). They contain functional groups at the interface between the polymer and water: the interfacial energy is minimized in this manner. The distance between water pools reaches 3 nm. Their isolation provides extremely low ionic conductivity of the membranes (about 1×10^{-3} S cm^{-2}). In the case of cationic polymer materials, there are positively charged fixed ions on the walls of the water pools, negatively charged counter-ions are in the water.

When the water content increases, water pools are consolidated into larger clusters, their diameter increases from 2 to 4 nm, the distance between them increases insufficiently. Narrow bridges are formed between clusters resulting in the organization of their continuous network and rapid increase of ionic conductivity. As supposed, the percolation is due to the effect of the interfacial energy on the one hand and the swelling limitations on the other hand. Clusters filled with water forma continuous network, the polymer islands are isolated.

Under higher water content, structure inversion occurs. In this case, the membrane is a network of the aggregates of the rods of polymer. The aggregates have spherical thickening, which isolates water pools. The hydrophobic phase (polymer rods) is solid, hydrophilic phase (water and ions inside it) is liquid.

Further increase of water content (volume fraction of 0.5–0.75) causes reorganization of polymer rods. Spherical shape is energetically beneficial due to the minimization of interfacial energy. However, it is impossible to reach this structure under high water content. The reason for this is a too large volume of the polymer, where functional groups are located, compared with the distance between functional groups. Conjoined cylindrical particles can be considered as the compromise option between the packing limitations and minimization of interfacial energy. The rod radius is 25 nm.

It should be noted that the experimental data are given only for the membranes, in which a volume fraction of water does not exceed 0.67. Note that the Gebel model is in agreement with the Gierke approach when the water fraction is 0.25–0.5. The polymer reorganization under high water content was studied with a SANS method later (Fumagalli et al. 2015). As opposed to water, which is bonded only with functional groups, methanol wets also the hydrophobic regions of ion exchange polymers (Vishnyakov and Neimark 2001).

According to all modern approaches to swollen ion exchange polymers (except Gebel views for the polymers containing extremely high amounts of water), water and polymer are different phases. It means, the voids filled with water can be considered as pores. They can be recognized with porosimetric methods, which involve water as the working liquid. These methods will be considered further.

3.5 METHODS FOR RESEARCH OF POROUS STRUCTURE OF IONIC POLYMERS

Several direct methods are applied to the investigation of porous polymers, such as electron microscopy, bubble point, gas or liquid adsorption-desorption, mercury intrusion, differential scanning calorimetry, and gas and liquid permeability. The bubble point technique is not suitable for ion exchange polymers, moreover, it can be applied only to monolithic materials, i.e. membranes. The scientific basis of the methods for pore recognition in ion exchange polymers is presented in detail (Kononenko et al. 2017; Tanis-Kanbur et al. 2021). Here we describe the results of measurements shortly.

3.5.1 Nitrogen Adsorption-Desorption

The method of nitrogen adsorption-desorption, which allows us to recognize only micro- and mesopores, provides the research of dry substances. The removal of free and bonded water from the samples is desirable during their pretreatment. Since micro- and mesoporosity of ion exchange polymers are realized only in swelling state, the technique gives inadequate information about pore size distribution. For this reason, the data for ion exchange materials are scarce and scattered. Figure 3.9 illustrates original data for strongly basic gel-like anion exchange resin Dowex SBR-P. These voids are evidently attributed only to the surface of glass-like granules. Further, the resin was modified with hydrated oxides of multivalent metals for ion exchange removal of As(V)-containing anions from water (Maltseva et al. 2019; Kolomiyets et al. 2016), and for electromembrane recovery of $HCrO_4^-$ ions from a multicomponent solution (Kolomiets et al. 2019).

Similar dV/dlogr values were obtained in (Saha and Streat 2005) for ion exchange resins produced by Purolite company. However, unexpected data of specific surface area are declared ($\approx$1000 m^2g^{-1}).

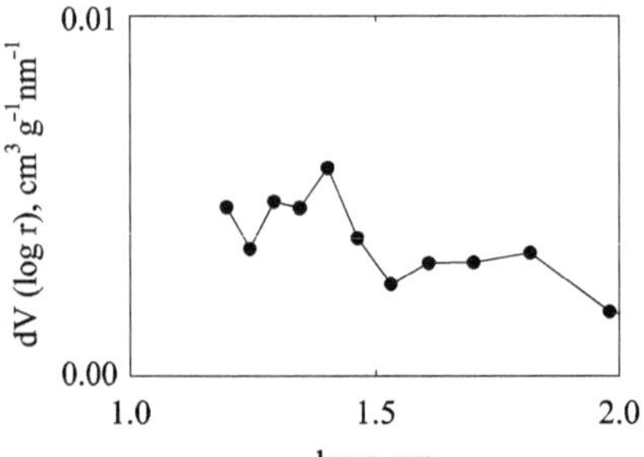

FIGURE 3.9 Pore size distribution for Dowex SBR-P anion exchange gel resin obtained with a method of nitrogen adsorption-desorption (original data).

TABLE 3.2
Evolution of Porous Structure of Homogeneous Anion Exchange Membranes Over Electrodialysis of Milky Whey

Life-time, %	S, m^2g^{-1}	V, cm^3g^{-1}	Fraction of Hydrophobic Pores
0	0	0	0.118
45	32	0.11	0.126
70	212	0.34	0.138
100	60	0.18	0.162

Source: Based on the data of (Garcia-Vasquez et al. 2013).

These unrealistic data are probably caused by incorrect calculations. For gel-like ion exchange resins, the specific surface area is less than 0.1 $m^2\ g^{-1}$ (Kun and Kunin 1967). In the case of macroporous resins, this value is 1.8–125 $m^2\ g^{-1}$. In general, the catalogs of the producers of ion exchange resins give the values up to 100 $m^2\ g^{-1}$.

Rather a large volume of pores (0.214 $cm^3\ g^{-1}$) has been reported by Pan et al. for polystyrenesulfone cation exchanger D-0.001 (Pan et al. 2007). Compared with other dry ion-exchange materials, this method gives a rather large specific surface area (0.214 $cm^3\ g^{-1}$). This resin was modified with zirconium hydrophosphate, the composite is selective toward Pb^{2+} ions.

The method of nitrogen adsorption-desorption was applied to investigations of AMX-SB homogeneous anion exchange membrane (Astom, Japan); the samples of the material were taken from an industrial electrodialysis apparatus for whey desalination. The membranes were depleted by 45–100% of their lifetime (Garcia-Vasquez et al. 2013). The evolution of the porous structure was studied in (Table 3.2). Regarding "fresh" membrane, no porosity was detected. Under electrodialysis conditions, the development of porosity occurs up to 70% of the life-time due to crack formation. Further, the newly formed pores are blocked with organic substances and insoluble inorganic salts. These pores are hydrophobic.

The data obtained with the method of nitrogen adsorption-desorption were represented for the heterogeneous cation exchange membrane obtained by assembling polyvinylchloride with strongly acidic resin (Hosseini et al. 2020). The specific surface area and average pore radius were estimated as 15 m^2g^{-1} and 14.73 nm respectively.

3.5.2 Mercury Intrusion

As compared with the method of nitrogen adsorption-desorption, mercury intrusion allows us to recognize larger pores. However, the requirement of this technique is dry material. Nevertheless, this approach was used in order to investigate a number of ion exchange polymers, especially macroporous resin and heterogeneous membranes.

A number of meso- and macroporous resins have been synthesized by means of polymerizing a mixed monomer solution of chloromethylstyrene and divinylbenzene followed by the treatment of polymer with trimethylamine (Takeda et al. 1986). The resulting resins are strongly basic anion exchangers, the capacity of which is 2.1–2.8 mmol g^{-1}. The samples, which possess higher exchange capacity, show smaller pores. Some resins show very small pore volume (up to 0.01 $cm^3\ g^{-1}$) similar to commercial resins produced by Dow Chemicals company. Certain samples demonstrate significant values of pore volume (0.25–0.4 $cm^3\ g^{-1}$). The volume of hydrophobic pores was determined by means of their filling with additionally sorbed highly concentrated electrolyte ($NdCl_3$). As found, the fraction of these pores relative to total porosity is 0.25–0.53. These results look unreliable since the total porosity was determined for dry samples, while hydrophobic pores were estimated for swollen resins (Figure 3.10).

Mercury intrusion was also applied to the investigations of other macroporous ion exchange resins (Kunin et al. 1962; Kun and Kunin 1967; Wood and Cooper 2001; Dunn et al. 1991). Especially this method is useful for the investigations of open-cell monolithic cation exchange materials (Inoue et al. 2004). As opposed to known ion exchange resins, which are obtained as granules, it is similar to polymer foam (Figure 3.11). The average pore radius was estimated as ≈10–11 μm. Very large pore volumes of 8.4–8.7 cm^3g^{-1} have been found. At the same time, the ion exchange capacity reaches 3.4–4.8 mmol g^{-1}: it is similar to that for strongly acidic resins.

Much lower pore size volumes (0.05–0.1 cm^3g^{-1}) have been found for heterogeneous ionic membranes, such as MA-40 and MK-40 produced by Schekinoazot, RF (Dobrevsky and Zvezdov 1979). However, the radius of dominant pores is 5 nm. These data are given for the materials before their usage in the electrodialysis process. However, the development of porosity after the operation (up to 0.25 cm^3g^{-1}) is suggested. The authors explain the change of porosity by the destruction of the membranes. As shown, the materials are destroyed unevenly in width and thickness. The porosity development depends on the membrane location in the electrodialysis apparatus (in the middle or from the side of the electrode). Nevertheless, no change in the size of dominant pores is reported.

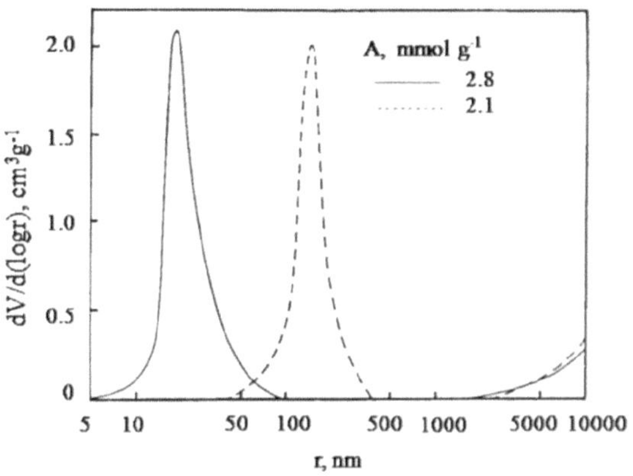

FIGURE 3.10 Pore size distribution for anion exchange resins. Plotted based on the data of (Takeda et al. 1986).

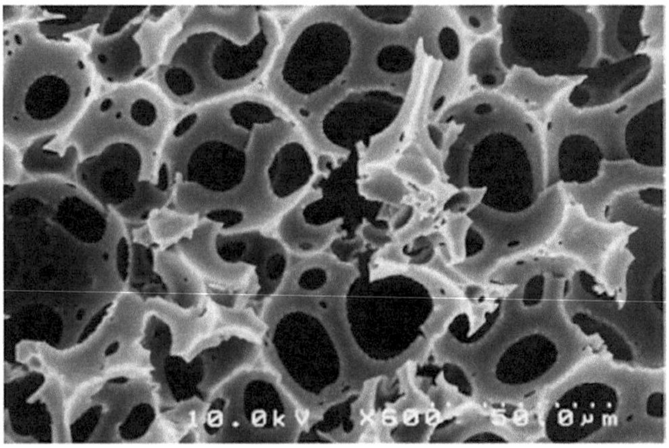

FIGURE 3.11 SEM image of open cell monolytic cation exchange resin (Inoue et al. 2004).

Reproduced according to the permission of Elsevier publisher.

3.5.3 Adsorption-Desorption of Water Vapor

Adsorption isotherms of vapor can be obtained using a vacuum spring Mac-Ben-Bakr scale or with isopiestic method. Water is usually used as a working liquid. In the case of isopiestic method, the needed pressure of water vapor is provided in a closed vessel under the solution of predetermined composition. Isopiestic method has been adapted to ion exchange polymers by Nandan and Gupta (1977). Isotherms of adsorption of water and methanol on polystyrenesulfonic acid type cation exchanger were obtained in this manner. Further, the isopiestic method was applied to ion exchange resins (Pushpa et al. 1988; Nandan et al. 1993; Toteja et al. 1997) or membranes (Krisilova et al. 2009; Zawodzinski et al. 1991) loaded with inorganic (Pushpa et al. 1988; Nandan et al. 1993; Toteja et al. 1997) and organic ions (Pushpa et al. 1988; Krisilova et al. 2009). No pore size distributions were calculated from the isotherms in mentioned works similar to the classical Helfferich monograph (Helfferich 1995).

Adsorption of water vapor provides recognition of micro- and mesopores. Porous structure of gel-ion exchange resin, which is strongly acidic, was estimated from the isotherms (Figure 3.12, Dzyazko et al. 2016). First of all, let us consider the isotherm of water adsorption-desorption. A hysteresis loop indicates the mesoporous structure of the resin. The loop can be divided into two regions. Region I is attributed to the loop of A-type (opened pores of cylindrical shape), the region

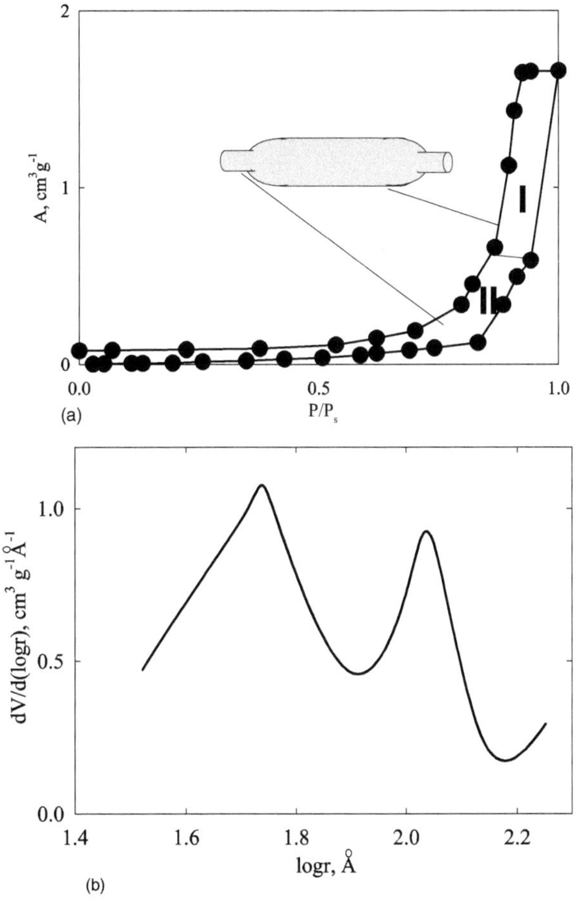

FIGURE 3.12 Isotherm of water vapor adsorption-desorption on-from Dowex HCR-S strongly acidic gel ion-exchange resin (a), pore size distribution (b). The graphs are plotted based on data (Dzyazko et al. 2016). Insertion: hydrophilic pores of ion exchange polymer.

II corresponds to E-type (pores with narrow constrictions) (Gregg and Sing 1991). This is in agreement with the cluster-channel model of ion exchange polymers (see Figure 3.4).

The sufficient disadvantage of the method is a narrow interval of pore size. It is possible to determine accurately only those pores with a radius that does not exceed 10 nm. The techniques, which allow us to recognize larger pores, are described further.

3.5.4 Differential Scanning Calorimetry

Differential scanning calorimetry (thermoporometry, DSC) gives a possibility to investigate the porosity of swollen ionic polymers in a wider interval of pore radius compared with the methods described above. As noted by Landry (2005), the DSC method recognizes the size of pores based on a decrease in the crystallization or melting of solvent in pores. This technique is sensitive to the liquid transitions (endothermic melting and exothermic freezing). In order to study ion exchange polymer materials, water is the most suitable working liquid, since their practical application requires contact with aqueous solutions in many cases. Another advantage of water is its fusion heat (334 J g^{-1}). It is an order of magnitude higher compared with the majority of organic liquids. Thus, DSC is sensitive to a small amount of water. The DSC technique allows us to recognize pores, which are no larger than 200 nm.

The application of the DSC technique to perfluorinated and hydrocarbon proton-conducting membranes has been studied by Kraemer et al. (2008). Compared with perfluorinatedpolymer (water uptake is 21% and 34% at the pretreatment temperature of 25°C and 80°C respectively), hydrocarbon membranes adsorb higher amount of water (up to 51% for the pretreatment at 25°C and 120% at 80°C). However, a larger volume of hydrophilic pores has been found namely for perfluorinated membrane (Nafion 117), which was pretreated at 25°C. Increasing the pretreatment temperature causes a larger pore volume for the hydrocarbon membrane than that for Nafion. About half of the total water content is located in channels, and the rest is in larger pores.

Later, the DSC method was applied to the investigation of heterogeneous ion exchange membranes (Figure 3.13, Kononenko et al. 2017). Here the intensive narrow peak at 20 nm is visible, it evidently corresponds to clusters. Larger hydrophobic pores are expressed as a wide shoulder.

Poly(ether ether ketone) membranes, which are characterized by various cross-linkage, were investigated (Al Lafi and Hay 2012). The radius of the clusters is independent of cross-linkage. Their volume increases with an increase of ion exchange capacity. This effect is stronger than the influence of cross-linkage.

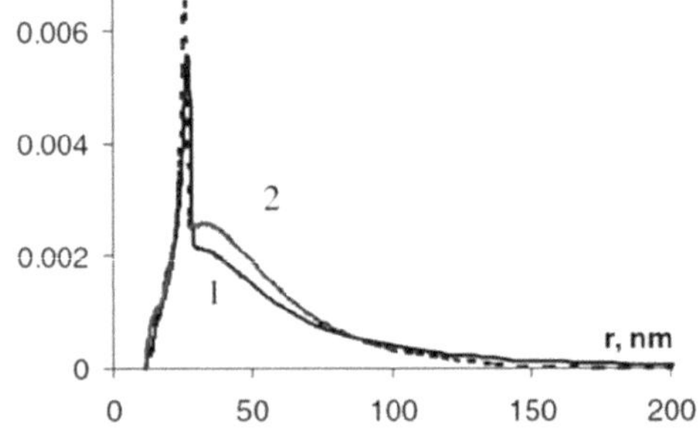

FIGURE 3.13 Pore size distributions obtained with DSC method for MK-40 heterogeneous cationic membrane.

Reproduced from (Kononenko et al. 2017) according to the permission of Elsevier publisher.

The disadvantage of DSC is the unknown effect of interactions of liquid with the surface of ionic polymers (Landry 2005). It is postulated that the temperature dependence of contact angle, surface tension, specific volume, and heat of fusion are defined. Nevertheless, the information given in the literature regarding this problem are very different. This makes the conversion of calorimetry data into pore size distributions difficult. As a result, some authors avoid this transformation. For instance, the DSC method was applied to the investigation of the materials, which were synthesized by sulfonation of polystyrene, polycarbonate, polysulfone, and poly(phenylene oxide) (Smitha et al. 2003). In all cases, sulfonation decreases the temperature of glass transition. As a result, the state of polymer can be changed from more amorphous to more crystalline. No transformation of the DSC data into pore size distribution was performed for the polyacrylonitrile-based composites containing sodium titanosilicate (Saberi et al. 2010), hexacyanoferrate of copper (Nilchi et al. 2011) or ammonium molybdophosphate (Nilchi et al. 2012), sulfonated poly(ether ether ketone) (Al Lafi and Hay 2015), and Nafion (Thompson et al. 2006; Lue and Shieh 2019, Voropaeva et al. 2018).

3.5.5 Standard Contact Porosimetry

The MSCP is related to untraditional techniques of pore recognition, however, it has been confirmed by the IUPAC in 2012 (Rouquerol et al. 2011). Theoretical basis of this method is represented in a number of reviews (Volfkovich and Sosenkin 2012; Volfkovich et al. 2001; Volfkovich and Bagotzky 1994; Kononenko et al. 2017). Similar to the mercury intrusion porosimetry, the method is valid in a very broad diapason of pore radius (from 1 nm to 100 μm). As opposite to this technique, the MSCP can be applied to swollen polymers. It is very important since the measurement conditions correspond to real processes involving ionic polymers. Moreover, no destruction of a sample occurs. Additionally, small pores can be determined correctly with MSCP. In the case of mercury intrusion, their walls are destroyed under high pressure.

Unlike the methods of DSC and liquid adsorption isotherms, the MSCP gives the information about porous structure in a wider interval of pore size –hydrophobic pores of micron size (structure defects) can be also recognized. Figures 3.14a and d illustrate typical pore size distributions for strongly acidic gel cation exchanger Dowex HCR-S (Dzyazko et al. 2012, 2014). All types of pores are seen. Pores, for which $\log r < 0.3$ nm, are related to channels, the maximum at $\log r = 1$ nm corresponds to clusters. At the same time, the peak at $\log r = 1.5$ nm is due to the pores between hydrophobic fragments of polymer chains. The structure defects are expressed as a broad peak at $\log r = 3.2$ nm. Larger pores are related to both structure defects and cavities between beads. As seen from Figure 3.14a, the pore volume of a swollen ion exchanger is much larger compared to that of dry material (in this case the method of nitrogen adsorption-desorption was used). The volume of structure defects is much smaller compared with hydrophilic pores. As opposite to gel-like resin, the macroporous ion exchanger shows a much larger fraction volume of structure defects relative to clusters and channels (Ponomarova et al. 2018).

Reorganization of the porous structure of gel-like resin during the treatment with polar organic solvent has been suggested in (Dzyazko et al. 2017a). In this case, the diffusion peak, which is attributed to clusters (see Figures 3.14a and d), becomes narrower and more intensive. It means ordering hydrophilic pores.

The hydration of structure defects leads to a growth of water amount in hydrophilic pores (compare the data of Figures 3.12b and 3.14a,d). In other words, the amount of free water increases. From the thermodynamical point of view, the value of water activity (a_{H2O}) becomes larger. In fact, this magnitude corresponds to the content of free water. Increasing the water activity causes a decrease of swelling pressure (π), which is equal to (Helfferich 1995):

$$\pi = -\frac{RT \ln a_{H_2O}}{v_{H_2O}} \tag{3.1}$$

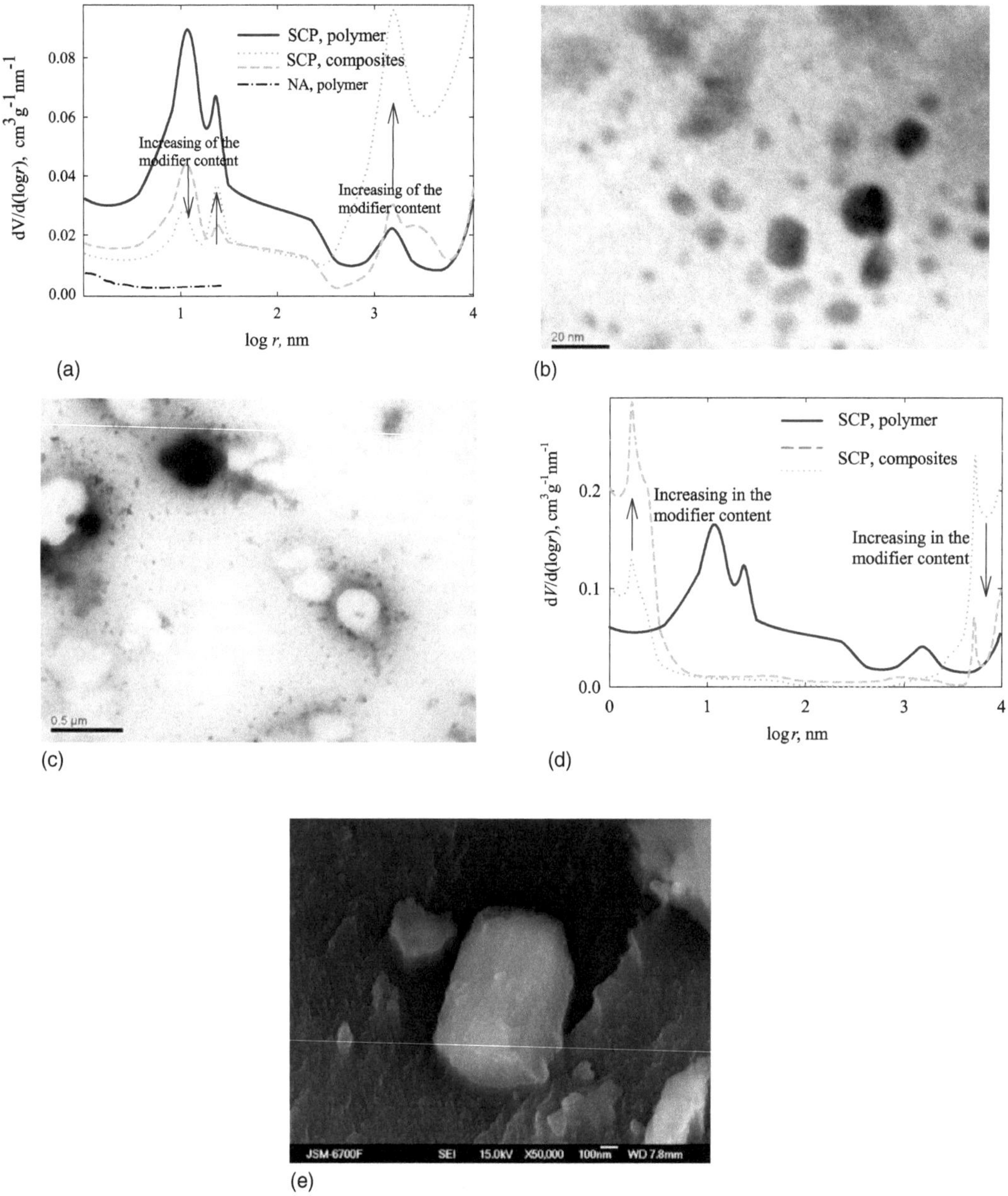

FIGURE 3.14 Dowex HCR-S strongly acidic gel resin and its composite containing primary, aggregated and agglomerated nanoparticles of zirconium hydrophosphate (a–c), only agglomerates (d, e). Pore size distributions were obtained with MSCP (a, d) and nitrogen adsorption-desorption technique (a), they are plotted based on data of: a –(Dzyazko et al. 2012), b –(Dzyazko et al. 2014). Non-aggregated nanoparticles and aggregates are seen in TEM image (b), agglomerates are visible in TEM (c) and SEM (e) images.

here T is the temperature, R is the gas constant, v_{H_2O} is the partial molar volume. A decrease of swelling pressure promotes the filling of hydrophilic pores with water.

The MSCP was applied to ionic polymers, namely to the polymer ion exchange sorbents (Kononenko et al. 1985) as well as to membranes (Berezina et al. 1994). First of all, heterogeneous membranes were investigated. The works devoted to the MSCP application to the study of ionic polymers are summarized in (Berezina et al. 2008). A very interesting aspect was stressed.

The indirect method, which allows us to determine the fraction pore volume, provides the measurements of electrical conductivity of the membrane. When the conductivities of the membrane (κ_m) and solution (κ_s) are equal, the following expression is valid:

$$\kappa_m = \kappa^a \kappa_s^{1-a} \tag{3.2}$$

where κ is the conductivity of a membrane impregnated with deionized water (without the contribution of surface conductivity), i.e. the conductivity due to hydrophilic regions, a and $1 - a$ are the fraction volumes of hydrophilic and hydrophobic pores, respectively. The solution concentration strongly affects the membrane conductivity. First of all, the κ_m-κ_s dependence is plotted on logarithmic coordinates. The concentration interval, where linearity is observed, is chosen. This diapason includes the point of $\kappa_m = \kappa_s$. The linear region is extrapolated to the ordinate axis, the κ value is obtained in this manner. The slope corresponds to $1 - a$. The fraction pore volumes obtained in this manner are in agreement with the data of MSCP.

Equation 3.2 is valid when the contribution of resistance at the membrane-solution interface is absent. When these conditions are not kept, the following expression is obtained (Dzyazko et al. 2015, 2017b):

$$\kappa_m = b_1 \kappa_s^{b_2} \tag{3.3}$$

Here b_1 and b_2 are the coefficients, which reflect screening of the conductivity of solution by the polymer matrix. The b_2 exponent is higher than 1 indicating the inapplicability of Equation 3.2. For preventing additional resistance of the interface of membrane-liquid, the membrane impregnated with a solution is placed between two mercury electrodes (Berezina et al. 1994). The MSCP advantage over this indirect method and mercury intrusion is a possibility to use water or less toxic working liquids such as octane, decane, etc.

Further, the MSCP was used for the investigations of a wide variety of ion exchange membranes. Perfluorinated membranes modified by polyaniline were investigated (Berezina et al. 2008): the maximum value of water content decreases by 8–15% compared with the pristine membrane depending on the aniline polymerization time. The volume of both hydrophilic and hydrophobic pores becomes smaller. Additional pores with size from hundreds of nanometers up to several microns, appear. Modifying causes a widening of the plateau of the volt-ammetric curves of the membrane. The potential of the overlimiting current for the composite is above 3 V, whereas this value is 0.85 V for the pristine membrane.

The absence of structure defects has been found for the AMX homogenous anion exchange membrane produced by Astom (Japan) (Sarapulova et al. 2019). When H^+ counter ions of this membrane are substituted by other ions, the pore volume is smaller compared with the H-form (Pismenskaya et al. 2019). However, the effect of counter ions on porosity is weaker than that for heterogeneous MA-41 membrane (Schekinoazot, RF). The conductivity of homogeneous membranes is higher in the electrolyte solutions. The pore structure of mosaic membranes, which includes both cation- and anion-exchange regions, is considered in (Volfkovich et al. 2018, 2020).

A very interesting field of MSCP application is polymer-inorganic ion exchange materials. Zirconium hydrophosphate (ZHP) is often used as a modifier to increase proton conductivity of membranes for fuel cell application. ZHP also enhances the selectivity of cation exchange resins toward toxic inorganic ions (Dzyazko et al. 2013a, 2014, 2017a) due to complex formation with phosphorus-containing functional groups of the modifier (Dzyazko et al. 2012). Hydrated oxides of zirconium and iron are applied to modify anion exchange materials (Perlova et al. 2018, Dzyazko et al. 2019).

The modifier nanopaticles are precipitated in different pores of polymer support. As a result, non-aggregated nanoparticles, aggregates, and agglomerates are formed. The type of particles strongly depends on the deposition conditions and the type of compound being deposited (Perlova et al. 2017, 2018, Maltseva et al. 2019).

Since the conditions of heating (≈80°C) before the measurements cannot provide the removal of water from the inorganic modifier, pore size distributions are related namely to an inorganic constituent. The size and volume of the polymer pores are affected by the type of embedded particles (see Figure 3.14). When ZHP nanoparticles, their aggregates, and agglomerates are present simultaneously in cation exchange resin (Figure 3.14b and c), the volume of hydrophilic pores decreases with increasing the modifier content (Figure 3.14a). This is due to their blockage with nanoparticles. However, the volume of hydrophobic pores becomes larger. Fixed ions and counter-ions of functional groups of ZHP in hydrophobic pores are the additional osmotic centers. As a result, osmotic pressure is realized in these voids. No change of cluster position has been found due to osmotic pressure caused by ZHP nanoparticles.

When the ion-exchanger contains only large agglomerates in structure defects, they squeeze hydrophilic and smaller hydrophobic pores, which are free from the modifier (Figure 3.14d). Increasing the modifier content causes gradual blockage of structure defects.

As known, an increase of cross-linkage of ion exchange polymer enhances its swelling pressure (Helfferich 1995). This value can be approximately estimated from the water adsorption isotherms by comparison with theoretical value for the polymer of similar type. Taking into consideration the modifier effect on the pressure of swelling, the embedding primary ZHP nanoparticles into the ionic polymer can be formally related to increasing the amount of cross-linking agents (Dzyazko et al. 2016). Aggregates in agglomerates in hydrophobic pores decrease swelling pressure. This effect is similar to a decrease incross-linkage.

Pore size distributions obtained with MSCP allow us to estimate the following parameters: α value, which is the relation of the volume of hydrophilic pores to the volume of hydrophobic pores, and γ magnitude, which is the relation of volumes of channels and clusters (Dzyazko et al. 2013a). A slightly modified α parameter has been proposed in (Dzyazko et al. 2013c): the volume of micropores was neglected. In the case of the composite modified with only agglomerates, the γ parameter was not determined, since all hydrophobic pores cannot be recognized exactly.

The α and γ parameters are changed due to the effect of incorporated particles. These values are represented as functions of the modifying cycle in Figure 3.15a. An increase in the α value and decrease of the γ parameter has been found for the composites after the first cycle of modifying. The sample only agglomerates shows lower α magnitude in comparison with the composites containing all types of particles. After the third modifying cycle, these parameters are not changed sufficiently. It means that the porous structure of the polymer constituent is not transformed. This change of porosity reflects on the electrochemical behavior of the composites. Figure 3.15b illustrates their electrical conductivity *vs* the concentration of mobile charge carriers, $\overline{C}$, for the composites containing all types of particles. Within the interval of the first–third modifying cycle, the conductivity decreases. This is caused by the transformation of the structure of the ionic polymer. Further, the conductivity increases practically linearly due to the contribution of the counter-ions of the functional groups of inorganic ionic modifier. Since ion transport is realized through hydrophilic pores, namely non-aggregated nanoparticles cause the growth of conductivity. Modifying ion exchange resin allows us to increase the conductivity of an ionic polymer three times (Dzyazko et al. 2013b). The influence of the porous structure of the polymer constituent on the conductivity of polymer-inorganic membranes is considered also in (Petrova et al. 2019; Kononenko et al. 2015). Thus, the MSCP method is useful for the clarification of the modifier effect on the functional properties of polymer-inorganic composites.

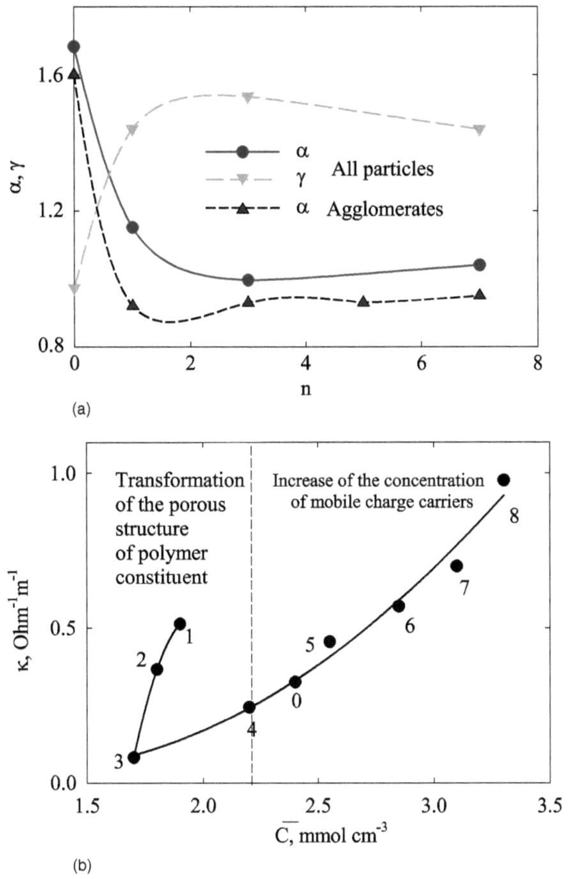

FIGURE 3.15 α and γ parameters *vs* the number of the modifying cycle (a), electrical conductivity of ion-exchangers *vs* concentration of mobile charge carriers (b). A Dowex HCR strongly acidic resin containing zirconium hydrophosphate was investigated. Numerals correspond to the modifying cycle (b).

3.6 CONCLUSIONS

As opposed to the majority of solids, the porous structure of ionic polymers is formed during swelling. A continuous system of hydrophilic pores provides their functional properties: the ability to sorb ionic species and electrical conductivity. Nowadays, the model of alternate clusters and channels is conventional despite some elaborations and even criticism. These hydrophilic pores are originated from heterogeneities, which exist in dry ionic polymers. These heterogeneities are caused by dipole-dipole interaction of polar functional groups. The polymers contain also hydrophobic pores, which are larger compared with hydrophilic pores. When ionic polymers are modified by an inorganic ion-exchanger, the modifier can be precipitated in different types of pores.

The porous structure of ionic polymers can be recognized with a number of methods. However, the MSCP looks most attractive, since it allows one to recognize pores in a diapason of five decades. Another advantage of this method is the possibility to estimate porous structure under the real conditions of processes, where ionic polymers are used.

REFERENCES

Allahyarov, E., and P.L. Taylor. 2007. Role of electrostatic forces in cluster formation in a dry ionomer. *J. Chem. Phys.* 127, no. 15 (October):154901. https://doi.org/10.1063/1.2793784

Al Lafi, A.G., and J.N. Hay. 2015. State of the water in crosslinked sulfonated poly(ether ether ketone). Two-dimensional differential scanning calorimetry correlation mapping. *Thermochim. Acta* 612: 63–69.

Al Lafi, A.G., and J.N. Hay. 2012. State of the water in crosslinked sulfonated poly(ether ether ketone). *J. Appl. Polym. Sci.* 128: 3000–3009.

Arnold, R., and D.F.A. Koch. 1966. Electrical conductivity of cation-exchange membranes in the hydrogen ion form. *Austral. J. Chem.* 19: 1299–1313.

Berezina, N.P., N.A. Kononenko, O.A. Dyomina, et al. 2008. Characterization of ion-exchange membrane materials: Properties vs structure. *Adv. Colloid. Interface Sci.* 139: 3–28.

Berezina, N.P., N.A. Kononenko, and Yu M. Volfkovich. 1994. Hydrophilic properties of heterogeneous ion-exchange membranes. *Russ. J. Electrochem.* 30: 329–335.

Borodin, P.M., M.K. Nikitin, and E.N. Sventitskii. 1965. Study of the structure of an electrolyte in the ion-exchange resin phase by the nuclear magnetic resonance method. *J. Struct. Chem.* 6: 176–178.

Dobrevsky, J., and A. Zvezdov. 1979. Investigation of pore structure of ion exchange membranes. *Desalination* 28: 283–289.

Dorenbos, G. 2020. How hydrophobic side chain design affects water cluster connectivity in model polymer electrolyte membranes: Linear versus Y-shaped side chains. *J. Hydrogen Energy* 45(58): 33906–33924.

Dreifus, B. 1986. Clustering and hydration in ionomers. In: *Coulombic interaction in macromolecular systems*, ed. A. Eisenberg and F.E. Bailey, 103–119. *ACS Symposium. Series* no. 302. Washington: American Chemical Society.

Dunn, L., M. Aboulezz, L. Cummings, et al. 1991. Characterization of synthetic macroporous ion-exchange resins in low-pressure cartridges and columns. *J. Chromatography A* 548: 165–178.

Dzyazko, Yu S., L.N. Ponomareva, Yu M. Volfkovich, et al. 2012. Effect of the porous structure of polymer on the kinetics of Ni^{2+} exchange on hybrid inorganic-organic ionites. *Russ. J. Phys. Chem.* 86: 913–919.

Dzyazko, Yu S., L.N. Ponomaryova, Yu M. Volfkovich, et al. 2013a. Polymer ion-exchangers modified with zirconium hydrophosphate for removal of Cd^{2+} ions from diluted solutions. *Sep. Sci. Technol.* 48: 2140–2149.

Dzyazko, Yu S., V.V. Trachevskii, L.M. Rozhdestvenskaya, et al. 2013b. Interaction of sorbed Ni(II) ions with amorphous zirconium hydrogen phosphate. *Russ. J. Phys. Chem.* 87: 840–845.

Dzyazko, Yu S., L.N. Ponomareva,Y.M. Volfkovich, et al. 2013c. Conducting properties of a gel ionitemodied with zirconium hydrophosphate nanoparticles. *Russ. J. Electrochem.* 49: 209–215.

Dzyazko, Yu S., L.N. Ponomaryova, Yu M. Volfkovich, et al. 2014. Ion-exchange resin modified with aggregated nanoparticles of zirconium hydrophosphate. Morphology and functional properties. *Micropor. Mesopor. Mater.* 198: 55–62.

Dzyazko, Yu, L. Rozhdestveskaya, Yu Zmievskii, et al. 2015.Heterogeneous membranes modified with nanoparticles of inorganic ion-exchangers for whey demineralization. *Mater. Today: Proc.* 2: 3864–3873.

Dzyazko, Yu, L. Ponomarova, Yu Volfkovich, et al. 2016. Influence of zirconium hydrophosphate nanoparticles on porous structure and sorption capacity of the composites based on ion exchange resin. *Chem. Chem. Technol.* 10: 329–335.

Dzyazko, Yu S., O.V. Perlova, N.A. Perlova, et al. 2017a. Composite cation-exchange resins containing zirconium hydrophosphate for purification of water from U(VI) cations. *Desalin. Water Treat.* 69: 142–152.

Dzyazko, Y.S., L.M. Rozhdestvenska, S.L. Vasilyuk, et al. 2017b. Composite Membranes Containing Nanoparticles of Inorganic Ion Exchangers for Electrodialytic Desalination of Glycerol. *Nanoscale Res. Let.* 12 (June): 438. https://doi.org/10.1186/s11671-017-2208-4

Dzyazko,Yu, E. Kolomyets, Yu Borysenko, et al. 2019. Organic-inorganic sorbents containing hydrated zirconium dioxide for removal of chromate anions from diluted solutions. *Mater. Today: Proc.* 6: 260–269.

Eisenberg, A. 1970.Clustering of ions in organic polymers. A theoretical approach. *Macromolecules.* 3: 147–154.

Eikerling, M., A.A. Kornyshev, and U. Stimming. 1997. Electrophysical properties of polymer electrolyte membranes: A random network model. *J. Phys. Chem. B* 101: 10807–10820.

Elliott, J.A., S. Hanna, H.H. Wills, et al. 2000. Interpretation of the small-angle X-ray scattering from swollen and oriented perfluorinated ionomer membranes. *Macromolecules* 33: 4161–4171

Fumagalli, M., S. Lyonnard, G. Prajapati, et al. 2015. Fast water diffusion and long-term polymer reorganization during nafion membrane hydration evidenced by time-resolved small-angle neutron scattering. *J. Phys. Chem. B* 119: 7068–7076.

Garcia-Vasquez, W., L. Dammak, C. Larchet, et al. 2013. Evolution of anion-exchange membrane properties in a full scale electrodialysis stack. *J. Membrane Sci.* 446: 255–265.

Gebel, G. 2000. Structural evolution of water swollen perfluorosulfonated ionomers from dry membrane to solution. *Polymer* 41: 5829–5838.

Gierke, T.D., G.E. Munn, and F.C. Wilson 1981. The morphology in nafion perfluorinated membrane products, as determined by wide-and small-angle x-ray studies. *J. Polym. Sci.: Polym. Phys. Ed.* 19: 1687–1704.

Glueckauf, E. 1962. A new approach to ion exchange polymers. *Proc. R. Soc. Lond. A* 268: 350–370.

Gregg, S.J., and K.S.W. Sing.1991. *Adsorption, surface area and porosity*. London: Academic Press.

Guo, D., C.X. Lin, E.N. Hu, et al. 2017. Clustered multi-imidazolium side chains functionalized alkaline anion exchange membranes for fuel cells. *J. Membrane Sci.* 541: 214–223.

Halim, J., F.N. Büchi, O. Haas, et al. 1994. Characterization of perfluorosulfonic acid membranes by conductivity measurements and small-angle X-ray scattering. *Electrochim. Acta* 39: 1303–1307.

Helfferich, F. 1995. *Ion exchange*. 2nd ed. New York: Dover.

Hodgdon, R.B., and J.R. Boyack. 1965. Study of swelling in two new ion exchange membranes. *J. Polymer Sci. Part A: General Papers* 3: 1463–1472.

Hosseini, S.M., H. Alibakhshi, E. Jashni, et al. 2020. A novel layer-by-layer heterogeneous cation exchange membrane for heavy metal ions removal from water. *J. Hazard. Mater.* 381: 120884.

Hsu, W.Y., and T.D. Gierke. 1983. Ion transport and clustering in nafion perfluorinated membranes. *J. Membrane Sci.* 13: 307–326.

Inoue, H., K. Yamanaka, A. Yoshida, et al. 2004. Synthesis and cation exchange properties of a new porous cation exchange resin having an open-celled monolith structure. *Polymer* 45: 3–7.

Ioselevich, A.S., A.A. Kornyshev, and J.H.G. Steinke. 2004. Fine morphology of proton-conducting ionomers. *J. Phys. Chem. B* 108: 11953–11963.

James, P.J., J.A. Elliott, T.J. McMaster, et al.2000. Hydration of Nafion studied by AFM and X-ray scattering. *J. Mater. Sci.* 35: 5111–5119.

Karlsson, L.E., B. Wesslen, and P. Jannasch. 2002. Water absorption and proton conductivity of sulphonated acrylamide copolymers. *Electrochem. Acta* 47: 3269–3275.

Kolomiets, E.O., T.V. Maltseva, and Yu S. Dzyazko. 2019. Organic-inorganic ion-exchangers for selective electrodeionization of diluted aqueous solutions. *Voprosy Khimii i Khimicheskoi Tekhnologii*. 1: 18–25.

Kolomiyets, Y., V. Belyakov, A. Palchik, et al. 2016.Effect of incorporation of hydrated oxides of Sn (IV), Zr (IV) and Fe (III) in a matrix of the anion exchanger dowex SBR-P on the sorption capacity towards the arsenic (V) anions. *Int. J. Water Wastewater Treat.* 2. http://dx.doi.org/10.16966/2381–5299.120

Kononenko, N.A., N.P. Berezina, Yu M. Volfkovich, et al.1985. Investigation of ion-exchange materials structure by standard porosimetry method, *J. Appl. Chem. USSR* 58: 2029–2033.

Kononenko, N.A., M.A. Fomenko, and Y.M. Volfkovich. 2015. Structure of perfluorinated membranes investigated by method of standard contact porosimetry. *Adv. Colloid. Interface Sci.* 222: 425–435.

Kononenko, N., V. Nikonenko, D. Grande, et al. 2017. Porous structure of ion exchange membranes investigated by various technique. *Adv. Colloid. Interface Sci.* 246: 196–216.

Kraemer, S., A.I. Sagidullin, G. Lindbergh, et al. 2008. Pore size distribution and water uptake in hydrocarbon and perfluorinated proton-exchange membranes as studied by NMR cryoporometry. *Fuel Cells*: 262–269.

Krisilova, E.V., T.V. Eliseeva, and V.F. Selemenev. 2009. The hydration of cation exchange membranes after sorption of basic amino acids. *Russ. J. Phys. Chem.* 83: 1954–1956.

Kun, K.A., and R. Kunin. 1967. The pore structure of macroreticular ion exchange resins. *J. Polymer Sci. Part A: General Papers*. 16: 1457–1469.

Kunin, R., E. Meitzner, and N. Bortnick. 1962. Macroreticular ion exchange resins. *J. Am. Chem. Soc.* 84: 305–306.

Landry, M.R. 2005. Thermoporometry by differential scanning calorimetry: Experimental considerations and applications. *Thermochim. Acta* 433: 27–50.

Lee, N., D.T. Duong, and D. Kim. 2018. Cyclic ammonium grafted poly (arylene ether ketone) hydroxide ion exchange membranes for alkaline water electrolysis with high chemical stability and cell efficiency. *Electrochim. Acta* 271: 150–157.

Lowry, S.R., and K.A. Mauritz. 1980. An investigation of ionic hydration effects in perfluorosulfonate ionomers by Fourier transform infrared spectroscopy. *J. Am. Chem. Soc.* 102: 4665–4667.

Lue, S.J., and Shieh, S.-J. 2019. Water states in perfluorosulfonic acid membranes using differential scanning calorimetry. *J. Macromolec. Sci. Part B: Phys.* 48: 114–127.

Maltseva, T.V., E.O. Kolomiets, Yu S. Dzyazko, et al. 2019. Composite anion-exchangers modified with nanoparticles of hydrated oxides of multivalent metals. *Appl. Nanosci.* 9: 997–1004.

Manohar, M., P.P. Sharma, and D. Kim. 2021. Intercalated poly (2-acrylamido-2-methyl-1-propanesulfonic acid) into sulfonated poly (1,4-phenylene ether-ether-sulfone) based proton exchange membrane: Improved ionic conductivity. *Molecules* 26, no. 1 (December): 161. https://doi.org/10.3390/molecules26010161
Mauritz, K.A. 1995.A water sorption isotherm model for ionomer membranes with cluster morphologies. *Macromolecules* 18: 483–491.
Mauritz, K.A., and R.B. Moore. 2004. State of Understanding of Nafion. *Chem. Rev.* 104 (10): 4535–4585.
Millar, J.R., D.G. Smith, W.E. Marr et al. 1964. Solvent-modified polymer networks. Part III. Cation-exchange equilibria with some univalent inorganic and organic ions. *J. Chem. Soc.* 0: 2740–2746.
Mosdale, R., G. Gebel, and M. Pineri. 1996. Water profile determination in a running proton exchange membrane fuel cell using small-angle neutron scattering. *J. Membr. Sci.* 118: 269–277.
Nandan, D., and A.R. Gupta. 1977. Solvent sorption isotherms, swelling pressures, and free energies of swelling of polystyrenesulfonic acid type cation exchanger in water and methanol. *J. Phys. Chem.* 81:1174–1179.
Nandan, D., B. Venkataramani, and A.R. Gupta. 1993. Ionic hydration and water sorption isotherms of ion exchange resins. *Langmuir* 9:1786–1793.
Nilchi, A., R. Saberi, M. Moradi, et al. 2012. Evaluation of AMP–PAN composite for adsorption of Cs+ ions from aqueous solution using batch and fixed bed operations. *J. Radioanal. Nucl. Chem.* 292: 609–617.
Nilchi, A., R. Saberi, M. Moradi, et al. 2011. Adsorption of cesium on copper hexacyanoferrate–PAN composite ion exchanger from aqueous solution. *Chem. Eng. J.* 172: 572–580.
Pan, B.C., Q.R. Zhang, W.M. Zhang et al. 2007. Highly effective removal of heavy metals by polymer-based zirconium phosphate: A case study of lead ion. *J. Colloid. Interface Sci.* 310: 99–105.
Petrova, D.A., A.N. Filippov, N.A. Kononenko, et al. 2019. Perfluorinated hybrid membranes modified by metal decorated clay nanotubes. *J. Membrane Sci.* 582: 172–181.
Perlova, N., Y. Dzyazko, O. Perlova, et al. 2017. Formation of zirconium hydrophosphate nanoparticles and their effect on sorption of uranyl cations, *Nanoscale Res. Lett.* 12 (March): 209.doi.org/10.1186/s11671–017–1987-y.
Perlova, O., Yu Dzyazko, I. Halutska, et al. 2018. Anion exchange resin modified with nanoparticles of hydrated zirconium dioxide for sorption of soluble U(VI) compounds. *Springer Proc. Phys.* 210: 3–15.
Pismenskaya, N., V. Sarapulova, E. Nevakshenova, et al. 2019. Concentration dependencies of diffusion permeability of anion-exchange membranes in sodium hydrogen carbonate, monosodium phosphate, and potassium hydrogen tartrate solutions. *Membranes* 9, no.12 (December): 170. https://doi.org/10.3390/membranes9120170
Pogorenko, Y.V., R.M. Pschenychnyi, A.O. Omelchuk, et al. 2019. Conductivity of aliovalent substitution solid solutions $Pb_{1-x}R_xSnF_{4+x}$ (R = Y, La, Ce, Nd, Sm, Gd) with β-$PbSnF_4$ structure. *Solid State Ionics* 338: 80–86.
Pogorenko, Y.V., R.M. Pschenychnyi, A.O. Omelchuk, et al. 2016.Electric conductivity of heterovalent substitution solid solutions of the $(1-x)PbF_2$–xYF_3–SnF_2 system. *Rus. J. Electrochem.* 52: 374–384.
Ponomarova, L., Yu Dzyazko, Yu Volfkovich, et al. 2018. Effect of incorporated inorganic nanoparticles on porous structure and functional properties of strongly and weakly acidic ion exchangers. *Springer Proc. Phys.* 214: 63–77.
Pushpa, K.K., D. Nandan, and R.M. Iyer. 1988. Thermodynamics of water sorption by perfluorosulphonate (Nafion-117) and polystyrene–divinylbenzene sulphonate (Dowex 50W) ion-exchange resins at 298 ± 1 K. *J. Chem. Soc., Faraday Trans.* 84:2047–2056.
Rouquerol, J., G. Baron, R. Denoyel, et al. 2011. Liquid intrusion and alternative methods for the characterization of macroporous materials (IUPAC Technical Report). *Pure Appl. Chem.* 84: 107–136.
Saberi, R., A. Nilchi, S. Rasouli Garmarodi, et al. 2010. Adsorption characteristic of 137Cs from aqueous solution using PAN-based sodium titanosilicate composite. *J. Radioanal. Nucl. Chem.* 284: 461–469.
Saha, B., and M. Streat. 2005. Adsorption of trace heavy metals: Application of surface complexation theory to a macroporous polymer and a weakly acidic ion-exchange resin. *Ind.* Chem. *Res.* 44: 8671–8681.
Samsonov, G.V., and V.A. Pasechnik. 1969. Ion exchange and the swelling of ion-exchange resins. *Russ. Chem. Rev.* 38: 547–566.
Sarapulova, V., I. Shkorkina, S. Mareev, et al. 2019. Transport characteristics of fujifilm ion-exchange membranes as compared to homogeneous membranes AMX and CMX and to heterogeneous membranes MK-40 and MA-41. *Membranes* 9, no.7 (July): 84. https://doi.org/10.3390/membranes9070084
Schmidt-Rohr, K., and Q. Chen. 2008. Parallel cylindrical water nanochannels in Nafion fuel-cell membranes. *Nature Mater.* 7: 75–83.
Smitha, B., S. Sridhar, and A.A. Khan. 2005. Solid polymer electrolyte membranes for fuel cell applications—A review*J. Membrane Sci.* 259: 10–26.

Smitha, B., S. Sridhar, and A.A. Khan. 2003. Synthesis and characterization of proton conducting polymer membranes for fuel cells. *J. Membr. Sci.* 225: 63–76.

Soldatov, V.S., E.G. Kosandrovich, and T.V. Bezyazychnaya. 2018. Hydration of ion exchangers: Thermodynamics and quantum chemistry calculations. IV. The state of ions and water molecules in alkali forms of sulfostyrene resins. *React. Funct. Polym.* 131: 219–229.

Soldatov, V.S., S. Pristavko, V. Zelenkovskii, et al. 2013. Hydration of ion exchangers: Thermodynamics and quantum chemistry calculations. *React. Funct. Polym.* 73: 737–744.

Soldatov, V., V. Zelenkovskii, and E. Kosandrovich. 2016a. Hydration of ion exchangers: Thermodynamics and quantum chemistry calculations. II an improved variant of the predominant hydrates model. *React. Funct. Polym.* 102: 147–155.

Soldatov, V., V. Zelenkovskii, and E. Kosandrovich. 2016b. Hydration of ion exchangers: Thermodynamics and quantum chemistry calculations. III. The state of the proton and water molecules in hydrogen form of sulfostyrene ion exchangers. *React. Funct. Polym.* 102: 156–164.

TakedaK., F. Kawakami, and M. Sasaki. 1986. Analysis of the pore structure of ion-exchange resins and the adsorption equilibrium of Fe^{3+}/Cl- Complex ions. *J. Macromolec. Sci. Part A* 23: 1137–1154.

Tanis-Kanbur, M.B., R.I. Peinador, J.I. Calvo, et al. 2021. Porosimetric membrane characterization techniques: A review. *J. Membrane Sci.* 619: 118750.

Thompson, E.L., T.W. Capehart, T.J. Fuller, et al. 2006. Investigation of low-temperature proton transport in Nafion using direct current conductivity and differential scanning calorimetry. *J. Electrochem. Soc.* 153: A2351–A2362.

Toteja, R.S.D., B.L. Jangida, M. Sundaresan, et al.1997. Water sorption isotherms and cation hydration in Dowex 50W and Amberlyst-15 ion exchange resins. *Langmuir* 13:2980–2982.

Vishnyakov, A., and A.V. Neimark. 2001. Molecular dynamics simulation of nafion oligomer solvation in equimolar methanol-water mixture. *J. Phys. Chem. B* 105: 7830–7834.

Volfkovich, Y.M., and V.E. Sosenkin, 2012. Porous structure and wetting of fuel cell components as the factors determining their electrochemical characteristics. *Russ. Chem. Rev.* 81: 936–959.

Volfkovich, Y.M., V.S. Bagotzky, V.E. Sosenkin et al. 2001. The standard contact porosimetry. *Colloids Surf. A:Physicochem. Eng. Aspects* 187–188: 349–365.

Volfkovich, Y.M. and V.S. Bagotzky. 1994. The method of standard porosimetry: 1. Principles and possibilities. *J. Power Sources* 48 (3): 327–338.

Volfkovich, Yu M., A. Yu Rychagov, A.A. Mikhalin, et al. 2018. Capacitive deionization of water using mosaic membrane. *Desalination* 426: 1–10.

Volfkovich, Yu M., N.A. Kononenko, A.A. Mikhalin, et al. 2020.Capacitive deionization of water involving mosaic membranes based on fibrous polymer matrices. *Desalin. Water Treat.* 182: 77–87.

Voropaeva, D. Yu, S.A. Novikova, T.L. Kulova, et al.2018. Solvation and sodium conductivity of nonaqueous polymer electrolytes based on Nafion-117 membranes and polar aprotic solvents. *Solid State Ionics* 324: 28–32.

Wescott, J.T., Y. Qi, L. Subramanian, and T.W. Capehart. 2006. Mesoscale simulation of morphology in hydrated perfluorosulfonic acid membranes. *J. Chem. Phys.* 124: 134702.

Wood, C.D., and A.I. Cooper. 2001. Synthesis of Macroporous Polymer Beads by Suspension Polymerization Using Supercritical Carbon Dioxide as a Pressure-Adjustable Porogen. *Macromolecules* 34: 5–8.

Zawodzinski, T.A., M. Neeman, L.O. Sillerud, et al. 1991. Determination of water diffusion coefficients in perfluorosulfonate ionomeric membranes. *J. Phys. Chem.* 95: 6040–6044.

4 Analysis of Qualitative and Quantitative Criteria of Porous Plastics

Figen Balo
Fırat University, Turkey

Lutfu S. Sua
School of Entrepreneurship and Business Administration, AUCA, Turkey

CONTENTS

4.1 INTRODUCTION

The worldwide production of plastics has dramatically increased with plastics gaining widespread application in almost every aspect of economies. Plastics have provided many advantages for communities with their ever-expanding implementations. Food packaged with plastic lasts longer, decreasing waste. Utilization of plastic as pipe materials enables the supply of safe drinking water, while plastic helps medical instruments like drips and surgical equipment to save lives. Plastic utilization in cars has lowered CO_2 emissions in transportation because of its lightweight [1]. Due to their big potential in a wide range of implementations in biomedicine, separation, energy storage, catalysis, optoelectronics, and much more, porous polymers have gained substantial interest in recent times [2–4]. They stand out for their distinctive characteristics, like scaffolds, excellent stability, well-defined porosities, and impressive superficies fields that can easily be functionalized. In general, the porous plastic term describes a porous material, molded, open-cell, made from any number of thermoplastic polymers. Through casted, it means an inflexible body that has three-dimensional properties. Through open-cells, it provides a substance that allows air (or fluid) to flow from one side of the component to the other as water flows through a sponge. The porous structure that activates fluidic contact over the course of the thermoplastic material is constituted in a fundamental sintering operation [5]. The porous polymer family contains poly-acetylenes [6], hyper-crosslinked polymers [4], intrinsic microporosity polymers [7, 8], conjugated microporous polymers [9], covalent organic frameworks [10], porous aromatic frameworks [11], and covalent triazine frameworks [12], which is also the development order.

DOI: 10.1201/9781003169604-4

With intrinsic microporosity polymers being an exception, porous polymers vary from other linear polymers because cross-linked structures are displayed in the former. In widespread organic solvents, they can be classified as insoluble or soluble in relation to their solubility. There are a few instances of intrinsic microporosity polymers that are easily soluble while hyper-crosslinked polymers, covalent organic frameworks, conjugated microporous polymers, covalent triazine frameworks, and porous aromatic frameworks are commonly insoluble in nature. Hyper-crosslinked polymers, which are predated by other porous polymers, are shaped through too distinct chemistry, have distinct features, and are comparatively more scalable, cheaper, although lacking in a constructional variety [13, 14]. Intrinsic microporosity polymers are amorphous polymers with inherent microporosity arising from the extremely contorted and rigid polymer chains' inefficient packaging [15, 16]. It should be noted that poly-acetylene [17], especially poly (trimethylsilylpropylene) accomplishes the constructional plan of intrinsic microporosity polymers and has been synthesized approximately 20 years prior to the acknowledged intrinsic microporosity polymers. Nonetheless, they have not been called on intrinsic microporosity polymers.

Although intrinsic microporosity polymers and hyper-crosslinked polymers are amorphous in nature, covalent organic frameworks are enlarged crystalline systems structured into well-defined constructions linked by robust covalent bonds from the molecular structure blocks' covalent assembly [18–20]. As they are crystalline, covalent organic frameworks are very different from other porous polymers. Natural metal-organic structures are more similar to them. The expertness of the networks of conjugated microporous polymers is their enlarged π-conjugation, which gives options for optoelectronic implementations [21, 22]. Covalent triazine frameworks have firm aromatic covalent bonds (C and N), high stability, and rich nitrogen ingredient [23, 24], made via aromatic nitriles' trimerization. Porous aromatic frameworks are unique because the extended frameworks' stability limitations (metal-organic frameworks and covalent organic frameworks) and the surface limitations of covalently bound porous polymer networks can be overcome (hyper-crosslinked polymers, conjugated microporous polymers, covalent triazine frameworks, and covalent organic frameworks) [25, 26]. The porous material used for CO_2 adsorption by physisorption is an effective way of reducing CO_2 grades [27]. CO_2 adsorption on porous adsorbents involves the weak bond creation that depletes lower energy than the liquid amines' use in which CO_2 is connected chemically to the amines [28, 29]. At lower pressure levels, porous adsorbents should have good CO_2 adsorption selectivity and high adsorption potential [30].

For industries that utilize renewable power resources, like H_2 storage and natural gas adsorbents, the production of such porous materials is very important [31, 32]. Chemically and physically, porous organic polymers are extremely constant and can be utilized under tough circumstances, like acidity, humidity, heat, and high temperatures [33]. They can be utilized for the storage and capture of gases, molecular separation, adsorption of toxins, and heterogeneous catalysis [34, 35]. Additionally, greatly covalent porous organic polymers have been planned to contain different functionalities on the basis of monomeric units or molecular building blocks, and their pore extents permit the CO_2 selective capture in natural gas (around 95 percent CH_4) or flue gas (75 percent N_2) over other gases [36, 37]. A few microporous organic polymers have elevated stability, small distributions of pore size, and wide superficies fields [38, 39]. Four widespread kinds of microporous organic polymers are available: intrinsic microporosity polymers, organic covalent frameworks, hyper-crosslinked polymers, and crosslinked polymers [40]. The most engaging materials for gas seize are called microporous organic polymers, but more changes can be made [41]. For the development of porous organic polymers, many synthetic processes have been utilized. The most popular ones include nitrile trimerization, condensation of boronic acid, acetylene trimerization catalyzed through cobalt, imine formation, and Sonogashira-Hagihara cross-coupling [42]. In the event of porous organic polymers, which indicate extreme adsorption efficiency at little pressure with adsorption through pore plenish, the micropore composition is more important than the superficies field [43]. On the other hand, microporous organic polymers have effective adsorption of CO_2 and big polar pore superficies [44–46], but their chemical imbalance makes them inappropriate for utilization in significant implementations. In selective adsorption [47], gas storage and capture [48–50], heterogeneous catalysis [51], and organic vapor capture [50] porous organic polymers with enduring microporosity can be utilized. For utilization in gas separation and storage [52], heterogeneous catalysis [53, 54], sensors [55], and electricity [56], the pores inside porous organic frameworks can be regulated throughout synthesis. In the presence of strong Lewis acids as catalysts, synthesis of porous organic frameworks requires the commercial monomer use [57]. As porous polymer networks with considerable efficiency, Schiff base networks with elevated superficies fields were improved [58]. In the CO_2 selective adsorption over other gases, like CH_4 and N_2, heteroatoms inside porous organic polymer compositions play an important role [59]. In low-pressure selective adsorption of CO_2, for instance, porous polymer bonds including sulfonate-grafted [60] and microporous organic polymer heterocycles [61] have been utilized.

4.2 SORTING OF POROUS POLYMERS

A wide range of studies has been carried out on polymeric materials with diverse grades of porosity. The porous polymers' attractiveness is largely related to their sole characteristics, like elevated superficies field and other status resulting from material spatial dimensions [62]. The porous materials are generally illustrated through the distribution of their size, pore size, total porosity amount, interconnectivity extent, and shape. The characteristics and dimensions of the pores vary depending on the porous material's implementation [63]. The pores have been graded as mesopores (lengths from 50 to 2 nm), macropores (lengths $\geq$ 50 nm), and micropores (lengths $\leq$ 2 nm), with respect to the description of the International Union of Pure and Applied Chemistry (IUPAC) [64]. As "nano" points to different extents, it is possible to designate any of the three types of porous materials above as nanoporous materials. Nonetheless, nanoporous materials apply to microporous or mesoporous materials in the majority of the literature [65].

4.2.1 Macroporous Polymers

In separation [66], catalysis [67], and gas storage [68], macroporous polymers [69] can be utilized. For the production of such polymers, nontemplating or templating treatments are utilized [70]. Utilizing emulsion templating treatments that include the creation of a very high internal-phase emulsion subsequently continual phase polymerization to fabricate polymer foams with low density, polymers with high porous can be synthesized [66, 71]. Photo-polymerization, which has a fast curing duration, enables the synthesis of high-internal polyphase monomer emulsions creating greatly undecided emulsions [72]. Macroporous polymers have pores between 1 μm and 50 nm, as described through IUPAC. They are created through vinyl-divinyl monomer copolymerization with the inert diluent's presence via free radical crosslinking. This diluent, which moves as a pore-creating agent, is on the basis of the nature of the resulting macroporosity and the pore structure. A linear polymer, a solvent, or a nonsolvent may be the diluent [73]. For example, the diluent is miscible with the monomers in the course of the macroporous polymer formation through the divinylbenzene and styrene's suspension polymerization, but it is also a comparatively weak polymer solvent. As the polymer shapes, it concentrates as a separate stage in the polymerizing droplet owing to its low resolubility in the diluent, and the residuary space is refilled through the diluent. A bead with a very high interior superficies field impregnated with big pores is the conclusion. A limp aggregate of diminutive submicron-sized granules with the transition volume between those porosity-constituting particles can be defined as the ultimate polymer grain [74]. In addition, diverse types are generated, each of which is utilized in a specialized implementation, depending on the kind of methodology utilized to shape the macroporous polymer. Other than phase segregation, various methods combine, gas-blowing, microemulsion formation, porogenation, and freeze-drying techniques to make macroporous materials. For instance, gels with macroporous are utilized to extract smaller molecules from material solutions suspended, monolithic polymers with macroporous are used as chromatographic matrices, beads with macroporous are utilized to diversify plasmid DNA chromatographically, and polymer scaffolds with macroporous are involved in network engineering implementations [75].

4.2.2 Microporous Polymers

Microporous materials have surface areas and high porosity and can be utilized for the development of renewable energy [76]. Due to their polymeric rigid chains, microporous polymers have excessive porosity with sub-nano and nanopores [53, 77]. Additionally, their little pore diameters (smaller than 2 nm) and wide superficies fields, low skeleton density, and high stability are found in organic polymers with micropores [76]. With products over other gases like N_2, they can be utilized selectively to store and transport molecules and small ions like CO_2, CH_4, and H_2 [78–80]. It is significant to establish synthesis methodologies for organic microporous polymers that utilize easy polymerization ways and low-expense precursors. In the nonexistence of catalysts, one-point synthesis of microporous organic polymers has been researched utilizing basis reactions of Schiff [65]. By coupling reactions, nickel-catalyzed cross-coupling of Ullmann, oxidative coupling, cross-coupling of Suzuki–Miyaura, cross-coupling of Sonogashira–Hagihara, cross-coupling of Buchwald–Hartwig, and Friedel–Crafts alkylation, several polymers with micropores have been synthesized effectively [49, 64]. In the soluble polymer synthesis with innate microporosity and other indissoluble microporous organic polymers, tetrahedral and spirocyclic monomers have been utilized effectively as firm structure blocks [81].

Microporous organic polymers have different chemical characteristics incorporated into their pores [82]. Through the integration of interstitial metal areas into microporous organic polymers, the second generation of the porous materials having beneficial physical and chemical features could be improved [82]. In thiol oxidation [83], coupling of Suzuki–Miyaura [84, 85], hydrogenation reactions [47], and other reactions, these materials can behave as heterogeneous catalysts [86]. There are different benefits to microporous organic polymers with a metal catalytic piece, like homogeneous dispersions of energetic metal nanogranules which suppress metal leakage and improve both selectivity and catalytic activity [87–89]. Thus, the processing of diverse microporous organic polymers with diverse characteristics is desirable [90, 91]. The polymers with micropores like amorphous fluoropolymers and polyacetylenes can be utilized in membranes of gas separation due to their good free volume components [92]. A few porous organic polymers, containing hyper-crosslinked polymers [93–95], intrinsic microporosity polymers [87], and microporous conjugated polymers [94] have been utilized in CO_2 seize. Additionally, porous organic frameworks including triazine [95, 96] and aromatic frameworks with pores [97] have been used in core cross-linked star operations [97, 98]. Generally, the polymers have abundant rotational and conformational elasticities that allow them to effectively optimize packing space and intramolecular cohesive interactions in the solid case.

In conclusion, the release volume within a polymer, which is the area within the substance not activated through the polymer molecules, can vary notably owing to specific elements (recent history, molecular structure, physical state, and the like). Therefore, if the substance has a big enough release volume, there will be substantial presentation and interconnectivity of microporosity [63]. It is plausible to discriminate between three groups of permanent polymers with micropores: hyper-crosslinked polymers, intrinsic microporosity polymers, and crystalline organic covalent frameworks [65]. In hyper-crosslinked polymers, the excess free volume of a big amount can be found through the structure of what is named a hyper-crosslinked system polymer within a solvent swollen polymer liquid, which, afterward the solvent's removal, constructs a material with microporous. The most examined group of microporous polymers made utilizing this notion is on the basis of polystyrene resins hyper-crosslinked [63]. Utilizing the reaction of Friedel–Crafts' between the polymer's phenyl rings and a crosslinker, the polymer chain's crosslink is accomplished [65]. Implementations for chromatography substrates, hyper-crosslinked polymers contain adsorption, storage of methane-hydrogen, and catalytic metal nanoparticle supports [63]. As for polymers of inherent microporosity, which constitute a new group of organic microporous materials based on polymers, microporosity derives from their free volume interconnected [65]. A polymer's specific volume is a mixture of both the material's free volume and the volume occupied through the polymer molecules. It is possible to measure the volume occupied through the polymer molecules as the unique van-der-Waals volume. In addition, because the molecules do not completely fill the uniform areas in totally sequenced crystals, a portion of the polymer's volume is brought as the release volume. As a consequence, with temperature, the specific volume raises directly proportional for a polymer. Whenever the glass transition temperature is reached, a slope shift takes place in such a way that the expansion coefficient in the rubbery state is higher than in the glassy state. In the glass passage temperature, this passage is strongly influenced through kinetics, where a slight rise in the ratio of cooling conclusions is observed as a very high increase. Having said that, it was obtained that if a polymer specimen is quickly cooled and held at a temperature subtly below the temperature of the transition glass, the effect would be the creation of an excessive released volume that is locked owing to quick cooling and that is eventually lost by physical aging. In other words, a microporous material can be shaped if abundant excessive volume can be confined inside an amorphous polymer. Implementations of polymers and molds innovations encompass hydrogen storage, heterogeneous catalysis, adsorption, and gas separation membranes [60–62].

The research on organic-sourced modular metal structures was precursored by Yaghi and his colleagues. Their success was a turning period that helped create alternative materials with pores for a wide range of implementations in an organic framework [66]. Through organic connection molecules including physical categories, metal-organic frameworks are structured of metal oxide or metal vertices bound. Although metal-organic frameworks have apparent benefits compared to other materials with pore due to their greater interior superficies fields, they show a necessary disadvantage in terms of the asset of heavy metal atoms inside their systems raising the effectiveness of their composition. Organic covalent frameworks on the other side put right this deficiencies because their molecular systems consist of light elements (oxygen, hydrogen, boron, and carbon) that decrease the structure's relative weight [67]. The primarily synthesized organic covalent frameworks, called Organic covalent frameworks, are found from the benzene-diboronic acid's molecular dehydration [68]. Organic covalent frameworks show big abilities for implementation in the area of carbon dioxide capture and hydrogen storage [69].

4.2.3 Mesoporous Polymers

This new group's members of polymers are made either by a firm molding strategy utilizing inorganic firm substance as formworks (such as silica nanoparticles or mesoporous silica) or through a soft molding strategy utilizing organic surfactant molecules as formworks (such as surfactant micelles or block copolymer) [59–61]. There are plenty of favorable implementations for mesoporous polymers with the inclusion of their utilization as catalyst backs, or even metal-free catalysts, in purification, membrane synthesis, and separation. For example, phenolic resins with mesoporous shaped through soft molding have been often utilized for the making ready of carbons with mesoporous by petrifaction. On the other side, polymeric C nitrides with mesoporous made via heating pioneers in the silica template's presence have been displayed to be organic-based semiconductors, which can be used as metal-free photocatalysts for other photocatalytic reactions and all water splitting [58, 65]. Resolutions to the difficulties faced through porous materials constructed as hierarchic need novel concepts and novel discoveries, which supply those entering this area with stimulant occasions. Such fascinating and exciting search aspects are the pore channels' physical alteration to minimize the frequency of side reactions, regulating pore crystallinity to enhance quantitative matching and thermal stability of pore sizes to the other side of poly length measures to maximize mass transfer. With the awareness of the relationship electrochemical efficiency obtained from mentioned review and between the structural, geometric, and organizational characteristics of pores at diverse length scales, it is plausible to develop a set of rules for material plan to synthesize as hierarchical organized porous materials.

4.3 METHODOLOGY

Utilization of multi-criteria quantitative assessment methodologies for resolving some engineering troubles has outgrown substantially. One of the two primary ingredients of quantitative multi-criteria assessment methodologies is related to the criteria weighing. Practically, weights of the criteria are identified in evaluating the potential and activity of the problem, and the like. The criteria's pairwise comparison is broadly implemented. A few practical and theoretical methodologies of evaluating the weights of criteria through specialists are understood. These different methodologies of incorporating certain criteria explaining the contemplated objective into a sole generalization criterion have been proposed and several diverse multi-criteria assessment methodologies have been generated. The most renowned, broadly implemented, and numerically grounded technic is the Analytic Hierarchy Process. Analytic Hierarchy Process that is formed under multi-criteria decision making is constituted of proper techniques for sorting of critical administration problems [99]. The Analytic Hierarchy Process was presented by Saaty in the 1970s and it has been applied in various fields [100]. The Analytic Hierarchy Process methodology is a sorting operation that is utilized in making group decisions and is broadly utilized around the world in a variety of fields like engineering, industry, government, business, health, education, and others. The methodology also permits for coherence tests in decision-making room to control and decrease discrepancies in judgments or opinions. The measurements of rate and coherence indicator are obtained from the paramount eigenvalue and eigenvectors, respectively. The methodology concentrates on prioritizing the choice criteria and differentiating the more significant criteria from the lesser significant ones. The Analytic Hierarchy Process is a methodology with concentrate sited on peer to spouse similarities that are appropriate to assess both quantitative and qualitative plan [101].

4.3.1 AHP Analysis

The materials to be evaluated within the scope of this study and the criteria which form the basis of comparison are provided in Table 4.1.

However, the listed criteria are not expected to have the same impact on the overall results. Thus, the relative priorities for these criteria have to be obtained on expert evaluations. Experts are asked to rate the relative importance of these criteria through pair-wise comparison based on the scale of comparison provided in Table 4.2.

Based on the pair-wise comparison, the decision matrix presented in Table 4.3 is constructed.

Pair-wise comparison values obtained from Table 4.3 result in the relative priorities of the selected criteria. Figure 4.1 presents these priorities.

Upon calculation of the criteria priorities, the final step of the methodology involves calculating the scores of each material alternative based on each of the criteria and finding the total score for the decision alternatives. It should be noted that *Relative Raw Material Price, Air Filtration, and Water Filtration* criteria are negatively correlated with the attractiveness of the material alternatives. In other words, lower values of these criteria are desired. Resulting scores for the 11 material alternatives that are used within this study are listed in Table 4.4.

The table exhibits individual scores of the alternatives for each criterion while the last row presents the total scores.

Figure 4.2 indicates that the material alternative with the highest score is *PE150* with a score of *0,068* is the best alternative based on the selected criteria used in the study.

TABLE 4.1
Properties of the Selected Materials

Material Grade	PTFE30	PVDF30	PP150	PP100	PP50	PE-HV	PE150	PE100	PE60	PE25	PE10
Material Options	-	-	H20/SS	H20/SS	H20/SS	H20/SS	H20/SS	H20/SS	H20/SS	H20/SS	H20/SS
Polymer Type	PTFE	PVDF	PP	PP	PP	PE	PE	PE	PE	PE	PE
Relative Raw Material Price	4	3	2	2	2	2	1	1	1	1	4
Nominal Pore Volume (%)	40	45	45	45	45	60	50	50	50	50	50
Nominal Pore Size (μ)	30	30	150	100	50	.	150	100	60	25	10
Air Filtration[a]	3 μ	3 μ	75 μ	50 μ	10 μ	1 μ	30 μ	15 μ	5 μ	1 μ	5 μ
Air Permeability[b]	5–25	10–50	70–200	60–150	40–120	40–80	70–120	50–100	30–90	10–50	<10
H_2O Intrusion Pressure (mBar)	80	60	10	12	35	65	16	42	70	152	175
Water Filtration[c]	N/A	15 μ	150 μ	100 μ	40 μ	10 μ	90 μ	20 μ	15 μ	10 μ	5 μ

Source: https://www.thermoporecorp.com/material-properties.

[a] ($\varepsilon > 98\%$ @ 50 ft/min).

[b] (ft/min @ 1.2″ H20 ΔP, material thickness of .125″).

[c] ($\varepsilon > 95\%$ @ 4 ft/min).

TABLE 4.2
The Comparison Scale [102]

Intensity	Definition	Explanation
1	Equal significance	2 events take part evenly to the target
2	Slight or weak	
3	Moderate significance	Judgment and experience insignificantly favor 1 event over another
4	Moderate positive	
5	Strong significance	Judgment and experience firmly favor one event over another
6	Strong positive	
7	Too firm or indicated significance	One event is favored too firmly over another; its superiority indicated practically
8	Very, very strong	
9	Extreme significance	The proof favoring 1 event over another is of the maximum feasible order of assertion
Reciprocals referred to above	If event *i* as compared to activity *j*, has one of the above non-zero numbers assigned to it, then *j* has the reciprocal value compared to *i*	An acceptable assertion
1.1–1.9	If the events are too close	

TABLE 4.3
The Decision Matrix

Decision Matrix		RRM. Price	N.Pore Vol.	N.Pore Size	Air Filtration	Air Permeability	H2O Intrusion P.	Water Filtration
		1	2	3	4	5	6	7
RRM.Price	1	1	1/5	1/3	1/2	1/3	1	1
N.Pore Vol.	2	5	1	2	3	5	7	7
N.Pore Size	3	3	1/2	1	2	3	5	5
Air Filtration	4	2	1/3	1/2	1	2	3	3
Air Permeability	5	3	1/5	1/3	1/2	1	2	2
H_2O Intrusion P.	6	1	1/7	1/5	1/3	1/2	1	1
Water Filtration	7	1	1/7	1/5	1/3	1/2	1	1

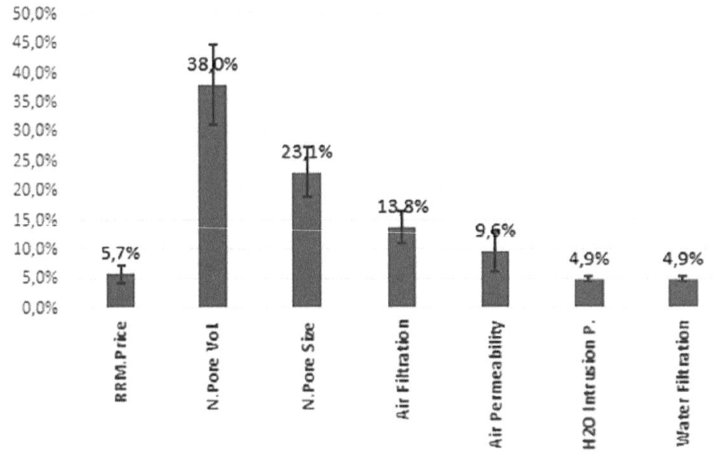

FIGURE 4.1 Relative priorities of the selected criteria.

TABLE 4.4
Resulting Scores for Material Alternatives

	PTFE30	PVDF30	PP150	PP100	PP50	PE-HV	PE150	PE100	PE60	PE25	PE10
RRM.Price	–0,010	–0,007	–0,005	–0,005	–0,005	–0,005	–0,002	–0,002	–0,002	–0,002	–0,010
N.Pore Vol.	0,029	0,032	0,032	0,032	0,032	0,043	0,036	0,036	0,036	0,036	0,036
N.Pore Size	0,009	0,009	0,045	0,030	0,015	0,021	0,045	0,030	0,018	0,007	0,003
Air Filtration	–0,002	–0,002	–0,052	–0,035	–0,007	–0,001	–0,021	–0,010	–0,003	–0,001	–0,003
Air Permeability	0,002	0,004	0,018	0,014	0,010	0,008	0,019	0,010	0,008	0,004	0,001
H2O Intrusion P.	0,005	0,004	0,001	0,001	0,002	0,004	0,001	0,003	0,005	0,010	0,012
Water Filtration	–0,004	–0,001	–0,015	–0,010	–0,004	–0,001	–0,009	–0,002	–0,001	–0,001	0,000
Total	**0,028**	**0,038**	**0,023**	**0,027**	**0,044**	**0,070**	**0,068**	**0,063**	**0,059**	**0,053**	**0,037**

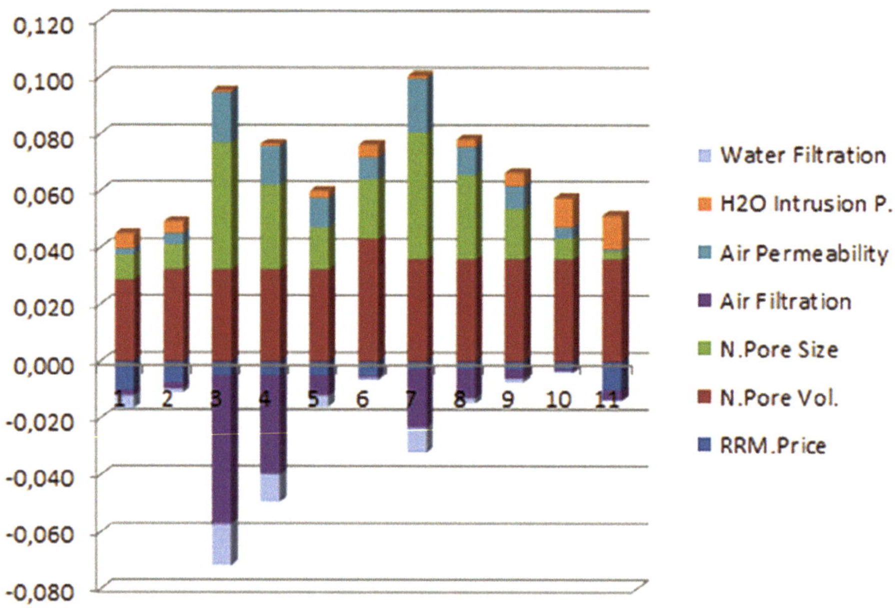

FIGURE 4.2 Resulting scores for material alternatives.

4.4 CONCLUSIONS

The success and progress in the synthesis of porous materials constructed hierarchically have given a perspective to adjust their chemical kinetics and diffusion behavior. These superiorities have encouraged the operation of different power storage devices and materials. With the improvement of society and economy, high-efficiency power storage devices and materials with high power and energy densities, excellent stability, and long lifetime are desperately needed. As a biomimetic material kind, constructed porous materials possess innate constructional benefits that track nature's laws, displaying extraordinary potency in the energy storage area. Particulary through designing and adjusting the pore parameters, like pore's geometric configuration, pore size, and the scale grades' number, the materials' transfer effectiveness and electrons/ions can be optimized, and the materials' electrochemical features can be maximized. The synergy impact from hierarchic porosity throughout multi-length measures, matching pore sizes in quantitative terms varying from micro levels to macro levels, and the geometric configuration's fine control of the pores, has been approved to be versatile and effective policies for planning sophisticated electrode materials. With this thought, it is very important to properly adjust the pore characteristics through the utilization of multidisciplinary expertise from chemistry, material science, biology, and physics for power storage implementations. In spite of considerable progress in the improvement of porous materials constructed as hierarchical for electrochemical power storage, there are still some important drawbacks that hinder their commercial implementations. In general, porous constructs as hierarchical give materials with high superficies field and rapid mass move. However, this raises the probability of side reactions in a particular way. Particularly, these undesirable side reactions give rise to irreversible loss of capacity, electrolyte consumption, and a reduction in the use of energetic materials, which is destructive to the features of the battery. The other obstruction is the little volumetric power density owing to the excessive porosity of materials with pores constructed as hierarchical. Miniaturization and integration are the trends of electronic products which seriously rely on new electrode materials with good volume power density. In the near future, this poses a big difficulty to materials with porous constructed as hierarchical for power storage implementation.

REFERENCES

[1] Andrady A, Neal M (2009) Applications and societal benefits of plastics. *Philos Trans* 364(1526): 1977–1984.
[2] Dawson R, Cooper AI, Adams DJ (2012) *Prog Polym Sci* 37:530–563.
[3] Das S, Heasman P, Ben T, Qiu S (2017) *Chem Rev* 177:1515–1563.
[4] Rogozhin SV, Davankov VA, Tsyurupa MP (1969) *Patent USSR* 299165.
[5] Saikat D, Ben T, Qiu S (2020) Shaping of porous polymers. *Polymer* 207:122928.
[6] Masuda T, Isobe E, Higashimura T (1983) *J Am Chem Soc* 105:7473–7474.
[7] McKeown NB, Hanif S, Msayib K, Tattershall CE, Budd PM (2002) *Chem Commun* 2782–2783.
[8] McKeown NB, Makhseed S, Budd PM (2002) *Chem Commun* 2780–2781.
[9] Jiang J-X, Su F, Trewin A, Wood CD, Campbell NL, Niu H, Dickinson C, Ganin AY, Rosseinsky MJ, Khimyak YZ (2007) *Angew Chem Int Ed* 46:8574–8578.
[10] Côté AP, Benin AI, Ockwig NW, O'Keeffe M, Matzger AJ, Yaghi OM (2005) *Science* 310:1166–1170.
[11] Ben T, Ren H, Ma S, Cao D, Lan J, Jing X, Wang W, Xu J, Deng F, Simmons JM, Qiu S, Zhu G (2009) *Angew Chem Int Ed* 48:9457–9460.
[12] Kuhn P, Antonietti M, Thomas A (2008) *Angew Chem Int Ed* 47:3450–3453.
[13] Tan L, Tan B (2017) *Chem Soc Rev* 46:3322–3356.
[14] Davankov VA, Tsyurupa MP (1990) *Chem Rev* 13:27–42.
[15] McKeown NB, Budd PM (2006) *Chem Soc Rev* 35:675–683.
[16] McKeown NB (2017) *Sci China Chem* 60:1023–1032.
[17] Nagai K, Masuda T, Nakagawa T, Freeman BD, Pinnau I (2001) *Prog Polym Sci* 26:721–798.
[18] Guan X, Chen F, Fang Q, Qiu S (2020) *Chem Soc Rev* 49:1357–1384; Kandambeth S, Dey K, Banerjee R (2019) *J Am Chem Soc* 141:1807–1822.
[19] Wang Z, Zhang S, Chen Y, Zhang Z, Ma S (2020) *Chem Soc Rev* 49:708–735.
[20] Lohse MS, Bein T (2018) *Adv Funct Mater* 28:1705553.
[21] Xu Y, Jin S, Xu H, Nagai A, Jiang D (2013) *Chem Soc Rev* 42:8012–8031.
[22] Lee MJ-S, Cooper AI (2020) *Chem Rev* 120:2171–2214.
[23] Krishnaraj C, Jena HS, Leus K, Van Der Voort P (2020) *Green Chem* 22:1038–1071.
[24] Liu M, Guo L, Jin S, Tan B (2019) *J Mater Chem* 7:5153–5172.
[25] Tian Y, Zhu G (2020) *Chem Rev* https://doi.org/10.1021/acs.chemrev.9b00687.
[26] Ben T, Qiu S (2013) *CrystEngComm* 15:17–26.
[27] Liebl MR, Senker J (2013) Microporous functionalized triazine based polyimides with high CO2 capture capacity. *Chem Mater* 25(6):970–980.
[28] D'Alessandro DM, Smit B, Long JR (2010) Carbon dioxide capture: prospects for new materials. *Angew Chem Int Ed* 49(35):6058–6082.
[29] Li G, Wang Z (2013) Microporous polyimides with uniform pores for adsorption and separation of CO_2 gas and organic vapors. *Macromolecules* 46(8):3058–3066.
[30] Wilmer CE, Farha OK, Bae Y-S, Hupp JT, Snurr RQ (2012) Structure–property relationships of porous materials for carbon dioxide separation and capture. *Energy Environ Sci* 5(12):9849–9856.
[31] Wang Z, Yuan S, Mason A, Reprogle B, Liu D-J, Yu LP (2012) Nanoporous porphyrin polymers for gas storage and separation. *Macromolecules* 45(18):7413–7419.
[32] Liu L, Wang X, Zhang Q, Li Q, Zhao Y (2013) Distinct interpenetrated metal–organic frameworks constructed from crown ether-based strut analogue. *CrystEngComm* 15(5):841–844.
[33] Rabbani MG, El-Kaderi HM (2012) Synthesis and characterization of porous benzimidazole-linked polymers and their performance in small gas storage and selective uptake. *Chem Mater* 24(8):1511–1517.
[34] McKeown NB, Budd PM (2006) Polymers of intrinsic microporosity (PIMs): organic materials for membrane separations, heterogeneous catalysis and hydrogen storage. *Chem Soc Rev* 35(8):675–683.
[35] Zhang Y, Wei S, Liu F, Du Y, Liu S, Ji Y, Yokoi T, Tatsumi T, Xiao F-S (2009) Superhydrophobic nanoporous polymers as efficient adsorbents for organic compounds. *Nano Today* 4(2):135–142.
[36] Islamoğlu T, Rabbani MG, El-Kaderi HM (2013) Impact of postsynthesis modification of nanoporous organic frameworks on small gas uptake and selective CO_2 capture. *J Mater Chem A* 1(35):10259–10266.
[37] El-Kaderi HM, Hunt JR, Mendoza-Cortés JL, Côté AP, Taylor RE, O'Keeffe M, Yaghi OM (2007) Designed synthesis of 3D covalent organic frameworks. *Science* 316(5822):268–272.
[38] Liu L, Zhang J (2013) Triptycene-based microporous polymer with pending tetrazole moieties for CO_2-capture application. *Macromol Rapid Commun* 34(23–24):1833–1837.

[39] Weston MH, Farha OK, Hauser BG, Hupp JT, Nguyen ST (2012) Synthesis and metalation of catechol-functionalized porous organic polymers. *Chem Mater* 24(7):1292–1296.
[40] Liu L, Li P-Z, Zhu L, Zou R, Zhao Y (2013) Microporous polymelamine network for highly selective CO_2 adsorption. *Polymer* 54(2):596–600.
[41] Zhu Y, Long H, Zhang W (2013) Imine-linked porous polymer frameworks with high small gas (H_2, CO_2, CH_4, C_2H_2) uptake and CO_2/N_2 selectivity. *Chem Mater* 25(9):1630–1635.
[42] Wu D, Xu F, Sun B, Fu R, He H, Matyjaszewski K (2012) Design and preparation of porous polymers. *Chem Rev* 112(7):3959–4015.
[43] Sevilla M, Fuertes AB (2011) Sustainable porous carbons with a superior performance for CO2 capture. *Energy Environ Sci* 4(5):1765–1771.
[44] Banerjee R, Phan A, Wang B, Knobler C, Furukawa H, O'Keeffe M, Yaghi OM (2008) High-throughput synthesis of zeolitic imidazolate frameworks and application to CO_2 capture. *Science* 319(5865):939–943.
[45] Furukawa H, Ko N, Go YB, Aratani N, Choi SB, Choi E, Yazaydin AÖ, Snurr RQ, O'Keeffe M, Kim J, Yaghi OM (2010) Ultrahigh porosity in metal-organic frameworks. *Science* 329(5990):424–428.
[46] Thomas A (2010) Functional materials: from hard to soft porous frameworks. *Angew Chem Int Ed* 49(45):8328–8344.
[47] Jin Y, Voss BA, Jin A, Long H, Noble RD, Zhang W (2011) Highly CO_2-selective organic molecular cages: what determines the CO_2 selectivity. *J Am Chem Soc* 133(17):6650–6658.
[48] Zhao Y-C, Cheng Q-Y, Zhou D, Wang T, Han B-H (2012) Preparation and characterization of triptycene-based microporous poly(benzimidazole) networks. *J Mater Chem* 22(23):11509–11514.
[49] Chen Q, Luo M, Hammershøj P, Zhou D, Han Y, Laursen BW, Yan C-G, Han B-H (2012) Microporous polycarbazole with high specific surface area for gas storage and separation. *J Am Chem Soc* 134(14):6084–6087.
[50] Furukawa H, Yaghi OM (2009) Storage of hydrogen, methane, and carbon dioxide in highly porous covalent organic frameworks for clean energy applications. *J Am Chem Soc* 131(25):8875–8883.
[51] Ding S-Y, Gao J, Wang Q, Zhang Y, Song W-G, Su C-Y, Wang W (2011) Construction of covalent organic framework for catalysis: Pd/COF-LZU1 in Suzuki–Miyaura coupling reaction. *J Am Chem Soc* 133(49):19816–19822.
[52] Li JR, Sculley J, Zhou HC (2012) Metal-organic frameworks for separations. *Chem Rev* 112(2):869–932.
[53] Filer A, Choi H-J, Seo J-M, Baek J-B (2014) Two and three dimensional network polymers for electro-catalysis. *Phys Chem Chem Phys* 16(23):11150–11161.
[54] Zhang Y, Riduan SN (2012) Functional porous organic polymers for heterogeneous catalysis. *Chem Soc Rev* 41(6):2083–2094.
[55] Xiang Z, Cao D (2012) Synthesis of luminescent covalent-organic polymers for detecting nitroaromatic explosives and small organic molecules. *Macromol Rapid Commun* 33(14):1184–1190.
[56] Feng X, Chen L, Honsho Y, Saengsawang O, Liu L, Wang L, Saeki A, Irle S, Seki S, Dong Y, Jiang D (2012) An ambipolar conducting covalent organic framework with self-sorted and periodic electron donor-acceptor ordering. *Adv Mater* 24(22):3026–3031.
[57] McKeown NB, Budd PM (2010) Exploitation of intrinsic microporosity in polymer-based materials. *Macromolecules* 43(12):5163–5176.
[58] Schwab MG, Fassbender B, Spiess HW, Thomas A, Feng X, Müllen K (2009) Catalyst-free preparation of melamine-based microporous polymer networks through Schiff base chemistry. *J Am Chem Soc* 131(21):7216–7217.
[59] Jin Y, Voss BA, McCaffrey R, Baggett CT, Noble RD, Zhang W (2012) Microwave-assisted syntheses of highly CO_2-selective organic cage frameworks (OCFs). *Chem Sci* 3(3):874–877.
[60] Lu W, Yuan D, Sculle J, Zhao D, Krishna R, Zhou H-C (2011) Sulfonate-grafted porous polymer networks for preferential CO_2 adsorption at low pressure. *J Am Chem Soc* 133(45):18126–18129.
[61] Luo Y, Li B, Wang W, Wu K, Tan B (2012) Hypercrosslinked aromatic heterocyclic microporous polymers: a new class of highly selective CO_2 capturing materials. *Adv Mater* 24(42):5703–5707.
[62] Cetinkaya M (2008) Synthesis and Characterization of Nanostructured Poly (p-xylylene) Films, *ProQuest*.
[63] Manjooran NJ, Pickrell GR (2005) Biologically self-assembled porous polymers. *J Mater Process Technol* 168(2):225–229.
[64] Michael S, Silverstein NRC, Hillmyer MA (2011) *Porous Polymers*, Wiley.
[65] Zhao D, Wan Y, Zhou W (2013) *Ordered Mesoporous Materials*, Wiley.

[66] Nischang I, Causon TJ (2016) Porous polymer monoliths: from their fundamental structure to analytical engineering applications. *TrAC-Trends Anal Chem* 75:108–117.
[67] Svec F, Fréchet JMJ (1996) New designs of macroporous polymers and supports: from separation to biocatalysis. *Science* 273(5272):205–211.
[68] Kimmins SD, Cameron NR (2011) Functional porous polymers by emulsion templating: recent advances. *Adv Funct Mater* 21(2):211–225.
[69] Okay O (2000) Macroporous copolymer networks. *Prog Polym Sci* 25(6):711–779.
[70] Wu D, Xu F, Sun B, Fu R, He H, Matyjaszewski K (2012) Design and preparation of porous polymers. *Chem Rev* 112(7):3959–4015.
[71] Silverstein MS (2014) Emulsion-templated porous polymers: a retrospective perspective. *Polymer* 55(1):304–320.
[72] Pierre SJ, Thies JC, Dureault A, Cameron NR, van Hest JCM, Carette N, Michon T, Weberskirch R (2006) Covalent enzyme immobilization on to photopolymerized highly porous monoliths. *Adv Mater* 18(14):1822–1826.
[73] Bo Mattiasson AK, Galaev IY Eds. (2010) *Macroporous Polymers Production Properties and Biotechnological/Biomedical Applications*, CRC Press, Taylor & Francis Group.
[74] Small H (1989) *Ion Chromatography*, Springer, Boston, MA.
[75] Mattiasson B, Kumar A, Galaev I (2010) *Macroporous Polymers: Production Properties and Biotechnological/Biomedical Applications*, CRC Press/Taylor & Francis Group, 2010.
[76] Xie Y, Wang T-T, Liu X-H, Zou K, Deng W-Q (2013) Capture and conversion of CO_2 at ambient conditions by a conjugated microporous polymer. *Nat Commun* 4:1960.
[77] Dawson R, Cooper AI, Adams DJ (2012) Nanoporous organic polymer networks. *Prog Polym Sci* 37(4):530–563.
[78] Xiang Z, Cao D (2013) Porous covalent-organic materials: synthesis, clean energy application and design. *J Mater Chem A* 1(8):2691–2718.
[79] Suh MP, Park HJ, Prasad TK, Lim D-W (2012) Hydrogen storage in metal-organic frameworks. *Chem Rev* 112(2):782–835.
[80] Budd PM, McKeown NB, Fritsch D (2005) Free volume and intrinsic microporosity in polymers. *J Mater Chem* 15(20):1977–1986.
[81] Ben T, Ren H, Ma S, Cao D, Lan J, Jing X, Wang W, Xu J, Deng F, Simmons JM, Qiu S, Zhu G (2009) Targeted synthesis of a porous aromatic framework with high stability and exceptionally high surface area. *Angew Chem Int Ed* 121(50):9621–9624.
[82] Jiang J-X, Wang C, Laybourn A, Hasell T, Clowes R, Khimyak YZ, Xiao J, Higgins SJ, Adams DJ, Cooper AI (2011) Metal–organic conjugated microporous polymers. *Angew Chem Int Ed* 50(5):1072–1075.
[83] Chen L, Yang Y, Jiang D (2010) CMPs as scaffolds for constructing porous catalytic frameworks: a built-in heterogeneous catalyst with high activity and selectivity based on nanoporous metalloporphyrin polymers. *J Am Chem Soc* 132(26):9138–9143.
[84] Zhang P, Weng Z, Guo J, Wang C (2011) Solution-dispersible, colloidal, conjugated porous polymer networks with entrapped palladium nanocrystals for heterogeneous catalysis of the Suzuki–Miyaura coupling reaction. *Chem Mater* 23(23):5243–5249.
[85] Li B, Guan Z, Wang W, Yang X, Hu J, Tan B, Li T (2012) Highly dispersed Pd catalyst locked in knitting aryl network polymers for Suzuki–Miyaura coupling reactions of aryl chlorides in aqueous media. *Adv Mater* 24(25):3390–3395.
[86] Xie Z, Wang C, de Krafft KE, Lin W (2011) Highly stable and porous cross-linked polymers for efficient photocatalysis. *J Am Chem Soc* 133(7):2056–2059.
[87] Hasell T, Wood CD, Clowes R, Jones JTA, Khimyak YZ, Adams DJ, Cooper AI (2010) Palladium nanoparticle incorporation in conjugated microporous polymers by supercritical fluid processing. *Chem Mater* 22(2):557–564.
[88] Yang Y, Ogasawara S, Li G, Kato S (2013) Water compatible Pd nanoparticle catalysts supported on microporous polymers: their controllable microstructure and extremely low Pd-leaching behavior. *J Mater Chem A* 1(11):3700–3705.
[89] Lu Y-M, Zhu H-Z, Li W-G, Hu B, Yu S-H (2013) Sizecontrollable palladium nanoparticles immobilized on carbon nanospheres for nitroaromatic hydrogenation. *J Mater Chem A* 1(11):3783–3788.
[90] Katsoulidis AP, Kanatzidis MG (2011) Phloroglucinol based microporous polymeric organic frameworks with –OH functional groups and high CO_2 capture capacity. *Chem Mater* 23(7):1818–1824.

[91] Katsoulidis AP, He J, Kanatzidis MG (2012) Functional monolithic polymeric organic framework aerogel as reducing and hosting media for Ag nanoparticles and application in capturing of iodine vapors. *Chem Mater* 24(10):1937–1943.
[92] Bernardo P, Drioli E, Golemme G (2009) Membrane gas separation: a review/state of the art. *Ind Eng Chem Res* 48(10):4638–4663.
[93] Qian H, He Q, Zheng J, Li S, Zhang S (2014) Catecholfunctionalized microporous organic polymer as supported media for Pd nanoparticles and its high catalytic activity. *Polymer* 55(2):550–555.
[94] Wood CD, Tan B, Trewin A, Niu H, Bradshaw D, Rosseinsky MJ, Khimyak YZ, Campbell NL, Kirk R, Stöckel E, Cooper AI (2007) Hydrogen storage in microporous hypercrosslinked organic polymer networks. *Chem Mater* 19(8):2034–2048.
[95] Rose M, Böhlmann W, Sabo M, Kaskel S (2008) Element–organic frameworks with high permanent porosity. *Chem Commun* 44(21):2462–2464.
[96] Zhu X, Tian C, Mahurin SM, Chai S-H, Wang C, Brown S, Veith GM, Luo H, Liu H, Dai S (2012). A superacid-catalyzed synthesis of porous membranes based on triazine frameworks for CO_2 separation. *J Am Chem Soc* 134(25):10478–10484.
[97] Chan-Thaw CE, Villa A, Katekomol P, Su D, Thomas A, Prati L (2010) Covalent triazine framework as catalytic support for liquid phase reaction. *Nano Lett* 10(2):537–541.
[98] Kiskan B, Antonietti M, Weber J (2012) Teaching new tricks to an old indicator: pH-switchable, photoactive microporous polymer networks fromphenolphthalein with tunable CO_2 adsorption power. *Macromolecules* 45(3):1356–1361.
[99] Cabala P (2010) Using the analytic hierarchy process in evaluating decision alternatives. *Operat Res Dec* 1:5–23.
[100] Saaty TL (1980) *The Analytic Hierarchy Process*. McGraw-Hill International, New York, NY.
[101] Zamani R, Yousefi P (2013) Optimal decision making approach for selecting effort estimation model. *Int J Mach Learn Comput* 3(4):1–4.
[102] Vaidya OS, Kumar S (2006) Analytic hierarchy process: an overview of applications. *Eur J Oper Res* 169:1–29.

5 Novel Research on Porous Polymers Using High Pressure Technology

Fernanda Wariss Figueiredo Bezerra, Jhonatas Rodrigues Barbosa, Lucas Cantão Freitas, Marielba de Los Angeles Rodriguez Salazar, Rafael Henrique Holanda Pinto, Giselle Cristine Melo Aires, and Raul Nunes de Carvalho Junior
Federal University of Pará, Belém, Brazil

Jorddy Neves da Cruz
Museu Paraense Emílio Goeldi, Belém, Brazil

CONTENTS

5.1 BACKGROUND

The interest in porous polymers has grown in recent years due to the global search for reducing the consumption of products derived from non-renewable sources, expanding the exploration of materials on a sustainable basis. Porous polymers simulate inorganic materials such as zeolites, porous carbons and silicates through monoliths, films and granules. Porous materials can be used as filters or for industrial applications such as the transport of substances in the pharmaceutical field in drug and active compound delivery systems, as well as in food packaging, tissue repair, reaction cartridges for transporting enzymes and catalysts, in microelectronics, in gas separation and storage, among other applications (Aizawa 2018; Xu et al. 2018; Taylor et al. 2020; Silverstein 2014; Zhuang et al. 2014).

DOI: 10.1201/9781003169604-5

Porous polymers belong to a class of highly open materials, with microstructure that plays a fundamental role in determining their mechanical properties. These properties depend both on intrinsic microscopic measurements (e.g., pore size and cell wall diameter) and on macroscopic extrinsic measurements, such as sample size (Rege, Hillgärtner, and Itskov 2019). Supercritical carbon dioxide ($scCO_2$) is the main solvent used in the synthesis of nanostructured materials due to its favorable characteristics such as being non-flammable, non-toxic, environmentally friendly, having diffusivity comparable to that of gaseous substances and almost zero surface tension (Sanli, Bozbag, and Erkey 2012; Cardea and Reverchon 2019). During the supercritical process, the formation of polymeric particles goes through sequential stages of nucleation, formation of the polymeric nucleus and growth that produce the desired particles. Thus, changes in operational conditions such as temperature, pressure and solution flow result in different forms of polymeric composites, such as microspheres and microfibers (Kumar et al. 2017).

Supercritical technology combined with the formation of particles, especially porous polymers, can be used in a variety of processes. Therefore, in this chapter, we will approach obtaining porous polymers from natural polysaccharides as raw material, discussing aspects related to synthesis through mechanisms at high pressure. In addition, we will highlight topics such as the general properties involved in processing, techniques related to the use of supercritical fluid in drying, foaming, scaffolds, food packaging, emulsion templating, supports for catalysts materials and metal-organic frameworks production.

5.2 POROUS POLYMERS BASED ON NATURAL POLYSACCHARIDES

Although there is a multitude of raw materials from which porous polymers can be synthesized, in this topic we will focus on polymers based on natural polysaccharides and supercritical fluid technology for processing. The main reason for choosing natural polysaccharides as a polymeric base for the development of porous polymers is based on the premise that in the last decade, the use of petroleum-derived materials has been seen as an impasse for sustainable development. Porous polymers, synthesized from inorganic materials (such as polymers of silica, graphene, oxides and others) or derived from petroleum products are exhaustible, not sustainable and still have restricted applications. The porous polymers of natural polysaccharides are quite the opposite of porous polymers of inorganic origins and petroleum derivatives. Since they are from renewable sources, in some cases even from agro-industrial residues, are reusable, have a wide range of applications, are biodegradable, biocompatible and non-toxic (Barbosa et al. 2021).

Polysaccharides have a wide range of applications, from food to pharmaceutical products, and today these polymers have been explored as potential materials for the synthesis of porous polymers and countless other classes of biomaterials. Polysaccharides also present intrinsic aspects of their composition and origins related to the evolution of species, and the evolution of sex between species, which explain much of their variability between species of living organisms (Barbosa and de Carvalho Junior 2020). Natural polysaccharides have incredible properties, such as gelatinization power and formation of complex polymeric networks, which are directly linked to their chemical structure, including monosaccharide composition, molecular weight, types of glycosidic bonds and degree of branching. It can also be obtained from different natural sources, including agro-industrial residues, applying innovative techniques such as pressurized fluids, eutectic solvents and associated technologies (Barbosa et al. 2020; Freitas et al. 2021). Although there are a variety of polysaccharides with incredible properties, we will only address the polysaccharides most used in the synthesis of polymeric materials, including cellulose, starch and pectin (Figure 5.1).

Cellulose is a high–molecular-weight biopolymer formed by repeated units of D-glucose linked by β-(1$\rightarrow$4) glycosidic bonds. It is one of the most abundant resources in the world, it is present in the cell wall of plants and can also be synthesized by some microorganisms (Embrechts et al. 2018). Porous cellulose-based polymers are divided into three types: porous polymers of nanofibrillated

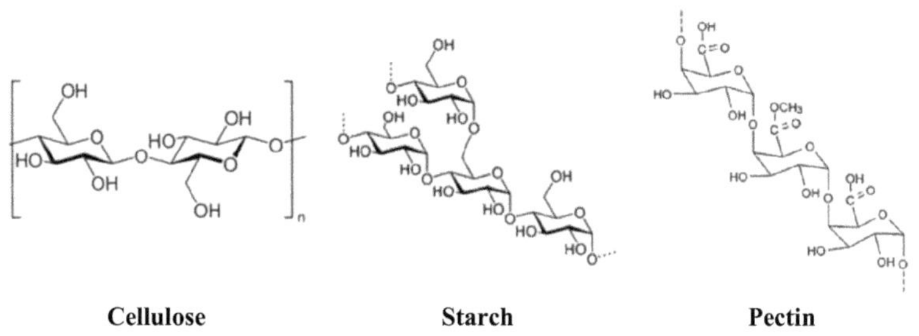

FIGURE 5.1 Basic chemical structure of cellulose, starch and pectin molecules.

cellulose, cellulose nanocrystals and bacterial cellulose. Porous polymers based on nanofibrillated cellulose have pores of about 10 to 100 mm after supercritical drying (da Silva Martins et al. 2021). The high content of nanofibrillated cellulose is important in the formation of three-dimensional polymeric networks, as it hinders the growth of ice crystals and increases the formation of smaller pores. Porous polymers based on cellulose nanocrystals have improved mechanical properties, due to the more complex structure formed by cellulose crystals and fibers intertwined in a three-dimensional network. It is considered that porous materials based on cellulose nanocrystals can be used as coatings of high hydrophobicity for medical applications and also in environmental remediation. Finally, porous polymers based on bacterial cellulose such as those produced by the bacterial metabolism of *Acetobacter xylinum* are considered biomaterials with unique properties such as biocompatibility and low toxicity, indicating applications in the pharmacological and food fields. These porous polymers have high mechanical resistance, good permeability in liquids and gases, maintain mechanical and structural properties even in united environments, and inhibit infectious processes. Currently, they have been applied as materials for drug delivery, wound healing and, when combined in metal oxide nanocomposites such as graphene, they are applied as sensors (Amiryaghoubi et al. 2021; Cao et al. 2021).

Starch is the main energy storage polysaccharide in plants, essential for the processes of embryogenesis and development of plant cells. Starch is formed by two high-molecular-weight biopolymers: amylose and amylopectin. Porous polymers made from starch have varied properties that depend on the starch composition, source of the starch, and proportion of amylose and amylopectin. Starch is a branched structure and therefore forms gels in an aqueous solution while still at room temperature. The gels can undergo retrogradation at temperatures above 60°C, producing high viscosity solutions, which can be used for the synthesis of aerogels and other materials (Embrechts et al. 2018). In the case of aerogels, the amylose content can determine the mechanical properties. However, only the amylose molecule does not form aerogel, as the heterogeneous structure does not withstand the pressure of drying systems. The main problem with porous starch-based polymers is related to low hardness, which limits their applications. However, the manufacture of porous starch polymers with some modifications in the polymeric system, for example, crosslinking with trisodium citrate can increase hardness properties and reduce adhesion (Q. Zhang et al. 2020).

Finally, pectin, a biopolymer formed by different esterified residues of galacturonic acid, is found mainly in citrus fruits and some different food residues. Depending on the degree of esterification, pectin is divided into high methoxyl pectin and low methoxyl pectin and still forms high viscosity gels, even at room temperature (Pierce et al. 2020). Pectin can form an intermolecular junction zone through ionic bonds between divalent cations when the pK is equal to or greater than (3–3.5). Porous polymers based on pectin are biodegradable, have a diversified chemical structure and can be modified by various techniques such as oxidation, copolymerization and carboxymethylation, which help to improve structural and mechanical properties (Pierce et al. 2020).

5.3 PARAMETERS INVOLVED IN THE POROUS POLYMERS PROCESSING BY HIGH PRESSURE

The application of high-pressure technologies for the development of porous materials is a widely discussed subject, and major advances in the process have been achieved currently. In recent years, several process parameters have been identified, including pressure, temperature and gas flow, which are considered relevant and which directly influence the size and shape of porous polymers. High-pressure technologies have as their main property the control of secondary parameters that directly influence the pore size and that can be used as guides for the construction of experimental models, and thus assist in the development of projects with porous polymers. The main application of high-pressure technologies in the processing of porous polymers is associated with drying projects and controlled pores development (Vareda, Lamy-Mendes, and Durães 2018).

The main parameters involved in the processing of porous polymers using high-pressure technology are: system pressure, pressurized gas flow, temperature, solvent elimination, drying time and gas expansion rates. In high-pressure systems such as those in supercritical fluid technology, all of these parameters mentioned can be controlled and further evaluated as input parameters in scale-up projects. Supercritical fluid technology has been constantly used in the processing of porous materials, mainly polymers such as aerogel and foaming porous. In general, supercritical fluid technology can be used as a means for controlled porous development, including control of size and shape (H. Wang et al. 2020).

5.4 SUPERCRITICAL FLUID DRYING FOR POROUS POLYMERS PROCESSING

Supercritical fluid technology has some advantages when compared to other technologies for porous polymers drying. Firstly, it is a green technology, associated with the low rate generation of toxic or polluting residues, the CO_2 used can be reused throughout the process, the pores formed have unique quality and specificity, the process parameters can be controlled and still offer sufficient data for scale-up studies. Finally, we can evaluate the technology as being innovative, as it applies sophisticated techniques for the synthesis of porous materials with control of the size and shape of the pores (X. Zhang et al. 2020).

The main functionality of supercritical fluid usage is for drying polymeric materials, to remove solvents (e.g., from alcogels) and thus producing porous paths (H. Wang et al. 2020). In this regard, the process of manufacturing porous polymers (e.g., drying), with supercritical fluid technology, can be divided into two approaches: supercritical drying at low temperature and supercritical drying at high temperature. When drying at high temperatures, the water in the polymeric network needs to be removed and replaced with an organic solvent such as acetone or alcohol. Then, the polymeric material is placed in a steel cell, where it will be pressurized and heated. Under these conditions, the solvent present in the polymeric network reaches the conditions of fluid in the supercritical state. After depressurizing the system, the fluid is removed, and the path taken by the fluid during drying forms well-structured porous networks. On the other hand, the application of drying at low temperatures uses a supercritical fluid to remove organic solvents (Cao et al. 2021). In this case, supercritical carbon dioxide ($scCO_2$) is used as a substitute for organic solvents in polymeric gels, mainly due to the excellent characteristics of diffusivity and density. The principles are the same for high-temperature drying work. However, there are some advantages associated with $scCO_2$ drying work, such as the use of low temperature and pressure conditions, CO_2 is a non-flammable solvent and it can be completely removed from the system to be reused (Kalaj et al. 2020).

The main advantage associated with drying by supercritical fluid is related to the reduction of surface tension in the pores, helping to avoid capillary pressure and collapse of the pores. Now, let's assess the influence of process parameters on the formation quality of porous materials.

It is important to highlight that the excessive flow rate of ethanol, or even other organic solvents, increases the diffusion rate of these solvents between the pores, leading to cracks and fractures in the polymeric material structure, negatively affecting the formation of the polymeric network and pores (Kalaj et al. 2020). Another parameter that influences the production of porous polymers, more specifically the size and shape of the pores, is time. It is reported that the rates of resistance to mass transfer as a function of time, directly influence the structure of porous polymers. For example, polymeric starch materials, such as starch aerogels, have significant differences in density depending on the drying time, being larger in the beginning and drastically reducing after 2 hours of drying. Another important point is that the very long drying time using high-pressure technologies leads to the formation of large surface areas. However, it is important to consider that the ideal drying must evaluate economic aspects and, therefore, very long times should be avoided (Wang et al. 2019).

Finally, the drying pressure plays a crucial role in the formation of pore polymers and the structure and shape of the pores. The increase in pressure during the production of porous polymers contributes to reducing the surface area of the polymeric network, which contributes to the increase in density. It is important to note that high pressure leads to the formation of small pores, while the rate of depressurization increases the expansion of the polymeric network, leading to the formation of large pores. In general, the porous polymers produced by high-pressure techniques have the following characteristics, regardless of the material used: the pore size distribution is at the level of nanopores and still has a low density and high porosity (Sun et al. 2021).

5.5 POROUS POLYMERS FOR FOAMING AND SCAFFOLDS BY SUPERCRITICAL TECHNOLOGY

Foamings and scaffolds are materials with specific characteristics such as high porosity, ultralightweight, large surface area, uniform pores with interconnectivity in the polymeric structure, low density and low refractive index (Tsai and Wang 2019; Soorbaghi et al. 2019). The application of $scCO_2$ is an alternative in the foaming synthesis, potentiating the production of scaffolds due to the performance of processes with a low volume of solvent, in moderate temperatures, with evaporation of the fluid under ambient pressure, and operations free of residues (Tsai and Wang 2019). Foamings and scaffolds can be applied in the field of biomedicine for tissue repair and controlled drug delivery, as well as in the environmental area, in processes of separation in the treatment of water and purification of nonpolar substances (Li et al. 2021; S. Li et al. 2019; Rajanna et al. 2015).

Scaffolds are synthesized from the exchange of solvents trapped in gelatinous material with the use of $scCO_2$, allowing the formation of mesopores (95–99% of empty space) (Soorbaghi et al. 2019). In general, scaffolds are produced from organic substances (polysaccharides, lignocellulosic polymers of biomass, proteins and others) and metal oxides (Soorbaghi et al. 2019; Freitas et al. 2021). Foaming synthesis is based on the sorption and dissolution of CO_2 in the polymeric raw material for a definite period (immersion time), under a definite pressure and temperature, followed by pressure release (depressurization). The last step forces the expansion of the polymer, and consequently, the CO_2 is removed, inducing the formation of pores in the polymeric matrix. Due to the plastifying effect of CO_2, foaming synthesis with supercritical technology is streamlined at mild temperatures, enabling the preparation of products with thermosensitive bioactive compounds (Santos-Rosales et al. 2020).

The biocompatibility of a foaming/scaffold is of fundamental importance for tissue engineering in promoting the desired functionality and minimal toxicity (Ferreira et al. 2020). Foaming/scaffold biodegradability is desirable for the viability of green adsorption processes, using materials such as poly(lactic acid), cotton, polyvinyl alcohol, chitin, polycaprolactone, cellulose and others (Li et al. 2021). In Table 5.1, it is possible to observe the different materials used in the manufacture of foamings/scaffolds, the potential applications and the functional activities investigated.

TABLE 5.1
Applications of Foaming and Scaffolds Produced Via $scCO_2$

Type of Foamings/Scaffolds	Application	Status	Reference
poly(lactic-co-glycolic acid) particles	Tissue engineering	Bone regeneration/*in vitro*	S. Li et al. (2019)
Poly-ε-caprolactone/silk fibroin/ nanohydroxyapatite	Tissue engineering	Bone regeneration/*in vitro*	Diaz-Gomez et al. (2017)
Graphene/chitosan/hydroxyapatite	Tissue engineering	Assessment of MTT cell viability	Yilmaz et al. (2019)
Silk fibroin	Tissue engineering	Assessment of MTT cell viability	Z. H. Li et al. (2019)
PLGA/curcumin and PLGA/ geantamycin	Drug delivery	Encapsulation of bioactive compounds	Ong et al. (2018)
Poly-ε-caprolactone/ transglutaminase	Drug delivery	Protein release	Kravanja et al. (2020)
PLGA/gemcitabine	Drug delivery	Impregnation/drug release	Álvarez et al. (2020)
SAMS of rice husk	Drug delivery	Release of ibuprofen and eugenol	Rajanna et al. (2015)
Poly(lactic acid)/poly(butylene succinate)	Water treatment	Oil adsorption	Bo Li et al. (2021)

PLGA = poly (D,L-lactic-co-glycolic acid) and **SAMS** = silica aerogel microparticles.

5.6 SUPERCRITICAL CO_2 IMPREGNATION IN POROUS POLYMERS FOR FOOD PACKAGING

$ScCO_2$-assisted impregnation has been an innovative technology for adding bioactive compounds to porous polymeric materials. This green technology has been a clean alternative compared to other traditional methods such as extrusion, coating and casting that involve toxic solvents, being environmentally unfriendly (Goñi, Gañán, Barbosa, et al. 2017; Champeau et al. 2015). Impregnation with supercritical fluid can be carried out using different solvents. However, CO_2 has been widely used due to its attractive properties such as moderate critical temperature and pressure (31.1°C and 73.7 bar) that allow isolated organic molecules present in extracts to present a low risk of thermal degradation (Ameri, Sodeifian, and Sajadian 2020; Bezerra et al. 2020; Mir et al. 2017).

The high diffusivity and low surface tension capacity of $scCO_2$ allow compounds to be incorporated into porous polymeric matrices. In this process, there are three important sequential steps which are dissolution, mixing of compounds with $scCO_2$ and depressurization (Carlès 2010). In the first stage of the process, the compounds are dissolved in CO_2 and then impregnated in the polymeric matrix. For this to occur, high pressure CO_2 increases the plasticity and swelling of the matrix polymers, allowing the compounds to be able to diffuse through the polymeric structure. The diffusion capacity of the compounds, plasticity and expansion of the matrix are directly influenced by the solubility of the compounds in CO_2 and by the temperature and pressure adopted in the process (Derossi, De Pilli, and Severini 2012). As these characteristics depend mostly on the CO_2 used as a solvent, molecules that have affinity for the polymer can be impregnated in the polymer. Thus, the compounds are trapped in the matrix due to the chemical absorption that occurs through a phenomenon known as molecular dispersion (Ashitha and Prince 2018).

In the depressurization phase, the compounds are fixed in the polymer structure by a phenomenon known as physical entrapment, which consists in the deposition of the compound in the polymeric matrix. The phenomenon occurs in CO_2-assisted impregnations, as they depend on the swelling and solubility of the compound in CO_2 (Ashitha and Prince 2018; Carlès 2010). This last step is very important because it can influence the absorption of compounds in the matrix. According to the investigated system, the rate of depressurization must be analyzed, as a rapid depressurization can

TABLE 5.2
Porous Polymers as a Support for Active Compounds

Active Component	Type of Polymer	Function	Reference
Eugenol	Linear low density polyethylene (LLDPE) films	Antioxidant	Goñi et al. (2016)
Cinnamaldehyde	Poly lactic acid (*e*-PLA)	Antimicrobial	de Dicastillo et al. (2018)
Phytol and *C. nutans* extracts	Silica and alginate aerogels	Antioxidant	Mustapa et al. (2016)
Polyphenols	Multilayer PET/PP food-grade film	Antioxidant	Belizón et al. (2018)
Caffeic acid	Polyethylene terephthalate/ polypropylene (PET/PP)	Antioxidant	Bastante et al. (2017)
Vitamin K3 and D	Alginate aerogels	Nutraceutical	Pantić, Knez, and Novak (2016)
Terpene ketones	LDPE films	Biopesticides	Goñi, Gañán, Herrera, et al. (2017)

allow better absorption of compounds in a polymeric matrix for which the molecules do not have good affinity (Sihvonen et al. 1999; Alavi et al. 1999).

The food industry and research groups have used impregnation with supercritical fluid to develop active food packaging that works as a barrier against UV rays, gases and microorganisms. These packages have been produced using biopolymers, recommending the use of materials of natural origin in detriment to synthetic polymers (Bastante et al. 2017; Goñi, Gañán, Herrera, et al. 2017). Active packaging is planned to influence actively the product, by protecting it, increasing its shelf life and preserving its sensory properties. In this way, these modern packagings are closely related to product quality (Kedzierski et al. 2020; Haghighi et al. 2020). Table 5.2 shows some studies that evaluated the use of the supercritical impregnation methodology in polymeric matrices that were applied in the development of food polymeric packaging.

5.7 SYNTHESIS OF POROUS POLYMERS BY SUPERCRITICAL EMULSION TEMPLATING

When talking about emulsion, it refers to the dispersion of two immiscible fluids that are mixed by mechanical shear or by a surfactant, in order to reach no macroscopic separation of phases. In a common emulsion, one of the liquids (dispersed phase) is in the form of thin globules within the other liquid (continuous phase), forming a stable mixture with an outer part and a core (Tutek et al. 2021). The original application of porous polymers by emulsion templating was in the synthesis of hydrophobic cross-linked porous polymer monoliths known as "polyHIPEs." High internal phase emulsions (HIPEs) are polymers synthesized and stabilized by surfactants that have more than 74% of internal phase occupying the volume of the emulsion. Currently, polymerized HIPEs systems include a variety of new emulsion systems such as polymerized hydrogel HIPE, porous carbons, porous inorganic systems and encapsulation systems (He et al. 2014; T. Zhang et al. 2019).

ScCO_2 started to be used in the polymerization of emulsions, acting as a template for the internal phase in the elaboration of emulsions CO_2-in-Water (CO_2/Water) or as a continuous phase in the case of emulsions Water-in-CO_2 (Water/CO_2), producing well-defined porous polymers (T. Zhang et al. 2019; J. Chen et al. 2019; Boyère, Jérôme, and Debuigne 2014; Kale and Deore 2017). The nonpolar character and low CO_2 dielectric constant limit its ability to dissolve polar and macromolecular substances. Thus, CO_2/Water emulsions are dispersion systems that have characteristics of common emulsions but retain the advantages of scCO_2. Water can dissolve more polar substances, such as the continuous phase of the CO_2/Water emulsion. Therefore, these emulsions have been widely used as modeling agents to synthesize porous materials, thickening agents and extraction agents (Bao et al. 2018; Ma et al. 2020).

Bao et al. (2018) studied the synthesis of amphiphilic nonfluorinated polymers (PVAc-b-PDMAEMA (poly(vinyl acetate)-block-poly(dimethylaminoethyl methacrylate)) and (PVDBM)-b-PDMAEMA (poly(vinyl acetate-alt-diethyl maleate)) with different ratios of CO_2-philic and hydrophilic groups using RAFT/MADIX polymerization and ATRP. From the stability measurements, it was possible to infer that the $PVDBM_8$-b-$PDMAEMA_{22}$ could emulsify 80% of CO_2 in the water, with stability for more than 48 hours, 20 Mpa and 25°C, presenting a stabilization period 20 times greater than PVAc-b-PDMAEMA did. Both the experimental results and the simulation results demonstrated that the reinforcement of CO_2-philicty with the surfactant can improve the stability of CO_2/Water emulsions.

Ohde (Ohde, Ohde, and Wai 2005) studied the separation and recycling of Pd^0 nanoparticles synthesized by water microemulsions, sodium bis(2-ethylhexyl) sulfosuccinate as surfactant and 1-octanol as cosolvent in CO_2 (Water/CO_2). The products of the olefin catalytic hydrogenations were dissolved in $scCO_2$ and separated from the cosolvent. Thus, the Pd^0 nanoparticles could be reused as catalysts in other hydrogenation reactions, presenting a high capacity for stability, reuse and easy separation of reaction products. From the results, the authors inferred that the method used in the study of recycling catalysts solubilized in the water core of the Water/CO_2 microemulsions can be applied to other catalysts such as enzymes, organometallic compounds, ionic species and nanoparticles.

5.8 POROUS POLYMERS AS SUPPORTS FOR CATALYSTS MATERIALS BY SUPERCRITICAL FLUID

Catalysts are substances that decrease activation energy and increase the speed of a chemical reaction without modifying it (Benaglia and Puglisi 2020; Itsuno and Haraguchi 2020). Unlike homogeneous catalysts, which are difficult to separate from formed products, heterogeneous catalysts can be used in porous supports that provide greater activity in relation to the oxidation states of metal ions, making them more stable and with greater mechanical resistance (Aspromonte et al. 2020; Feng et al. 2021). A good catalytic support must present basic characteristics such as having a high specific surface area and porosity, low reactivity in the reaction and minimally affect the life span, selectivity and sensitivity of the catalyst (Mukhtar et al. 2020; Wu et al. 2019).

The use of polymeric supports attributes several advantages to the catalysts, such as: the high surface area guarantees a uniform distribution of the active species, increasing their reactivity and improving the cost-benefit, which are relevant factors for more noble and high-cost catalysts; avoid the sintering of the active phase due to the effect of high temperature and facilitate heat transfer avoiding the accumulation of heat in highly exothermic reactions; and prevent disintegration in industrial catalytic beds, ensuring a longer service life for catalysts, improving resistance to poisoning reactions, ensuring greater activity and selectivity to these materials (Sun et al. 2016; Itsuno 2013; Itsuno and Haraguchi 2020; Flores-López et al. 2021; He et al. 2014; Kundrat et al. 2021).

The production of polymeric supports using $scCO_2$ for depositing catalysts or even for their recovery or recycling has been the subject of recent research. The use of supercritical technology allows generating high metallic contents reached with high dispersion and without significantly affecting the textural properties of the support, in addition to having the production of ecofriendly catalysts (Nasrollahzadeh et al. 2019, 2021). In the study of Yuan et al. (2020), the redox reaction was thermally induced between polyvinyl alcohol (PVA) and silver nitrate ($AgNO_3$), combined with $scCO_2$ foaming technology, proposing an innovative idea for the preparation of the micron-sized Ag foam catalyst. The authors found a reaction rate ten times higher than that from the commercial catalyst when applied to catalyze the reduction of 4-nitrophenol. Aizawa (2018) demonstrated in his research a new method of producing porous polymers that use liquid CO_2 pressed by a piston. The author elucidated that liquid CO_2 was more advantageous than $scCO_2$, making the device configuration simpler at room temperature where the pressurization was performed by the piston. In this way, the system can serve to impregnate catalysts in the polymeric supports.

$ScCO_2$ was used as an alternative in the production of catalytic aerogel nanopores that were synthesized from the reduction of H_2PdCl_4 with sodium borohydride followed by drying with $scCO_2$, which when tested in formic acid oxidation showed excellent catalytic activity (Douk, Saravani, and Farsadrooh 2019). The authors report that obtaining the aerogel in a non-polar solvent (hexane) and assembling by nanosheets using carboxylic acid (RCOOH) with different alkyl groups are strategies that still face several limitations, such as long gelling time and use of surfactant materials in the synthesis process. The $scCO_2$ was also used in pulsed laser ablation techniques with the advantage of having greater penetration into porous materials compared to conventional liquids, due to its high diffusivity and negligible surface tension. Additionally, these characteristics can be easily controlled by temperature and pressure and can be optimized in the process (Labusch et al. 2021). In addition to the process of producing porous polymers as supports for catalysts, $scCO_2$ has been reported as a noble catalyst recycling technique, being more efficient than conventional processes such as pyrometallurgical or chemical corrosion. These conventional processes can destroy the catalytic support and emit a large amount of toxic fumes that need post-treatment before being released into the environment (Li et al. 2018).

5.9 POROUS METAL–ORGANIC FRAMEWORKS POLYMERS BY SUPERCRITICAL FLUID PROCESSING

Metal-organic frameworks (MOFs), also known as porous coordination polymers (PCPs) or covalent organic frameworks (COFs), are porous and crystalline materials based on the bond between metal ions and charged organic ligands, which may have structurally polymeric features, high thermal and chemical stability, excellent optical nature, large surface area, high porosity and geometrically controllable pores. MOFs can be used in several areas such as gas storage and separation, catalysis, encapsulation of active compounds, drug delivery, enzyme immobilization, optoelectronic, data storage, and so on (X. Zhang et al. 2020; Kumar et al. 2017; Li et al. 2014; Matsuyama 2018; Peng et al. 2014).

The synthesis can be accomplished through approaches using solvothermal synthesis (Sanada et al. 2019; Embrechts et al. 2018), electrochemical synthesis (Martinez Joaristi et al. 2012; Lu et al. 2014), mechanochemical synthesis (D. Chen et al. 2019; Z. Wang et al. 2020), ultrasonic or sonochemical synthesis (Esrafili et al. 2019; Masoomi, Bagheri, and Morsali 2017), layer-by-layer synthesis (Shi et al. 2018; Hinman et al. 2018), microwave-assisted synthesis (H. Zhang et al. 2019; Vakili et al. 2018) and supercritical route (Doan et al. 2020; López-Periago et al. 2015). Despite the efficiency presented, researches have been carried out to improve some limitations of MOFs, such as high processing costs, use of toxic solvents, low electrical conductivity, toxicity and non-biodegradability. The $scCO_2$ has been reported in the synthesis, drying and activation of porous structures as a promising route in obtaining these materials, as it can be performed without the usage of organic solvents, metallic complexing agents and catalysts. For the process, it is of fundamental importance that the solubility of metallic and organic binders in CO_2 can be sufficient for the complete formation of the reaction, requiring the use of CO_2-philic reagents due to the nonpolar nature of CO_2 (Marrett et al. 2018; Dapaah and Liu 2020; Kumar et al. 2017; Peng et al. 2014).

ScCO2 was used in the final extraction of solvents (dimethylformamide and ethanol), aluminum and unreacted chromium (M^{3+}) from intermediate aerogels in order to obtain PCPs of metal ions and carboxyl of porphyrins. According to the authors (Xu et al. 2018), the prepared porous materials showed a greater surface area, high thermal stability and high efficiency in the adsorptive capacity to rhodamine B (RhB) and methylene blue (MB) dyes. Song et al. (2019) also studied $scCO_2$ extraction of residual reagents and unreacted solvent molecules to obtain metalloporphyrins-based PCPs. The authors concluded that the method was effective in preventing the collapse of frame construction and obtaining a material with greater thermal stability, high specific surface area and with catalytic activity and stability for the ethylbenzene oxidation. The activation of microporous coordination polymers (MCPs) from N,N-dimethylformamide-solvated crystals by flowing $scCO_2$ was compared with the conventional solvent exchange/evacuation or solvent exchange/scCO_2 treatment. The mentioned study demonstrated that the method by flowing scCO_2 presented structures with larger surface areas, in addition to being a faster, scalable, economical and environmentally friendly method when compared to conventional activation by evacuation (Liu, Wong-Foy, and Matzger 2013).

5.10 CONCLUDING REMARKS

The growing demand for the use of sustainable and innovative materials in the various sectors of the industry has stimulated the search for materials that supply this need through technologies that also corroborate the search for these materials. Porous polymers are highly technological materials capable of producing composites with applications in the chemical, electronic, food, cosmetic and pharmaceutical industries. The use of supercritical technology can assist in the synthesis and processing of these materials due to their characteristics that make them environmentally friendly and, at the same time, competitive because they have a cost/benefit ratio similar to other techniques, in the long run.

ACKNOWLEDGMENTS

This study was financed in part by the Coordenação de Aperfeiçoamento de Pessoal de Nível Superior – Brasil (CAPES) – Finance Code 001.

REFERENCES

Aizawa, Takafumi. 2018. "A New Method for Producing Porous Polymer Materials Using Carbon Dioxide and a Piston." *Journal of Supercritical Fluids* 133 (July 2017): 38–41. doi:10.1016/j.supflu.2017.09.018.

Alavi, S. H., B. K. Gogoi, M. Khan, B. J. Bowman, and S. S. H. Rizvi. 1999. "Structural Properties of Protein-Stabilized Starch-Based Supercritical Fluid Extrudates." *Food Research International* 32 (2): 107–18. doi:10.1016/S0963-9969(99)00063-0.

Álvarez, I., C. Gutiérrez, J. F. Rodríguez, A. De Lucas, and M. T. García. 2020. "Production of Drug-Releasing Biodegradable Microporous Scaffold Impregnated with Gemcitabine Using a CO_2 Foaming Process." *Journal of CO_2 Utilization* 41 (January): 101227. doi:10.1016/j.jcou.2020.101227.

Ameri, Arezu, Gholamhossein Sodeifian, and Seyed Ali Sajadian. 2020. "Lansoprazole Loading of Polymers by Supercritical Carbon Dioxide Impregnation: Impacts of Process Parameters." *Journal of Supercritical Fluids* 164 (October): 104892. doi:10.1016/j.supflu.2020.104892.

Amiryaghoubi, Nazanin, Marziyeh Fathi, Abolfazl Barzegari, Jaleh Barar, Hossein Omidian, and Yadollah Omidi. 2021. "Recent Advances in Polymeric Scaffolds Containing Carbon Nanotube and Graphene Oxide for Cartilage and Bone Regeneration." *Materials Today Communications* 26 (January): 102097. doi:10.1016/j.mtcomm.2021.102097.

Ashitha, G. N., and M. V. Prince. 2018. "Vacuum Impregnation: Applications in Food Industry." *International Journal of Food and Fermentation Technology* 8 (2): 141–51. doi:10.30954/2277-9396.02.2018.3.

Aspromonte, Soledad Guadalupe, Federico Andrés Piovano, Esther Alonso, and Alicia Viviana Boix. 2020. "Synthesis of Supported Mesoporous Catalysts Using Supercritical CO_2." *Advances in Microporous and Mesoporous Materials*, 1–40. doi:10.5772/intechopen.92740.

Bao, Lei, Shuyi Fang, Dongdong Hu, Yuan Zong, Ling Zhao, Weikang Yuan, and Tao Liu. 2018. "Stabilization of CO_2-in-Water Emulsions by Nonfluorinated Surfactants with Enhanced CO_2-Philic Tails." *Journal of Supercritical Fluids* 133 (September 2017): 163–70. doi:10.1016/j.supflu.2017.10.007.

Barbosa, J. R., R. H. H. Pinto, H. S. Martins, and R. N. de Carvalho Junior. 2021. "Bioaerogels: Synthesis Approaches, Biomedical Applications and Cell Uptake." In *Materials Research Foundations*, 43–56. doi:10.21741/9781644901298-3.

Barbosa, Jhonatas Rodrigues, and Raul Nunes de Carvalho Junior. 2020. "Occurrence and Possible Roles of Polysaccharides in Fungi and Their Influence on the Development of New Technologies." *Carbohydrate Polymers* 246: 116613. doi:10.1016/j.carbpol.2020.116613.

Barbosa, Jhonatas Rodrigues, Maurício Madson Dos Santos Freitas, Luiza Helena da Silva Martins, and Raul Nunes de Carvalho Junior. 2020. "Polysaccharides of Mushroom *Pleurotus* Spp.: New Extraction Techniques, Biological Activities and Development of New Technologies." *Carbohydrate Polymers* 229: 115550. doi:10.1016/j.carbpol.2019.115550.

Bastante, C. Cejudo, L. Casas Cardoso, C. Mantell Serrano, and E. J. Martínez de la Ossa. 2017. "Supercritical Impregnation of Food Packaging Films to Provide Antioxidant Properties." *Journal of Supercritical Fluids* 128 (May): 200–207. doi:10.1016/j.supflu.2017.05.034.

Belizón, M., M. T. Fernández-Ponce, L. Casas, C. Mantell, and E. J. Martínez de la Ossa. 2018. "Supercritical Impregnation of Antioxidant Mango Polyphenols into a Multilayer PET/PP Food-Grade Film." *Journal of CO_2 Utilization* 25 (November 2017): 56–67. doi:10.1016/j.jcou.2018.03.005.

Benaglia, Maurizio, and Alessandra Puglisi. 2020. *Catalyst Immobilization. Catalyst Immobilization.* John Wiley & Sons. doi:10.1002/9783527817290.

Bezerra, F. W. F., M. S. De Oliveira, P. N. Bezerra, V. M. B. Cunha, M. P. Silva, W. A. Costa, R. H. H. Pinto, et al. 2020. "Extraction of Bioactive Compounds." In *Green Sustainable Processes for Chemical and Environmental Engineering and Science*, edited by Inamuddin, Abdullah M. Asiri, and Arun M. Isloor, 1st ed., 149–67. Cambridge: Elsevier Inc. doi:10.1016/B978-0-12-817388-6.00008-8.

Boyère, Cédric, Christine Jérôme, and Antoine Debuigne. 2014. "Input of Supercritical Carbon Dioxide to Polymer Synthesis: An Overview." *European Polymer Journal* 61: 45–63. doi:10.1016/j.eurpolymj.2014.07.019.

Cao, Jingsong, Chen Shen, Xiqing Wang, Yawei Zhu, Sansan Bao, Xiaodan Wu, and Yujie Fu. 2021. "A Porous Cellulose-Based Molecular Imprinted Polymer for Specific Recognition and Enrichment of Resveratrol." *Carbohydrate Polymers* 251 (September 2020): 117026. doi:10.1016/j.carbpol.2020.117026.

Cardea, Stefano, and Ernesto Reverchon. 2019. "Supercritical Fluid Processing of Polymers." *Polymers* 11 (10): 2–5. doi:10.3390/polym11101551.

Carlès, Pierre. 2010. "A Brief Review of the Thermophysical Properties of Supercritical Fluids." *Journal of Supercritical Fluids* 53 (1–3): 2–11. doi:10.1016/j.supflu.2010.02.017.

Champeau, M., J. M. Thomassin, T. Tassaing, and C. Jérôme. 2015. "Drug Loading of Polymer Implants by Supercritical CO_2 Assisted Impregnation: A Review." *Journal of Controlled Release*. doi:10.1016/j.jconrel.2015.05.002.

Chen, Dong, Jiahua Zhao, Pengfei Zhang, and Sheng Dai. 2019. "Mechanochemical Synthesis of Metal–Organic Frameworks." *Polyhedron* 162: 59–64. doi:10.1016/j.poly.2019.01.024.

Chen, Jian, Umair Azhar, Yongkang Wang, Jihong Liang, and Bing Geng. 2019. "Preparation of Fluoropolymer Materials with Different Porous Morphologies by an Emulsion Template Method Using Supercritical Carbon Dioxide as a Medium." *RSC Advances* 9 (20): 11331–40. doi:10.1039/C9RA00777F.

Dapaah, Malcom Frimpong, and Baojian Liu. 2020. "Recent Advances of Supercritical CO_2 in Green Synthesis and Activation of Metal–Organic Frameworks." *Journal of Inorganic and Organometallic Polymers and Materials* 30 (3): 581–95. doi:10.1007/s10904-019-01354-2.

Derossi, A., T. De Pilli, and C. Severini. 2012. "The Application of Vacuum Impregnation Techniques in Food Industry." In *Scientific, Health and Social Aspects of the Food Industry*, edited by Benjamin Valdez, 25–56. IntechOpen. doi:10.5772/31435.

Diaz-Gomez, Luis, Carlos A. García-González, Jiamian Wang, Fang Yang, Salvador Aznar-Cervantes, Jose Luis Cenis, Ricardo Reyes, et al. 2017. "Biodegradable PCL/Fibroin/Hydroxyapatite Porous Scaffolds Prepared by Supercritical Foaming for Bone Regeneration." *International Journal of Pharmaceutics* 527 (1–2): 115–25. doi:10.1016/j.ijpharm.2017.05.038.

de Dicastillo, Carol López, Carolina Villegas, Luan Garrido, Karina Roa, Alejandra Torres, María José Galotto, Adrián Rojas, and Julio Romero. 2018. "Modifying an Active Compound's Release Kinetic Using a Supercritical Impregnation Process to Incorporate an Active Agent into PLA Electrospun Mats." *Polymers* 10 (5). 479. doi:10.3390/polym10050479.

Doan, Huan V., Fei Cheng, Thandeka Dyirakumunda, Mark R. J. Elsegood, Jiamin Chin, Oliver Rowe, Carl Redshaw, and Valeska P. Ting. 2020. "Using Supercritical CO_2 in the Preparation of Metal-Organic Frameworks: Investigating Effects on Crystallisation." *Crystals* 10 (1). doi:10.3390/cryst10010017.

Douk, Abdollatif Shafaei, Hamideh Saravani, and Majid Farsadrooh. 2019. "Three-Dimensional Inorganic Polymer of Pd Aerogel as a Highly Active Support-Less Anode Catalyst toward Formic Acid Oxidation." *International Journal of Hydrogen Energy* 44: 18028–37. doi:10.1016/j.ijhydene.2019.05.084.

Embrechts, Heidemarie, Martin Kriesten, Kilian Hoffmann, Wolfgang Peukert, Martin Hartmann, and Monica Distaso. 2018. "Elucidation of the Formation Mechanism of Metal-Organic Frameworks via in-Situ Raman and FTIR Spectroscopy under Solvothermal Conditions." *Journal of Physical Chemistry C* 122 (23): 12267–78. doi:10.1021/acs.jpcc.8b02484.

Esrafili, Leili, Alireza Azhdari Tehrani, Ali Morsali, Lucia Carlucci, and Davide M. Proserpio. 2019. "Ultrasound and Solvothermal Synthesis of a New Urea-Based Metal-Organic Framework as a Precursor for Fabrication of Cadmium(II) Oxide Nanostructures." *Inorganica Chimica Acta* 484 (Ii): 386–93. doi:10.1016/j.ica.2018.09.025.

Feng, Lijuan, Lu Qiu, Shunli Wang, Zhaosen Chang, Yan Lu, Anwang Dong, Qi Chen, and Zhuyin Sui. 2021. "Metal Phenanthroline-Based Porous Polymeric Hybrid Catalysts for Direct Conversion of Methane." *Journal of Porous Materials* 28: 487–93. doi:10.1007/s10934-020-01013-9.

Ferreira, Filipe V., Caio G. Otoni, Kevin J. De France, Hernane S. Barud, Liliane M. F. Lona, Emily D. Cranston, and Orlando J. Rojas. 2020. "Porous Nanocellulose Gels and Foams: Breakthrough Status in the Development of Scaffolds for Tissue Engineering." *Materials Today* 37: 126–41. doi:10.1016/j.mattod.2020.03.003.

Flores-López, S. L., B. Karakashov, L. A. Ramírez-Montoya, J. A. Menéndez, V. Fierro, A. Arenillas, M. A. Montes-Morán, and A. Celzard. 2021. "Effect of the Porosity and Microstructure on the Mechanical Properties of Organic Xerogels." *Journal of Materials Science* 56: 10312–10325. doi:10.1007/s10853-021-05882-6.

Freitas, Lucas Cantão, Jhonatas Rodrigues Barbosa, Ana Laura Caldas da Costa, Fernanda Wariss Figueiredo Bezerra, Rafael Henrique Holanda Pinto, and Raul Nunes de Carvalho Junior. 2021. "From Waste to Sustainable Industry: How Can Agro-Industrial Wastes Help in the Development of New Products?" *Resources, Conservation and Recycling* 169 (January): 105466. doi:10.1016/j.resconrec.2021.105466.

Goñi, María L., Nicolás A. Gañán, Silvia E. Barbosa, Miriam C. Strumia, and Raquel E. Martini. 2017. "Supercritical CO_2-Assisted Impregnation of LDPE/Sepiolite Nanocomposite Films with Insecticidal Terpene Ketones: Impregnation Yield, Crystallinity and Mechanical Properties Assessment." *Journal of Supercritical Fluids* 130 (December): 337–46. doi:10.1016/j.supflu.2017.06.013.

Goñi, María L., Nicolás A. Gañán, Jimena M. Herrera, Miriam C. Strumia, Alfonsina E. Andreatta, and Raquel E. Martini. 2017. "Supercritical CO_2 iof LDPE Films with Terpene Ketones as Biopesticides against Corn Weevil (*Sitophilus zeamais*)." *Journal of Supercritical Fluids* 122 (April): 18–26. doi:10.1016/j.supflu.2016.11.017.

Goñi, María L., Nicolás A. Gañán, Miriam C. Strumia, and Raquel E. Martini. 2016. "Eugenol-Loaded LLDPE Films with Antioxidant Activity by Supercritical Carbon Dioxide Impregnation." *Journal of Supercritical Fluids* 111: 28–35. doi:10.1016/j.supflu.2016.01.012.

Haghighi, Hossein, Fabio Licciardello, Patrizia Fava, Heinz Wilhelm Siesler, and Andrea Pulvirenti. 2020. "Recent Advances on Chitosan-Based Films for Sustainable Food Packaging Applications." *Food Packaging and Shelf Life*. doi:10.1016/j.fpsl.2020.100551.

He, Hongkun, Wenwen Li, Melissa Lamson, Mingjiang Zhong, Dominik Konkolewicz, Chin Ming Hui, Karin Yaccato, et al. 2014. "Porous Polymers Prepared via High Internal Phase Emulsion Polymerization for Reversible CO_2 Capture." *Polymer* 55 (1): 385–94. doi:10.1016/j.polymer.2013.08.002.

Hinman, Joshua G., Jacob G. Turner, Daniel M. Hofmann, and Catherine J. Murphy. 2018. "Layer-by-Layer Synthesis of Conformal Metal-Organic Framework Shells on Gold Nanorods." *Chemistry of Materials* 30 (20): 7255–61. doi:10.1021/acs.chemmater.8b03341.

Itsuno, Shinichi. 2013. "Polymer Catalysts." In *Encyclopedia of Polymeric Nanomaterials*, edited by Shiro Kobayashi and Klaus Müllen, 1st ed., 1–9. Springer-Verlag Berlin. doi:10.1007/978-3-642-36199-9.

Itsuno, Shinichi, and Naoki Haraguchi. 2020. "Catalysts Immobilized onto Polymers." In *Catalyst Immobilization: Methods and Applications* 23–75. doi:10.1002/9783527817290.ch2.

Kalaj, Mark, Kyle C. Bentz, Sergio Ayala, Joseph M. Palomba, Kyle S. Barcus, Yuji Katayama, and Seth M. Cohen. 2020. "MOF-Polymer Hybrid Materials: From Simple Composites to Tailored Architectures." *Chemical Reviews* 120 (16): 8267–8302. doi:10.1021/acs.chemrev.9b00575.

Kale, Santosh Nemichand, and Sharada Laxman Deore. 2017. "Emulsion Micro Emulsion and Nano Emulsion: A Review." *Systematic Reviews in Pharmacy* 8 (1): 39–47. doi:10.5530/srp.2017.1.8.

Kedzierski, Mikaël, Benjamin Lechat, Olivier Sire, Gwénaël Le Maguer, Véronique Le Tilly, and Stéphane Bruzaud. 2020. "Microplastic Contamination of Packaged Meat: Occurrence and Associated Risks." *Food Packaging and Shelf Life* 24 (June): 100489. doi:10.1016/j.fpsl.2020.100489.

Kravanja, Gregor, Mateja Primožič, Željko Knez, and Maja Leitgeb. 2020. "Transglutaminase Release and Activity from Novel Poly(ε-Caprolactone)-Based Composites Prepared by Foaming with Supercritical CO_2." *Journal of Supercritical Fluids* 166. doi:10.1016/j.supflu.2020.105031.

Kumar, Pawan, Kowsalya Vellingiri, Ki Hyun Kim, Richard J. C. Brown, and Manolis J. Manos. 2017. "Modern Progress in Metal-Organic Frameworks and Their Composites for Diverse Applications." *Microporous and Mesoporous Materials* 253 (January 2016): 251–65. doi:10.1016/j.micromeso.2017.07.003.

Kundrat, V., R. Rosentsveig, O. Brontvein, R. Tenne, and J. Pinkas. 2021. "Synthesis and Characterization of WS_2/SiO_2 Microfiber." *Journal of Materials Science* 56: 10834–46. doi:10.1007/s10853-021-05979-y.

Labusch, Marc, Sissy Puthenkalam, Ernst Cleve, Stephan Barcikowski, and Sven Reichenberger. 2021. "Pore Penetration of Porous Catalyst Supports by *in-situ*-Adsorbed, Agglomeration-Quenched Nanoparticles from Pulsed Laser Ablation in Supercritical CO_2." *Journal of Supercritical Fluids* 169: 105100. doi:10.1016/j.supflu.2020.105100.

Li, Bin, Hui-Min Wen, Wei Zhou, and Banglin Chen. 2014. "Porous Metal–Organic Frameworks for Gas Storage and Separation: What, How, and Why?" *The Journal of Physical Chemistry Letters* 5: 3468–79. doi:10.1021/jz501586e.

Li, Bo, Guoqun Zhao, Guilong Wang, Lei Zhang, Jie Gong, and Zhanlin Shi. 2021. "Biodegradable PLA/PBS Open-Cell Foam Fabricated by Supercritical CO_2 Foaming for Selective Oil-Adsorption." *Separation and Purification Technology* 257 (September 2020). doi:10.1016/j.seppur.2020.117949.

Li, Shuang, Chaobo Song, Shengbing Yang, Weijun Yu, Weiqi Zhang, Guohua Zhang, Zhenhao Xi, and Eryi Lu. 2019. "Supercritical CO_2 Foamed Composite Scaffolds Incorporating Bioactive Lipids Promote Vascularized Bone Regeneration via Hif-1α Upregulation and Enhanced Type H Vessel Formation." *Acta Biomaterialia* 94: 253–67. doi:10.1016/j.actbio.2019.05.066.

Li, W. S. J., Fabien Gasc, Jordanna Pinot, Jérémy Causse, Hélène Poirot, Julien Pinaud, Cécile Bouilhac, Hervé Simonaire, Danielle Barth, and Patrick Lacroix-Desmazes. 2018. "Extraction of Palladium from Alumina-Supported Catalyst in Supercritical CO_2 Using Functional Fluorinated Polymers." *Journal of Supercritical Fluids* 138: 207–14. doi:10.1016/j.supflu.2018.04.013.

Li, Ze Hao, Li Wang, Hong Lian Dai, Xin Yu Wang, Jia Shen Li, and Zheng Zhao. 2019. "Fabrication, Characterization, and in Vitro Evaluation of Biomimetic Silk Fibroin Porous Scaffolds via Supercritical CO_2 Technology." *Journal of Supercritical Fluids* 150: 86–93. doi:10.1016/j.supflu.2019.04.006.

Liu, Baojian, Antek G. Wong-Foy, and Adam J. Matzger. 2013. "Rapid and Enhanced Activation of Microporous Coordination Polymers by Flowing Supercritical CO_2." *Chemical Communications* 49 (14): 1419–21. doi:10.1039/c2cc37793d.

López-Periago, A., O. Vallcorba, C. Frontera, C. Domingo, and J. A. Ayllón. 2015. "Exploring a Novel Preparation Method of 1D Metal Organic Frameworks Based on Supercritical CO_2." *Dalton Transactions* 44 (16): 7548–53. doi:10.1039/c5dt00521c.

Lu, Chunjing, Teng Ben, Shixian Xu, and Shilun Qiu. 2014. "Electrochemical Synthesis of a Microporous Conductive Polymer Based on a Metal-Organic Framework Thin Film." *Angewandte Chemie* 126 (25): 6572–76. doi:10.1002/ange.201402950.

Ma, Liru, Lei Bao, Dongdong Hu, Ling Zhao, and Tao Liu. 2020. "Effect of Interfacial Properties on the Stability of Ultra-Dry CO_2-in-Water (C/W) Foams Stabilized with Zwitterionic Surfactants and Nonionic/Anionic Polymers: Experimental and DPD Simulation." *Journal of Supercritical Fluids* 158: 104722. doi:10.1016/j.supflu.2019.104722.

Marrett, Joseph M., Cristina Mottillo, Simon Girard, Christopher W. Nickels, Jean Louis Do, Gandrath Dayaker, Luzia S. Germann, et al. 2018. "Supercritical Carbon Dioxide Enables Rapid, Clean, and Scalable Conversion of a Metal Oxide into Zeolitic Metal-Organic Frameworks." *Crystal Growth and Design* 18 (5): 3222–28. doi:10.1021/acs.cgd.8b00385.

Martinez Joaristi, Alberto, Jana Juan-Alcañiz, Pablo Serra-Crespo, Freek Kapteijn, and Jorge Gascon. 2012. "Electrochemical Synthesis of Some Archetypical Zn^{2+}, Cu^{2+}, and Al^{3+} Metal Organic Frameworks." *Crystal Growth and Design* 12 (7): 3489–98. doi:10.1021/cg300552w.

Masoomi, Mohammad Yaser, Minoo Bagheri, and Ali Morsali. 2017. "Porosity and Dye Adsorption Enhancement by Ultrasonic Synthesized Cd(II) Based Metal-Organic Framework." *Ultrasonics Sonochemistry* 37: 244–50. doi:10.1016/j.ultsonch.2017.01.018.

Matsuyama, Kiyoshi. 2018. "Supercritical Fluid Processing for Metal–Organic Frameworks, Porous Coordination Polymers, and Covalent Organic Frameworks." *Journal of Supercritical Fluids* 134 (December 2017): 197–203. doi:10.1016/j.supflu.2017.12.004.

Mir, Shabir Ahmad, Manzoor Ahmad Shah, Basharat Nabi Dar, Ali Abas Wani, Shaiq Ahmad Ganai, and Jyoti Nishad. 2017. "Supercritical Impregnation of Active Components into Polymers for Food Packaging Applications." *Food and Bioprocess Technology* 10 (9): 1749–54. doi:10.1007/s11947-017-1937-9.

Mukhtar, Ahmad, Sidra Saqib, Nurhayati Binti Mellon, Sikander Rafiq, Muhammad Babar, Sami Ullah, Nawshad Muhammad, et al. 2020. "A Review on CO_2 Capture via Nitrogen-Doped Porous Polymers and Catalytic Conversion as a Feedstock for Fuels." *Journal of Cleaner Production* 277: 123999. doi:10.1016/j.jclepro.2020.123999.

Mustapa, A. N., A. Martin, L. M. Sanz-Moral, M. Rueda, and M. J. Cocero. 2016. "Impregnation of Medicinal Plant Phytochemical Compounds into Silica and Alginate Aerogels." *Journal of Supercritical Fluids* 116 (October): 251–63. doi:10.1016/j.supflu.2016.06.002.

Nasrollahzadeh, Mahmoud, Reza Akbari, Solmaz Sakhaei, Zahra Nezafat, Saeed Banazadeh, Yasin Orooji, and Gurumurthy Hegde. 2021. "Polymer Supported Copper Complexes/Nanoparticles for Treatment of Environmental Contaminants." *Journal of Molecular Liquids* 330: 115668. doi:10.1016/j.molliq.2021.115668.

Nasrollahzadeh, Mahmoud, Mohaddeseh Sajjadi, Mohammadreza Shokouhimehr, and Rajender S. Varma. 2019. "Recent Developments in Palladium (Nano)Catalysts Supported on Polymers for Selective and Sustainable Oxidation Processes." *Coordination Chemistry Reviews* 397: 54–75. doi:10.1016/j.ccr.2019.06.010.

Ohde, Mariko, Hiroyuki Ohde, and Chien M. Wai. 2005. "Recycling Nanoparticles Stabilized in Water-in-CO_2 Microemulsions for Catalytic Hydrogenations." *Langmuir* 21 (5): 1738–44. doi:10.1021/la0482709.

Ong, Yi Xian Jolene, Lai Yeng Lee, Pooya Davoodi, and Chi Hwa Wang. 2018. "Production of Drug-Releasing Biodegradable Microporous Scaffold Using a Two-Step Micro-Encapsulation/Supercritical Foaming Process." *Journal of Supercritical Fluids* 133 (July 2017): 263–69. doi:10.1016/j.supflu.2017.10.018.

Pantić, Milica, Željko Knez, and Zoran Novak. 2016. "Supercritical Impregnation as a Feasible Technique for Entrapment of Fat-Soluble Vitamins into Alginate Aerogels." *Journal of Non-Crystalline Solids* 432 (January): 519–26. doi:10.1016/j.jnoncrysol.2015.11.011.

Peng, Li, Jianling Zhang, Zhimin Xue, Buxing Han, Xinxin Sang, Chengcheng Liu, and Guanying Yang. 2014. "Highly Mesoporous Metal-Organic Framework Assembled in a Switchable Solvent." *Nature Communications* 5: 1–7. doi:10.1038/ncomms5465.

Pierce, Aidan, Yifan Zheng, Willi L. Wagner, Henrik V. Scheller, Debra Mohnen, Maximilian Ackermann, and Steven J. Mentzer. 2020. "Visualizing Pectin Polymer-Polymer Entanglement Produced by Interfacial Water Movement." *Carbohydrate Polymers* 246 (March): 116618. doi:10.1016/j.carbpol.2020.116618.

Rajanna, Suresh Kumar, Dharmendr Kumar, Madhu Vinjamur, and Mamata Mukhopadhyay. 2015. "Silica Aerogel Microparticles from Rice Husk Ash for Drug Delivery." *Industrial and Engineering Chemistry Research* 54 (3): 949–56. doi:10.1021/ie503867p.

Rege, Ameya, Markus Hillgärtner, and Mikhail Itskov. 2019. "Mechanics of Biopolymer Aerogels Based on Microstructures Generated from 2-d Voronoi Tessellations." *Journal of Supercritical Fluids* 151: 24–29. doi:10.1016/j.supflu.2019.04.018.

Sanada, Tomoe, Satoshi Tominaka, Kazuo Kojima, and Anthony K. Cheetham. 2019. "Violet Luminescence from Zinc-Based Metal-Organic Frameworks Prepared by Solvothermal Synthesis." *Bulletin of the Chemical Society of Japan* 92 (2): 427–34. doi:10.1246/bcsj.20180256.

Sanli, D., S. E. Bozbag, and C. Erkey. 2012. "Synthesis of Nanostructured Materials Using Supercritical CO_2: Part I. Physical Transformations." *Journal of Materials Science* 47: 2995–3025. doi:10.1007/s10853-011-6054-y.

Santos-Rosales, Víctor, Marta Gallo, Philip Jaeger, Carmen Alvarez-Lorenzo, José L. Gómez-Amoza, and Carlos A. García-González. 2020. "New Insights in the Morphological Characterization and Modelling of Poly(ε-Caprolactone) Bone Scaffolds Obtained by Supercritical CO_2 Foaming." *Journal of Supercritical Fluids* 166. doi:10.1016/j.supflu.2020.105012.

Shi, Xiansong, Rui Wang, Ankang Xiao, Tianzhi Jia, Shi Peng Sun, and Yong Wang. 2018. "Layer-by-Layer Synthesis of Covalent Organic Frameworks on Porous Substrates for Fast Molecular Separations." *Applied Nano Materials* 1 (11): 6320–26. doi:10.1021/acsanm.8b01537.

Sihvonen, M., E. Järvenpää, V. Hietaniemi, and R. Huopalahti. 1999. "Advances in Supercritical Carbon Dioxide Technologies." *Trends in Food Science and Technology* 10 (6–7): 217–22. doi:10.1016/S0924-2244(99)00049-7.

da Silva Martins, Luiza Helena, João Moreira Neto, Jhonatas Rodrigues Barbosa, Gilson Celso Albuquerque Chagas Junior, Johnatt Allan Rochade Oliveira, Mahendra Rai, and Alessandra Santos Lopes. 2021. "Carboxymethyl Cellulose-Coated Polypropylene Films Containing Essential Oil for Food Preservation." In *Biopolymer-Based Nano Films*, 133–47. doi:10.1016/B978-0-12-823381-8.00008-9.

Silverstein, Michael S. 2014. "PolyHIPEs: Recent Advances in Emulsion-Templated Porous Polymers." *Progress in Polymer Science* 39 (1): 199–234. doi:10.1016/j.progpolymsci.2013.07.003.

Song, Dengmeng, Xiaolong Bai, Yong Yang, Xin Zhao, and Ning Wang. 2019. "Metalloporphyrins-Al^{3+} Porous Coordination Polymers: Preparations, Characterizations and Catalytic Properties." *Applied Organometallic Chemistry* 33 (9): e5055. doi:10.1002/aoc.5055.

Soorbaghi, Fatemeh Pashaei, Mojgan Isanejad, Sara Salatin, Milad Ghorbani, Samira Jafari, and Hossein Derakhshankhah. 2019. "Bioaerogels: Synthesis Approaches, Cellular Uptake, and the Biomedical Applications." *Biomedicine and Pharmacotherapy* 111 (January): 964–75. doi:10.1016/j.biopha.2019.01.014.

Sun, Minghui, Chen Chen, Lihua Chen, and Baolian Su. 2016. "Hierarchically Porous Materials: Synthesis Strategies and Emerging Applications." *Frontiers of Chemical Science and Engineering* 10 (3): 301–47. doi:10.1007/s11705-016-1578-y.

Sun, Yan, Youlu Chu, Weibing Wu, and Huining Xiao. 2021. "Nanocellulose-Based Lightweight Porous Materials: A Review." *Carbohydrate Polymers* 255 (September 2020): 117489. doi:10.1016/j.carbpol.2020.117489.

Taylor, Dominic, Scott J. Dalgarno, Zhengtao Xu, and Filipe Vilela. 2020. "Conjugated Porous Polymers: Incredibly Versatile Materials with Far-Reaching Applications." *Chemical Society Reviews* 49: 3981–4042. doi:10.1039/c9cs00315k.

Tsai, Wen Chyan, and Yadong Wang. 2019. "Progress of Supercritical Fluid Technology in Polymerization and Its Applications in Biomedical Engineering." *Progress in Polymer Science* 98: 101161. doi:10.1016/j.progpolymsci.2019.101161.

Tutek, Karol, Anna Masek, Anna Kosmalska, and Stefan Cichosz. 2021. "Application of Fluids in Supercritical Conditions in the Polymer Industry." *Polymers* 13 (5): 1–17. doi:10.3390/polym13050729.

Vakili, Reza, Shaojun Xu, Nadeen Al-Janabi, Patricia Gorgojo, Stuart M. Holmes, and Xiaolei Fan. 2018. "Microwave-Assisted Synthesis of Zirconium-Based Metal Organic Frameworks (MOFs): Optimization and Gas Adsorption." *Microporous and Mesoporous Materials* 260 (August 2017): 45–53. doi:10.1016/j.micromeso.2017.10.028.

Vareda, João P., Alyne Lamy-Mendes, and Luisa Durães. 2018. "A Reconsideration on the Definition of the Term Aerogel Based on Current Drying Trends." *Microporous and Mesoporous Materials* 258: 211–16. doi:10.1016/j.micromeso.2017.09.016.

Wang, Hong, Yue Shao, Shilin Mei, Yan Lu, Miao Zhang, Jian Ke Sun, Krzysztof Matyjaszewski, Markus Antonietti, and Jiayin Yuan. 2020. "Polymer-Derived Heteroatom-Doped Porous Carbon Materials." *Chemical Reviews* 120 (17): 9363–9419. doi:10.1021/acs.chemrev.0c00080.

Wang, Yixin, Yuehong Su, Weiling Wang, Ying Fang, Saffa B. Riffat, and Fatang Jiang. 2019. "The Advances of Polysaccharide-Based Aerogels: Preparation and Potential Application." *Carbohydrate Polymers* 226 (August): 115242. doi:10.1016/j.carbpol.2019.115242.

Wang, Zihao, Zongzhe Li, Marcus Ng, and Phillip J. Milner. 2020. "Rapid Mechanochemical Synthesis of Metal-Organic Frameworks Using Exogenous Organic Base." *Dalton Transactions* 49 (45): 16238–44. doi:10.1039/d0dt01240h.

Wu, Jinlun, Fei Xu, Shimei Li, Pengwei Ma, Xingcai Zhang, Qianhui Liu, Ruowen Fu, and Dingcai Wu. 2019. "Porous Polymers as Multifunctional Material Platforms toward Task-Specific Applications." *Advanced Materials* 31 (1802922): 1–45. doi:10.1002/adma.201802922.

Xu, Yujing, Qiong Yu, Dongbo Zhao, Wei Zhang, Ning Wang, and Jun Li. 2018. "Synthesis and Characterization of Porphyrin-Based Porous Coordination Polymers Obtained by Supercritical CO_2 Extraction." *Journal of Materials Science* 53 (14): 10534–42. doi:10.1007/s10853-018-2305-5.

Yilmaz, Pelin, Elif Öztürk Er, Sezgin Bakirdere, Kutlu Ülgen, and Belma Özbek. 2019. "Application of Supercritical Gel Drying Method on Fabrication of Mechanically Improved and Biologically Safe Three-Component Scaffold Composed of Graphene Oxide/Chitosan/Hydroxyapatite and Characterization Studies." *Journal of Materials Research and Technology* 8 (6): 5201–16. doi:10.1016/j.jmrt.2019.08.043.

Yuan, Bin, Shuangqiao Yang, Meng Wang, Xia Jiang, and Shibing Bai. 2020. "Preparation of Ag Foam Catalyst Based on *in-situ* Thermally Induced Redox Reaction between Polyvinyl Alcohol and Silver Nitrate with Supercritical CO_2 Foaming Technology." *Polymer* 206 (August): 122858. doi:10.1016/j.polymer.2020.122858.

Zhang, He, Xue Wang Gao, Li Wang, Xinsheng Zhao, Qiu Yan Li, and Xiao Jun Wang. 2019. "Microwave-Assisted Synthesis of Urea-Containing Zirconium Metal-Organic Frameworks for Heterogeneous Catalysis of Henry Reactions." *CrystEngComm* 21 (9): 1358–62. doi:10.1039/c8ce02153h.

Zhang, Qianyu, Zhigao Wang, Cheng Zhang, Rotimi E. Aluko, Jian Yuan, Xingrong Ju, and Rong He. 2020. "Structural and Functional Characterization of Rice Starch-Based Superabsorbent Polymer Materials." *International Journal of Biological Macromolecules* 153: 1291–98. doi:10.1016/j.ijbiomac.2019.10.264.

Zhang, Tao, Rajashekharayya A. Sanguramath, Sima Israel, and Michael S. Silverstein. 2019. "Emulsion Templating: Porous Polymers and Beyond." *Macromolecules* 52 (15): 5445–79. doi:10.1021/acs.macromol.8b02576.

Zhang, Xuan, Zhijie Chen, Xinyao Liu, Sylvia L. Hanna, Xingjie Wang, Reza Taheri-Ledari, Ali Maleki, Peng Li, and Omar K. Farha. 2020. "A Historical Overview of the Activation and Porosity of Metal-Organic Frameworks." *Chemical Society Reviews* 49 (20): 7406–27. doi:10.1039/d0cs00997k.

Zhuang, Xiaodong, Fan Zhang, Dongqing Wu, and Xinliang Feng. 2014. "Graphene Coupled Schiff-Base Porous Polymers: Towards Nitrogen-Enriched Porous Carbon Nanosheets with Ultrahigh Electrochemical Capacity." *Advanced Materials* 26 (19): 3081–86. doi:10.1002/adma.201305040.

6 Porous Polymer for Heterogeneous Catalysis

Vivek Mishra, Simran Aggarwal, and Shubham Pandey
Amity Institute of Click-Chemistry Research and Studies (AICCRS), Noida, India

CONTENTS

DOI: 10.1201/9781003169604-6

6.1 INTRODUCTION

Catalysis is important in today's world, particularly in industries where chemical reactions are used to turn raw materials into useful products. It serves as the foundation for many industrial processes, and advances in catalytic reactions not only enhance the efficacy of chemical reactions but also reduce waste and environmental impact.[1–3] For the advancement of environmentally friendly chemical processes, there is a need for highly selective, active, recyclable, and cost-effective catalysts. Previously, homogeneous catalysts were commonly used for chemical transformations, but they have several drawbacks. They are difficult to isolate and recycle, and their thermal stability is minimal. Heterogeneous catalysts have many advantages over homogeneous catalysts, including fast separation and recycling, high thermal and chemical stability, and so on. Continuous efforts are being made in making heterogeneous catalysts with easy recyclability and completely accessible and highly designable active sites.[4–8] In the past, zeolites were used as a porous heterogeneous catalyst in industries for adsorption, separation, and catalysis.[9, 10] After that metal-organic frameworks (MOFs) and covalent organic frameworks (COFs) came into the picture that serves application in gas adsorption and storage, sensing, catalysis, optoelectronics, etc. According to some researchers, MOFs are a part of porous organic polymers (POPs) while for some others they make a separate class. POPs have a large surface area, high thermal and chemical stability, and chemical composition and pore walls that are easily manageable. They have exceptional chemical tunability due to the wide range of functional groups used in their synthesis. Since they are insoluble in reaction media, they can be easily recycled. Furthermore, the pore size of POPs enriches the reactants, resulting in their increased catalytic activity. POPs with solid covalent bonds are classified as crystalline and amorphous. COFs are crystalline POPs produced by fusing light elements including oxygen, carbon, boron, and silicon. They have wide surface areas and porous structures that are uniform.[11–14] Amorphous POPs, on the other hand, have disordered porosity and a large pore size distribution.[15–21] Various methodologies for synthesizing POPs have been established due to the versatility of molecular design and the variety of organic chemical reactions. Sonogashira-Hagihara,[22] Suzuki–Miyaura,[23] olefin metathesis,[24] and other coupling reactions were effective in producing POPs with predefined structures and functions. Since all these coupling reactions require the use of metal catalysts, which raise reaction costs and pollute the environment, the creation of polymerization methods for the preparation of POPs is highly desirable to make POPs both environmentally and economically beneficial. Various metal-free methodologies such as free-radical polymerization,[25, 26] benzodioxane[27] and keto-enamine[28, 29] formation reaction, borosilicate condensations,[30] etc. have been well developed for synthesizing POPs. These approaches are successful in introducing target applications.

6.2 STABILITY AND FUNCTIONALIZATION OF POPs

POPs, as previously said, have a high chemical and thermal stability. This can be illustrated with the help of an illustration. Dong et al.[31] devised a simple technique for synthesizing polydivinylbenzene (PDVB) polymers under solvothermal conditions using free radical polymerization of a vinyl-functionalized monomer as the precursor (Figure 6.1).

The synthesized PDVB can be functionalized and a large number of SO_3H groups can be incorporated onto the porous PDVB structures by treating polydivinylbenzene (PDVB) with a powerful oxidizing agent like $ClSO_3H$ (Scheme 6.1).[32]

After examining nitrogen sorption isotherms, it was discovered that SO_3H-PDVB maintains its porous structure. Also, SO_3H-PDVB was found to show greater catalytic activity than Amberlyst-15 and Beta in the esterification and acylation reactions (Scheme 6.2).

The production of a Schiff base-modified Cu-coordinated PDVB (PDVB-SB-Cu) that preserves its porosity structure was achieved by a sequence of reactions on PDVB, including nitration, reduction, aldimine condensation, and coordination with Cu (Scheme 6.3).[33] The resultant polymer PDVB-SB-Cu shows an Ullmann biaryl ether coupling reaction (Scheme 6.4).

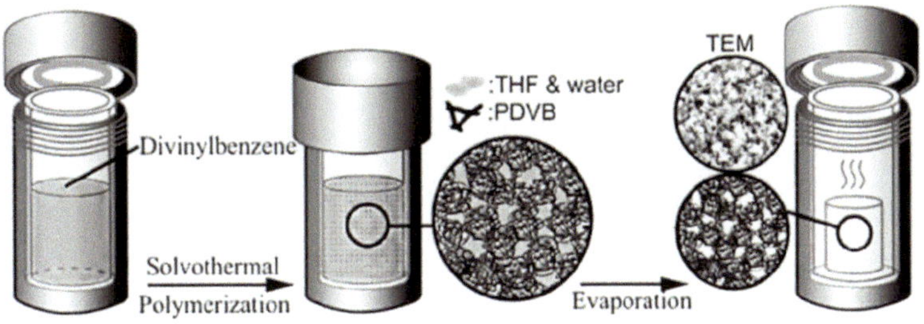

FIGURE 6.1 Sketch of preparation of porous polydivinylbenzene (PDVB) under solvothermal conditions. **Reprinted with permission from RSC (2015).**

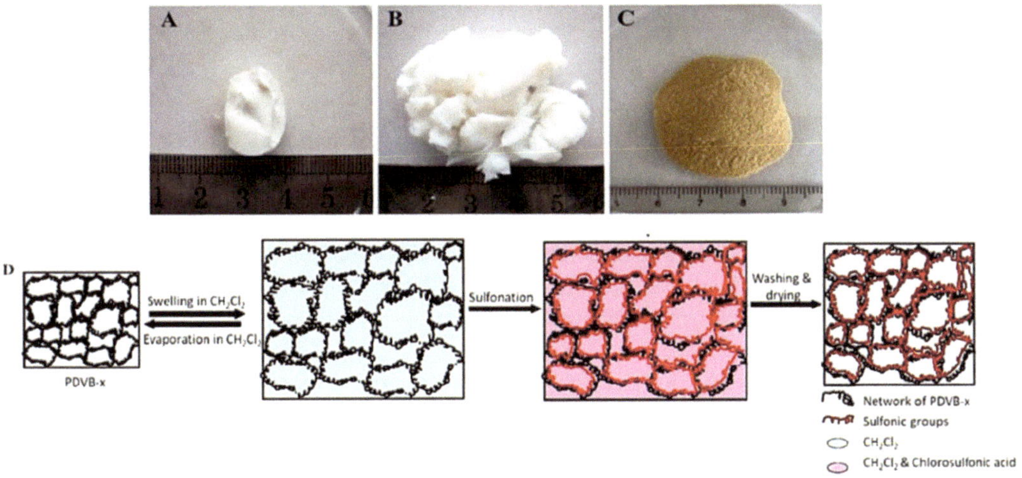

SCHEME 6.1 Images of PDVB-0.1 (a) before swelling in dichloromethane, (b) swollen PDVB-0.1 in dichloromethane, (c) sulfonated PDVB-0.1, and (d) sulfonation of PDVB-x by HSO_3Cl in CH_2Cl_2.

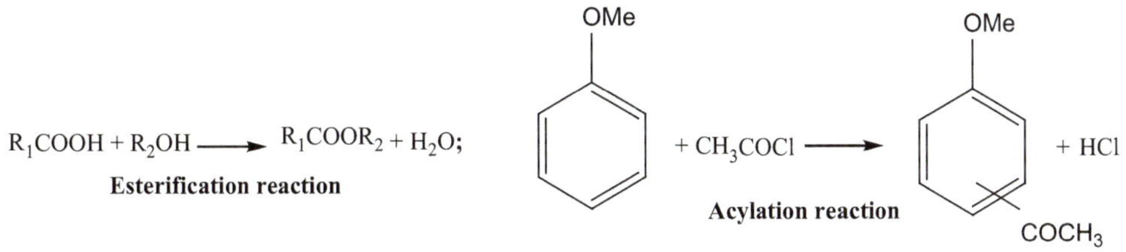

SCHEME 6.2 Catalytic esterification and acylation of anisole with acetyl chloride over PDVB-0.1-SO_3H catalyst.

On the other hand, the porous COFs get partially or destroyed by similar treatments.[34, 35] Because of the organic system's reactivity, range of organic reactions, and strength, it is crucial to highlight that POPs may be readily functionalized with various organic groups, which open up a lot of possibilities for designing heterogeneous catalysts with the appropriate active sites and ligands. Functional groups can also be inserted into POPs during the synthesis procedure by polymerization of the functionalized monomers, where both the functional groups and the polymer matrix can be precisely modified. These modifications may result in samples with hierarchical porosity and

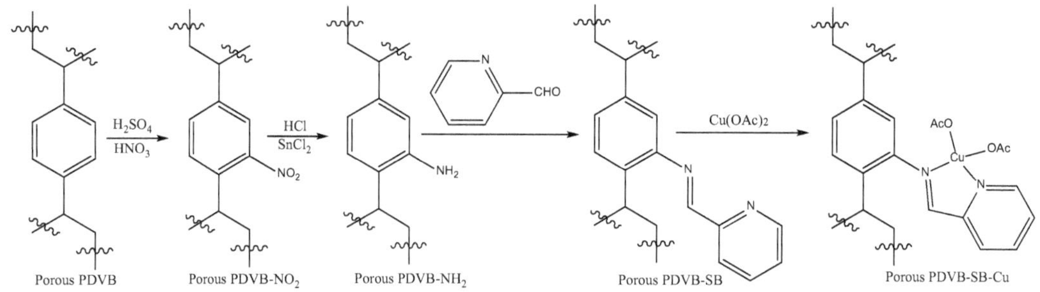

SCHEME 6.3 Synthesis of copper-coordinated porous organic polymer catalyst (PDVB-SB-Cu).

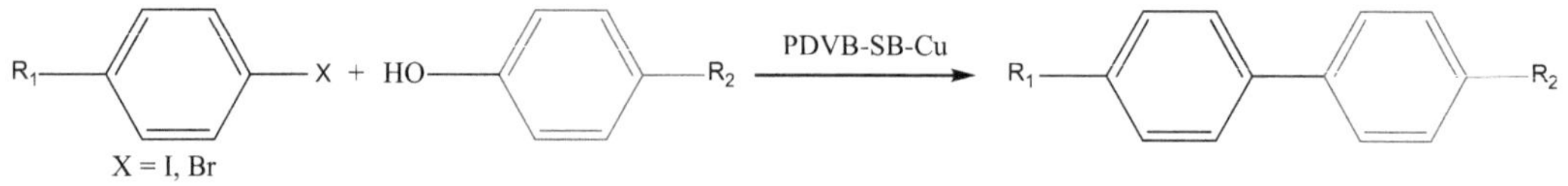

SCHEME 6.4 PDVB-SB-Cu as a catalyst in the Ullmann coupling reaction.

controllable wettability. The catalyst's wettability is advantageous for reactant entry, while the catalyst's hierarchical porosity is advantageous for reactant and product mass transfer.

For your convenience and better understanding, we will include a brief overview of the strategies for synthesizing and functionalizing POPs-based catalysts.

6.3 STRATEGIES FOR SYNTHESIZING POP CATALYST

In polymerization reactions, it is possible to change the porosity of the polymer and insert hierarchical porosity in the polymer. Hierarchical porosity is also useful in changing the wettability of the synthesized catalysts. Since the phenyl ring is very reactive, moiety-containing phenyl ring, namely PDVB can be readily derivatized with various functional groups via various chemical reactions to prepare active catalysts for heterogeneous catalysis. For the homogeneous dispersal of functional groups over the porous framework and for preventing the pores from blocking during post-synthesis, a co-polymerization strategy has been developed. Modification with vinyl groups, followed by co-polymerization with DVB, several catalytic functionalities may be integrated into the polymer framework. For increasing the density of the functional groups in the POP, a self-polymerization strategy has been developed. In this technique, vinyl-functionalized compounds can be directly incorporated into the POP.

6.3.1 Co-polymerization

6.3.1.1 Acidic and Basic Groups

In various acid-catalyzed reactions, water is obtained as a by-product that can initiate side reactions or can poison the active sites of the polymer.[36] This can be prevented by using a hydrophobic catalyst because superhydrophobicity helps in removing water from the active sites of the catalyst thus, shifting the equilibrium toward the product side. Therefore, Liu et al.[37] developed hierarchically porous solid acid catalysts with high hydrophobic nature and controllable acidic concentration by copolymerizing DVB and sodium p-styrene sulphonate in a variety of ratios and subsequent treatment with H_2SO_4 (Scheme 6.5).

From the N_2 adsorption-desorption isotherms and TEM images of the POP, it was found that the POP has a high BET surface area, high mesoporosity, and large pore volumes. Also, the water

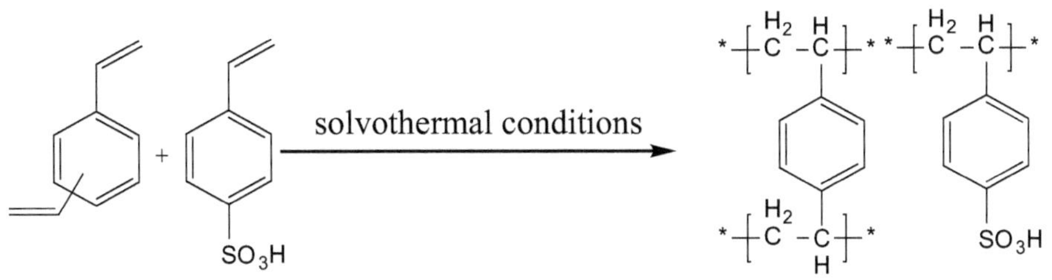

SCHEME 6.5 Synthesis of hierarchically POP via copolymerization of divinylbenzene and sodium p-styrene sulfonate (H-PDVB-x-SO_3H) in solvothermal conditions.

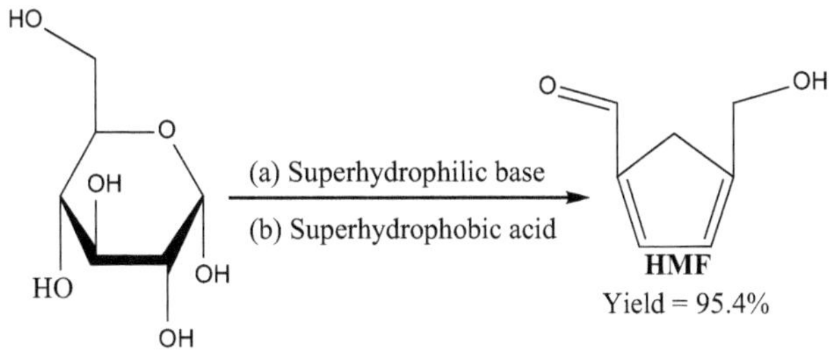

SCHEME 6.6 Formation of HMF using a superhydrophobic acid catalyst and super-hydrophilic base catalyst.

adsorption test and contact angle showed that POP has amazing hydrophobic properties. These acidic POPs have high thermal stability as compared to the acidic Amberlyst 15 resin. The catalytic activity of the synthesized POP was used in the esterification reaction of 1-butanol with acetic acid and the condensation reaction of ethylene glycol with benzaldehyde.

In the process of converting glucose to 5-hydroxymethylfurfural (HMF), an important intermediate in synthesizing various chemicals and fuels, a base catalyst is needed in the isomerization of glucose to fructose, and acid catalyst is required in the dehydration of fructose to finally give HMF. But in the acid-catalyzed conversion of fructose to HMF, HMF undergoes hydrolysis to yield levulinic acid (LA) and formic acid (FA) as the by-product, which reduces HMF yield.[38] Theoretical simulations have revealed that the superhydrophobic acid catalyst can significantly prevent hydrolysis of HMF, producing HMF as the only product. Acid catalysts based on porous PDVB-SO_3H groups gave HMF in high yields.[39] For the isomerization of glucose to fructose, a base catalyst having super hydrophilic properties was required which was prepared via copolymerization of 1-vinylimidazole, divinylbenzene, and N,N-methylenebisacrylamide. When super hydrophilic base catalyst (P-VI-0) and superhydrophobic acid catalyst (P-SO_3H-154) were used in the conversion of glucose to HMF, HMF was obtained in 95.4% yield (Scheme 6.6). These findings pave the way for the design and development of novel catalysts having high activities and superior selectivities for biomass conversion.

6.3.1.2 Ionic Groups

Ionic liquids (ILs) are useful in various fields. It is used in organic reactions to replace organic solvents to provide a green approach for synthesizing compounds and as a catalyst in chemical reactions. But ILs are highly viscous and therefore their diffusion is very slow. To overcome this problem, ILs are thought to be immobilized on porous materials.[40]

Liu et al.[41] synthesized mesoporous polymers (superhydrophobic) via copolymerizing DVB and 1-vinylimidazolate (vim) in ethyl acetate at 100 °C to give PDVB-vim. After this quaternary ammoniation with CH_3I and anion exchange treatment with HSO_3CF_3 was carried out to get ILs-functionalized polymers (Scheme 6.7).

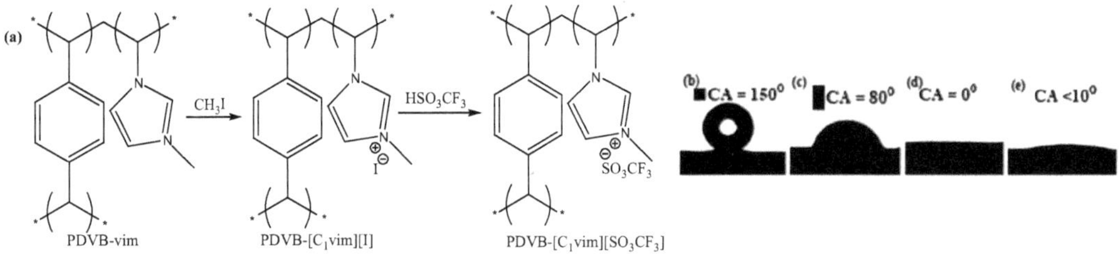

SCHEME 6.7 (a) Scheme for synthesizing PDVB-[C$_1$vim][SO$_3$CF$_3$]; (b,c) Contact angle for a water droplet on the surface of PDVB-vim; and PDVB-[C$_1$vim][SO$_3$CF$_3$]; (d,e) Contact angle of a methanol droplet and tripalmitin droplet on the surface PDVB-[C$_1$vim][SO$_3$CF$_3$]. Reprinted from ACS (2012).

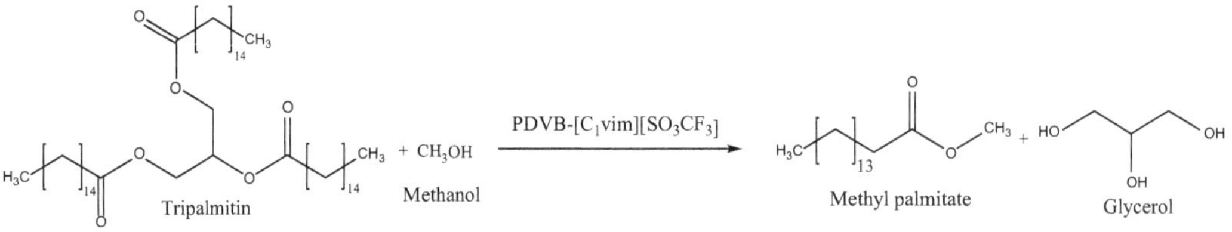

SCHEME 6.8 Transesterification of tripalmitin with methanol using PDVB-[C$_1$vim][SO$_3$CF$_3$].

From N_2 adsorption-desorption isotherms and TEM images, mesoporosity of the synthesized polymer was confirmed. The contact angle for a water droplet on PDVB-C$_1$vim and PDVB-[C$_1$vim][SO$_3$CF$_3$] surface was found to be 150 °C and 80 °C respectively, which suggests the hydrophobicity of the synthesized polymer. The contact angle for water is almost zero when a drop of tripalmitin or methanol came in contact with the water droplet, this shows the oleophilicity of PDVB-[C$_1$vim][SO$_3$CF$_3$].

The catalytic activity of the polymer was tested in the transesterification of tripalmitin with methanol to give methyl palmitate and glycerol (Scheme 6.8).

6.3.1.3 Ligand Groups

In many organic reactions, organic ligands control the catalytic performance of a catalyst, it is because of their electronic properties, steric effect, hydrophilic/hydrophobic nature, and acid/base/redox properties.[42] Organic ligands can be immobilized over POPs as they have advantages of both organic and inorganic supports. Since phenanthroline binds strongly with copper and can be used in copper-catalyzed reactions, it can be an effective polymeric ligand that has the advantage of recyclability and low leaching of metals. Sun et al.[43] synthesized a series of phenanthroline-based porous polymers (POP-Phen) via reaction of N-(1,10-phenanthroline-5-yl)acrylamide with divinylbenzene in solvothermal conditions, followed by coordination with Cu (Scheme 6.9). N_2 sorption, SEM and

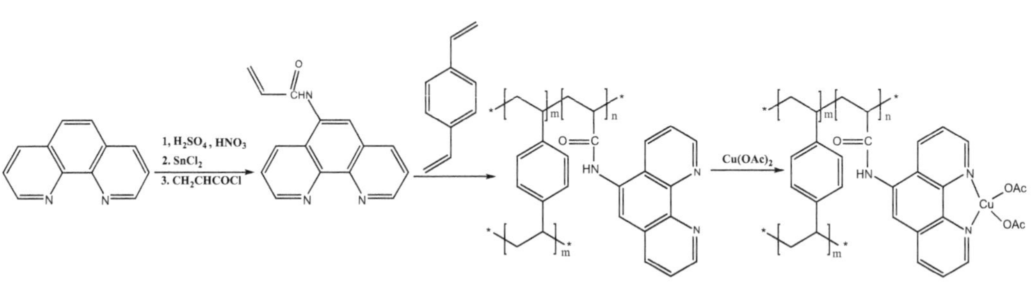

SCHEME 6.9 Synthesis of POP-Phen via copolymerizing DVB and N-(1,10-phenanthroline-5-yl)acrylamide.

TEM images, and thermogravimetric analysis verify that the polymer possesses large pore volume and surface area, hierarchical porosity, and outstanding thermal and chemical stability.

The synthesized polymer was found to have high activity in catalyzing oxidative homocoupling of alkynes (Glazer coupling) and Huisgen 1,3-dipolar cycloaddition (click reaction) of alkynes with benzyl azide in mild conditions (Scheme 6.10). When POP-Phen was coordinated with Pd instead of Cu, it was found to have high activity and selectivity in carbon-carbon bond formation reactions.

6.3.1.4 Chiral Groups

Asymmetric catalysis offers a way to make enantiomerically pure compounds.[44] However, few homogeneous chiral catalysts have complicated recycling processes and heterogeneous chiral catalysts have lower activity than their homogeneous counterparts.[45] Sun et al.[46] suggested a way of integrating chiral organic ligands with hierarchical POPs by copolymerization of divinylbenzene and vinyl-functionalized chiral ligands under solvothermal conditions. They incorporated R-BINAP [2,2′-bis(diphenylphosphino)-1,1′-binaphthyl] into PDVB via series of reactions as represented in Scheme 6.11.

N_2 sorption and SEM images show that the polymer is mesoporous. Also, the CD spectra of the polymers having reverse stereochemistry showed mirror signals, proving that the polymers are chiral. After metalation with Ru species, PCP-BINAP was found to have high activity in the asymmetric hydrogenation of methyl methacrylate (Scheme 6.12). Different ketoesters with Ru/PCP-BINAP can be transformed effectively to highly enantioselective chiral alcohols. The reason for high enantioselectivity is the insertion of BINAP into the polymer, instead of grafting it on the surface of the polymer that provides uniformly distributed active sites of the catalyst and the fact that Ru species were effectively coordinated with the chiral BINAP ligand in the polymer.

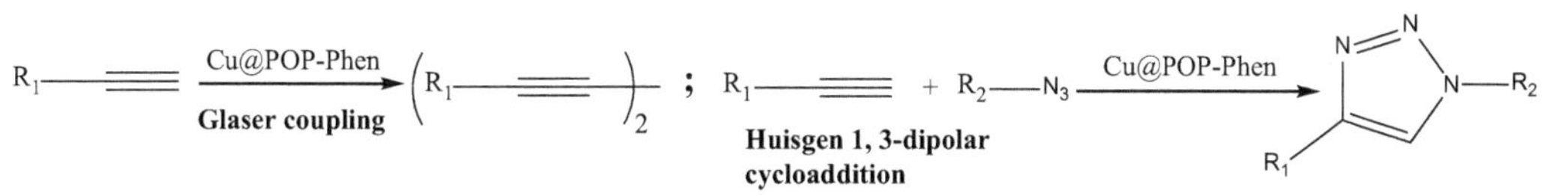

SCHEME 6.10 Glazer coupling and Huisgen 1,3-dipolar cycloaddition catalyzed by Cu-coordinated POP-Phen.

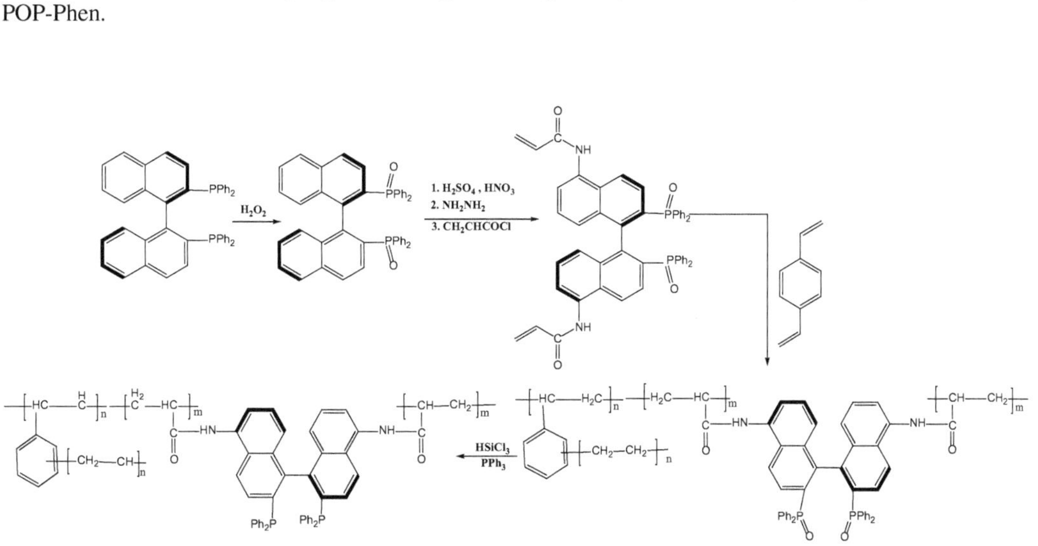

SCHEME 6.11 Synthesis of porous polymer of (R)-BINAP and DVB.

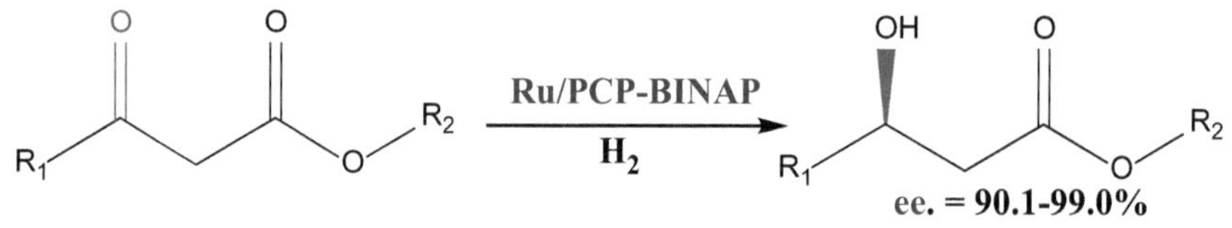

SCHEME 6.12 Asymmetric hydrogenation of methyl methacrylate using Ru/PCP-BINAP.

6.3.1.5 Porphyrin Group

Metalloporphyrin has piqued the attention of many researchers because of its ability to model biocatalysis.[47] Jiang et al.[48] demonstrated an effective method for incorporating metalloporphyrin units into POPs through a Suzuki coupling reaction catalyzed by Pd (Scheme 6.13).

FeP-CMP has dense active sites, intrinsic nanopores available to reactants, and a large surface area to promote chemical transformations. It has a BET surface area of 1270 $m^2\ g^{-1}$ and is chemically stable. FeP-CMP was found to be highly active and selective in catalyzing the oxidation of sulfides to sulfoxides using O_2 under mild conditions (Scheme 6.14).

FeP-CMP can catalyze the conversion of various sulfides to corresponding sulfoxides and possess chemoselectivity. It can be easily recycled and has a high turnover number (TON) 97320, and therefore, FeP-CMP acts as an artificial catalyst having activity similar to natural enzymes.

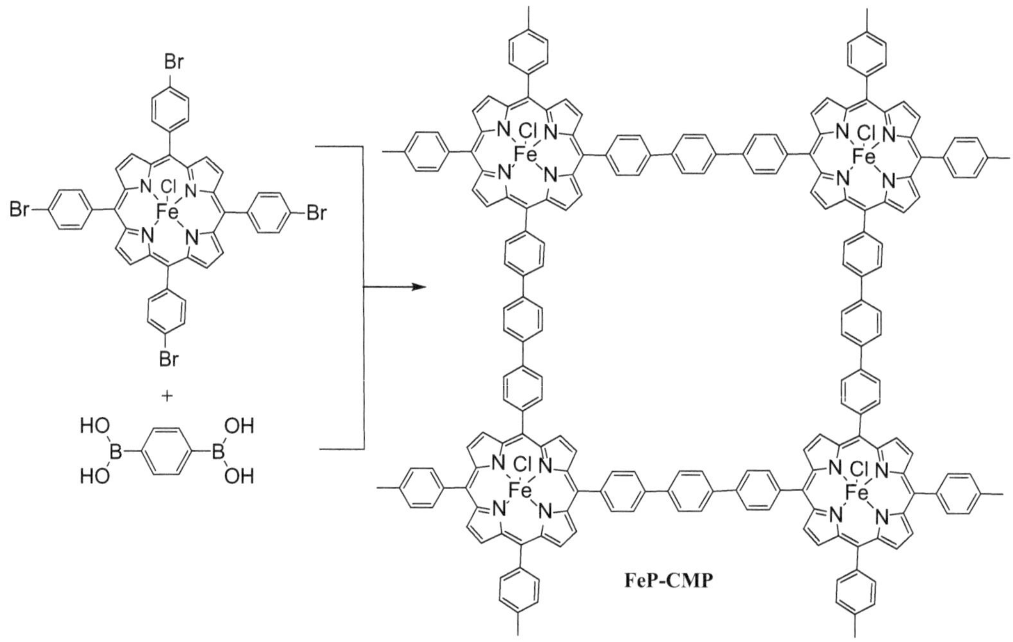

SCHEME 6.13 Synthesis of nanoporous FeP-CMP via Suzuki coupling reaction.

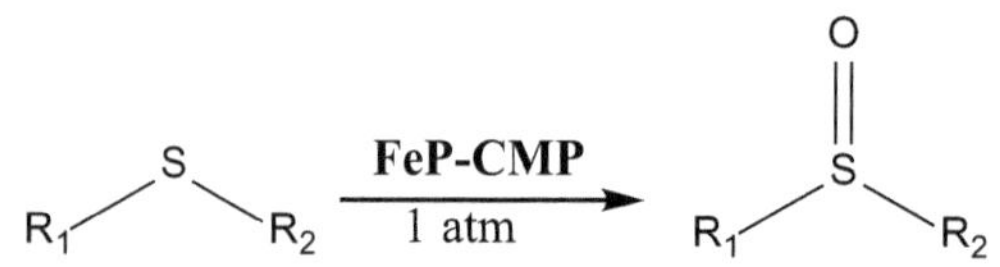

SCHEME 6.14 FeP-CMP catalyzing the oxidation of sulfides to sulfoxides.

6.3.2 Self-Polymerization

Porous polymers synthesized via self-polymerization have two advantages over the ones synthesized via co-polymerization. The former has a very high concentration and uniform distribution of functional groups. Based on the type of catalytically active site, they are classified into the following categories:

6.3.2.1 Organic Ligand Groups

Sun et al. suggested the preparation of porous organic ligands (POLs) by self-polymerization of vinyl-functionalized monomers under solvothermal conditions.[49] Firstly, vinyl-functionalized triphenylphosphine (v-PPh_3) was synthesized which undergoes self-polymerization to form POL-PPh_3. POL-PPh_3 was then metallated with Ru species to obtain Ru/POL-PPh_3 which shows great activity in the hydroformylation of aldehydes. Ru/POL-PPh_3 was more selective than the homogeneous counterparts $RhH(CO)(PPh_3)_3$ and $Rh(CO)_2(acac)$ for similar conversions, due to the large concentration of PPh_3 ligands in the catalyst. After this, Sun et al.[50] prepared various stable and flexible POLs using different vinyl-functionalized monomers (Figure 6.2). These were also found to have hierarchical porosity and large pore volume. Surprisingly, POL-dppe has swelling property in various organic solvents, which is advantageous for organometallic complexes. POL-dppe was then metallated with Rh to obtain Rh/POL-dppe which shows great chemoselectivity for aldehydes and regioselectivity for branched aldehydes in the hydroformylation of n-octene.

6.3.2.2 Organocatalyst Groups

Although there exist many homogeneous catalysts for the chemical reactions, difficulty in their separation has always been an issue. POPs offer new ways of linking porous polymer networks with organocatalysts to make the recycling of catalysts easy. Huangfu et al.[51] performed a polymerization reaction of vinyl-functionalized 2,2,6,6-tetramethylpiperidine (V-TEMP) in AIBN at 100 °C and then carried out its oxidation with m-CPBA at 0 °C to give polymerized TEMPO (Figure 6.3). N_2 adsorption-desorption isotherms, SEM images, and TEM images confirm the hierarchical porosity of the catalyst.

PPO-TEMPO is useful in the selective conversion of alcohols to corresponding ketones and aldehydes in a short span of time (Scheme 6.15). As compared to TEMPO in homogeneous catalysis, PPO-TEMPO possesses high activity and selectivity in the oxidation of 4-methoxybenzyl alcohol.

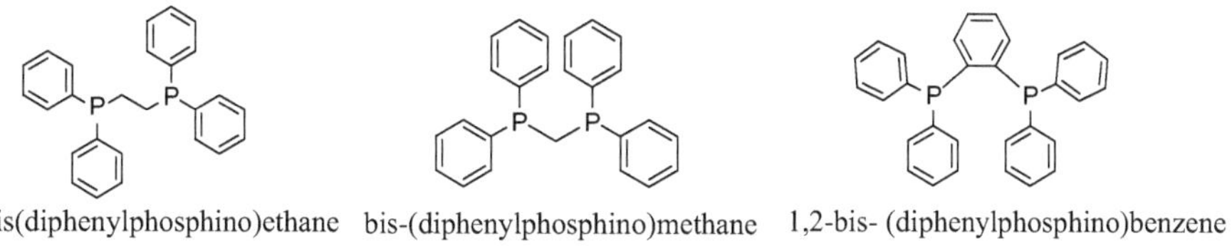

FIGURE 6.2 Structure of dppe, dppm, and dppb.

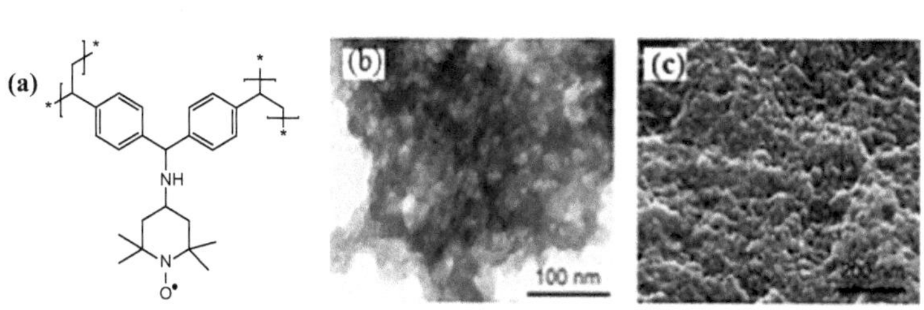

FIGURE 6.3 (a) Structure, (b) SEM, and (c) TEM images of PPO-TEMPO.

SCHEME 6.15 Oxidation of alcohol using PPO-TEMPO as a heterogeneous catalyst.

Interestingly, the hydrothermally treated PPO-TEMPO shows high selectivity (98%) and conversion (>99%) in oxidizing benzyl alcohol. Such remarkable hydrothermal stability of the catalyst will turn out to be very useful in the future.

6.3.2.3 Ionic Groups

v-PPh_3 can also act as the precursor to form a vinyl-functionalized quaternary phosphonium monomer which undergoes self-polymerization to yield porous ionic polymer (Scheme 6.16).[52]

N_2 sorption isotherm and SEM and TEM images verify the hierarchical porosity of the polymer.

The synthesized ionic polymer has excellent thermal stability which is evident from its high decomposition temperature of 400 °C observed from thermogravimetric analysis, and high chemical stability as the porous ionic organic polymer maintains its porous framework in a strongly acidic medium (10 M HCl).

The polymer shows excellent amphiphilicity with many solvents such as ethanol, water, hexane, and DMF, thus can be used in phase transfer catalysis such as in olefin epoxidation, dibenzothiophene oxidation with H_2O_2, etc. When PQP reacts with peroxotungstate, an anion-exchange reaction occurs, yielding W_2O_{11}/PQP. The high catalytic activity of W_2O_{11}/PQP is due to its amphiphilic nature that helps the hydrophobic and hydrophilic reactants to be miscible with each other and due to the highly porous framework that helps in better adsorption of reactants.

6.3.2.4 Chiral Ligand Groups

Lin et al.[53] produced various chiral crosslinked polymers (CCPs) with high porosities and BET surface areas of 689–974 $m^2\ g^{-1}$, as well as strong chemical strength since they retain their porosity following treatment with conc. HCl. Since the synthesized CCPs have many chiral centers, they can prove to be highly efficient in asymmetric catalysis in the future. When these CCPs were metallated with Ti(O-*i*-Pr)$_4$, the obtained catalysts were found to be highly active in the asymmetric diethylzinc addition to aldehydes. Also, the product was obtained in good enantiomeric excess using this method. Interestingly, the catalyst can be recycled and reused ten times without any significant loss of conversion and enantioselectivity. For synthesizing POP-based chiral catalyst, the focus is on the synthesis of the homochiral porous polymer frameworks under solvothermal conditions.

6.3.2.5 Porphyrin Groups

Since the concentration of CO_2 in the atmosphere is increasing day-by-day, therefore, it is highly necessary and of utmost interest to utilize this atmospheric CO_2 to synthesize valuable chemicals via an environment-friendly process.[54] Earlier metal-porphyrins were employed in transformation

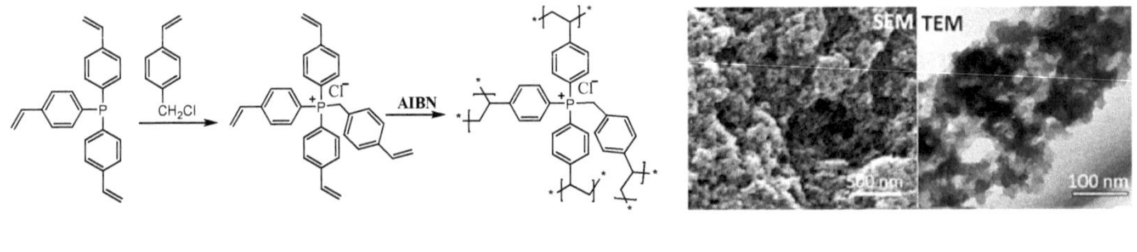

SCHEME 6.16 Synthetic methodology for the preparation of quaternary phosphonium salt polymer (PQP) along with their SEM and TEM images. Reprinted with permission from RSC, 2015.

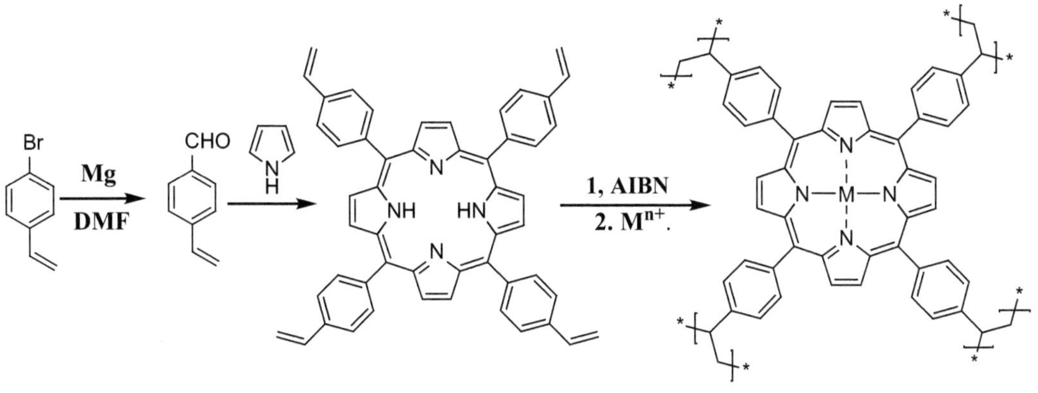

SCHEME 6.17 Synthetic methodology for the synthesis of POP-TPP via self-polymerization of tetraphenylporphyrin.

reactions of CO_2 but in this method, product separation from the catalyst was difficult. Therefore, porous heterogeneous catalysts were employed for CO_2 capture and conversion. Dai et al.[55] prepared porous organic polymer (POP-TPP) via self-polymerization of tetrastyrylporphyrin in the presence of AIBN as an initiator (Scheme 6.17).

The structure of the polymer was confirmed by ^{13}C and ^{1}H NMR spectra, and the hierarchical porosity of the polymer was confirmed by the N_2 sorption isotherm, SEM image, and TEM image. The polymer has excellent thermal stability up to 350 °C as determined by thermogravimetric analysis and high CO_2 adsorption capacity. When the polymer was metallated with Co, Mg, or Zn, there was a slight increase in the CO_2 adsorption capacity from 29.7 $cm^3\ g^{-1}$ to 32.6–37.5 $cm^3\ g^{-1}$ at 25 °C and 1 atm. The activity of Co/POP-TPP was tested in the cycloaddition reaction of epichlorohydrin and CO_2 to form cyclic carbonate at 29 °C and 1 atm and it was observed that the conversion efficiency of Co/POP-TPP is 95.6% that is near to the efficiency of Co/TPP homogeneous catalyst which is 97.4% when only CO_2 was used. And, when impure CO_2 i.e. 15% CO_2 + 85% N_2 by volume was used, the catalytic efficiency of Co/POP-TPP was higher than that of the Co/TPP. The high efficacy of Co/POP-TPP is due to the high concentration of CO_2 in its pores. Hence, heterogeneous catalysts are of potent application in the chemical fixation of CO_2 and therefore, will help in reducing global warming and the greenhouse effect.

6.4 APPLICATIONS OF VARIOUS POROUS POLYMERS

6.4.1 CO_2 Capture and Utilization

CO_2 is an abundant, non-flammable, non-toxic, cheap, and renewable source of carbon. Also, it leads to global warming to a great extent. For many years, many efforts are being made for the chemical fixation of CO_2. One such important transformation is the coupling of CO_2 and epoxides to yield cyclic carbonates as the latter can be used for various purposes such as electrolyte in Li-ion batteries, as aprotic solvents, and as a precursor for the synthesis of polycarbonates and polyurethanes. In the past, many efficient homogeneous catalysts have been developed for catalyzing the coupling reaction of CO_2 and epoxide but their complicated synthesis, metal-contaminated products, and complex recycling of the catalyst have limited their application. To overcome this problem, heterogeneous catalysts are being developed and POPs turned out to be the most famous among them. Owing to the high chemical and thermal stability, permanent porous structure, and alterable textural properties, POPs can be efficiently used for CO_2 conversion. The inclusion of CO_2-philic groups and metal-organic moieties in POPs results in the formation of materials that can capture CO_2 and can carry out its transformation reactions. Some of these examples are discussed below:

6.4.1.1 Ionic Liquid/Zn-PPh_3 Integrated POP

Ionic liquids exist in various structures with negligible volatility, stability, and solubility. Thus ionic liquids can be incorporated into porous polymers for CO_2 utilization. Porous organic polymers containing ionic liquids (ILs) have low skeletal density, are chemically stable, and have the capability of pioneering a wide range of functionality into the POP framework.[56] Wang et al.[57] designed a class of ILs and PPh_3 integrated POP named PPh_3-ILX@POPs (X= Cl, Br, I), followed by treatment with ZnX_2 in THF as the solvent to yield PPh_3-ILX-ZnX_2@POPs (Scheme 6.18).

The BET surface area of PPh_3-ILBr-$ZnBr_2$@POPs (482 $m^2\ g^{-1}$) was less than that of PPh_3-ILBr@POPs (591 $m^2\ g^{-1}$) which is due to more weight of Zinc salt in the former case. Both of these POPs are stable, and their structures and compositions were determined by spectroscopic techniques and ICP analysis. From SEM and TEM studies, it was found that the polymer has a rough surface and hierarchical porosity which is vital for the interfacial transfer of reactant and product in transformations involving gaseous species. Also, the elemental mapping of PPh_3-ILBr-$ZnBr_2$@POPs shows the outstanding homogenization of the active functional sites in the polymer. Such homogenization in the micropores generates suitable cooperativity for the CO_2 adsorption and conversion process. PPh_3-ILX@POPs selectively adsorbs CO_2 over N_2 which clearly shows that PPh_3 groups in POPs play an important role in CO_2 adsorption and activation. After the successful synthesis of the catalyst, a cycloaddition reaction of CO_2 and propylene oxide was performed to check the efficacy of the synthesized catalyst (Scheme 6.19).

The activity order of PPh_3-ILX@POPs in the cycloaddition reactions with varying ionic liquids was observed to be IL-Br > IL-I > IL-Cl due to the nucleophilicity and the leaving ability of the halide ions which play an essential role in ring-opening and ring-closing respectively and their TOFs range from 216 to 704 h^{-1}. After that, the catalytic activity of Zn-loaded POP i.e. PPh_3-ILBr-ZnX_2@POPs was evaluated in the cycloaddition reaction, and the activity order of various halides was observed to be $ZnBr_2$ > ZnI_2 > $ZnCl_2$, and TOF of up to 3520 h^{-1} was achieved. When PPh_3-$ZnBr_2$@POPs was used as the catalyst, only traces of the product were obtained which proves the importance of the ionic liquid for the catalytic system.

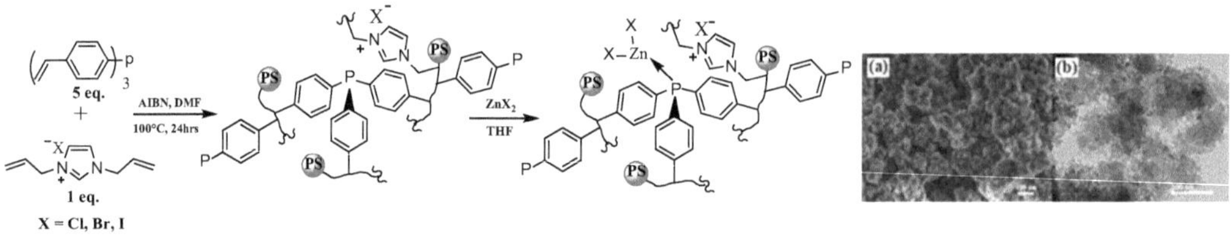

SCHEME 6.18 Synthesis of PPh_3-ILX@POPs under solvothermal conditions and its impregnation with ZnX_2 to get PPh_3-ILX-ZnX_2@POPs along with (a) SEM and (b) TEM images of PPh_3-ILBr@POPs. Reprinted from ACS, 2016.

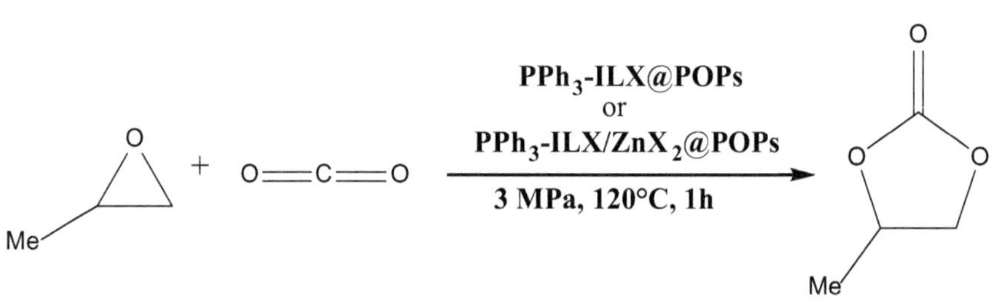

SCHEME 6.19 Reaction of CO_2 with propylene oxide.

6.4.1.1.1 *Mechanism of the Cycloaddition Reaction*

Initially, the activation of the epoxide takes place by the coulombic interactions of $ZnBr_2$-PPh_3 part of the POP followed by the nucleophilic attack of bromide ion to form alcoholate. This further attacks CO_2 generating an acyclic carbonate that underwent intramolecular substitution reaction resulting in the formation of cyclic carbonate with the release of catalyst (Scheme 6.20).

Also, the catalyst can be recycled and used up to four cycles without any significant reduction in the yield of the product, propene carbonate.

To check the versatility of the prepared PPh_3-ILX-ZnX_2@POPs catalysts, PPh_3-ILBr-$ZnBr_2$@POPs were utilized as the catalyst in the cycloaddition of CO_2 with various substrates and it was observed that when the substrates having larger dimensions were used, the yield of the product decreases. Also, the electron-donating nature of the alkane and phenyl group on the epoxide increases the electron density of the carbon atom on which bromide attacks, thereby deactivating the process of cycloaddition.

6.4.1.2 Triphenylphosphine-based POP

Realizing the importance of ionic liquids, Wu et al.[58] synthesized an ionic POP based on triphenylphosphine moiety. In this, firstly, the Friedel-Crafts reaction of PPh_3 and DB (α-dibromo-p-xylene) was carried out using $FeCl_3$ as Lewis acid catalyst to form porous polymer Triphenylphosphine-based POP (TPDB). Then, the functionalization of TPDB was done by using 3-Bromo-1-propanol (BP) to get TPDB-BP followed by reaction with triethylamine (TEA) to give TPDB-BP-TEA as the functional POP (Scheme 6.21). After the successful synthesis of the catalyst, it was used to carry out the cycloaddition reaction to form cyclic carbonate.

Initially, the cycloaddition reaction was carried out in the absence of any catalyst and no product formation took place. When TPDB was used as a heterogeneous catalyst, selectivity was found to be only 59% and 1,2-propanediol was formed as the major by-product. On using TPDB-BP as the

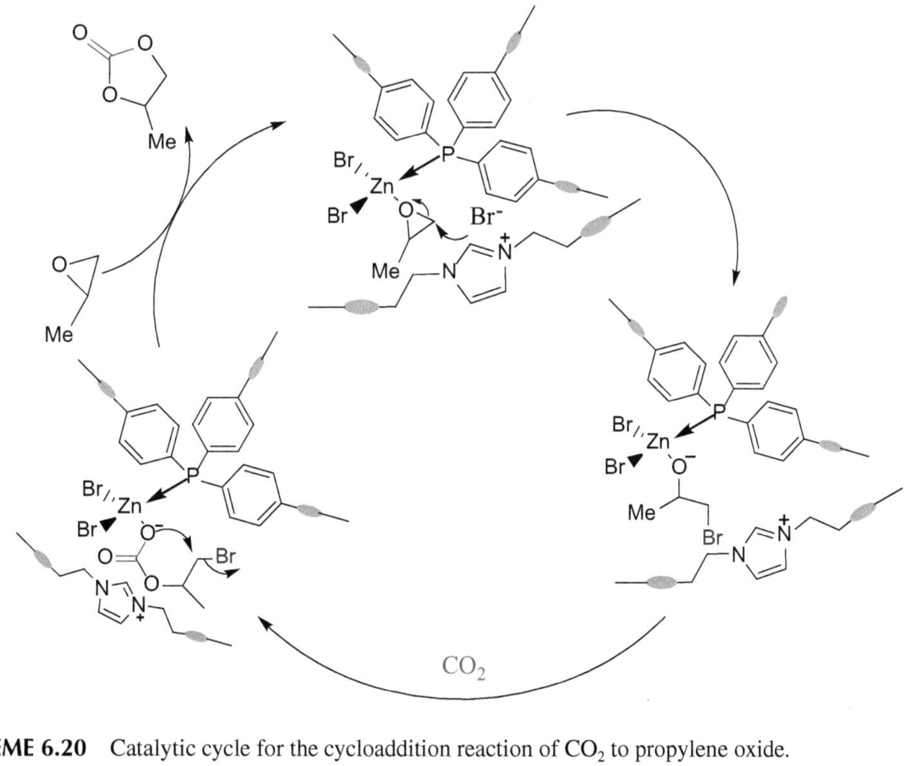

SCHEME 6.20 Catalytic cycle for the cycloaddition reaction of CO_2 to propylene oxide.

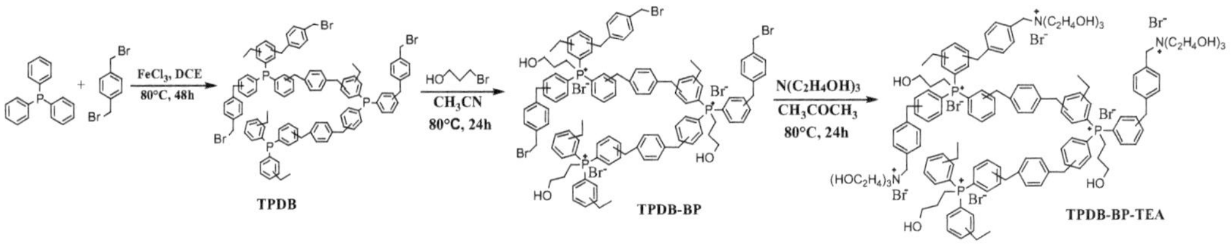

SCHEME 6.21 Synthesis of TPDB-BP-TEA from PPh_3 and DB.

catalyst, an increase in both yield and selectivity was observed. When TPDB-BP-TEA was used as the catalyst, a dramatic increase in the yield of the product and selectivity of the catalyst was observed. The presence of bromide ion and hydroxy groups is responsible for the drastic increase in the catalytic activity of TPDB-BP-TEA. Br^- acts as a main active center in the cycloaddition reactions.[59, 60] Also, the presence of hydroxy groups on the solid surface effectively accelerates the ring-opening of epoxides due to hydrogen bonding.[61, 62] Another reason for the excellent catalytic activity of TPDB-BP-TEA is its porous structure with a high surface area[63] that helps in the easy diffusion of reactants and products through the catalyst surface. TPDB-BP-TEA was recovered and recycled and was found to be effective up to five catalytic cycles.

6.4.2 Energy Storage

Energy storage is the capture of energy produced at one time for use at a later time to maintain a balance between energy demand and energy production. There are many energy storage devices, one among them is supercapacitors which have high power density, longer lifespan, and faster charge/discharge rate. They serve as a tool for storing electrical energy at an electrode-electrolyte interface. There is a wide range of available electrode materials, for example, activated carbon,[64] graphene,[65] transition metal oxides/sulfides,[66] etc., but they have certain limitations associated with them like low specific capacitance, poor recyclability, etc. POPs have high conductivity because of extended π-conjugation, high surface area, porous structure, and have good redox activity that provides significant pseudocapacitance which makes them an appropriate alternative to be used as an electrode material. Chen et al.[67] synthesized a crystalline POP named PyrOxin POP via a single-step polymerization reaction of 8-hydroxyquinoline, perylene, and $CHCl_3$ with anhydrous $AlCl_3$ as the catalyst[68] (Scheme 6.22). The synthesized POP was found to have a BET surface area of 221.4 $m^2\ g^{-1}$, while SEM and TEM images confirm the presence of microporosity in the polymer.

The presence of a large number of phenolic groups and redox-active pyridinic nitrogen enhances its polarity, electrolyte accessibility, electrical conductivity, and provides reversible faradaic capacitance. PyrOxin POP's electrochemical activity was tested in an aqueous 6M KOH. Figure 6.4 shows a cyclic voltammetric (CV) curve of the PyrOxin POP electrode at scan rates of 10–200 mV s^{-1}. With an increase in the scan rate, the CV curve changes a bit and shows redox peaks that indicate fast electron/ion diffusion pathways within the PyrOxin electrode.

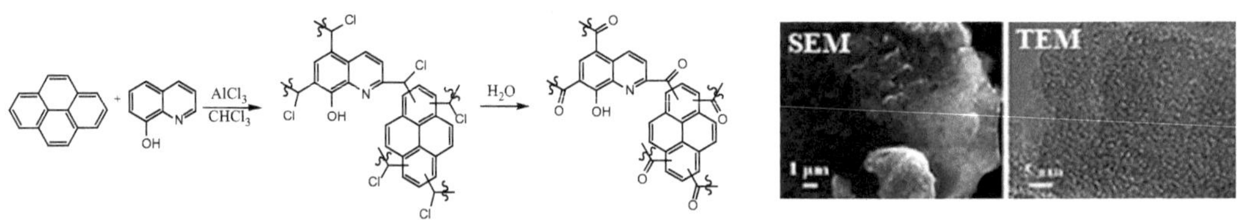

SCHEME 6.22 Synthesis of PyrOxin POP starting from 8-hydroxyquinoline and perylene and SEM and TEM images of PyrOxin.

Reprinted with permission from Elsevier, 2018.

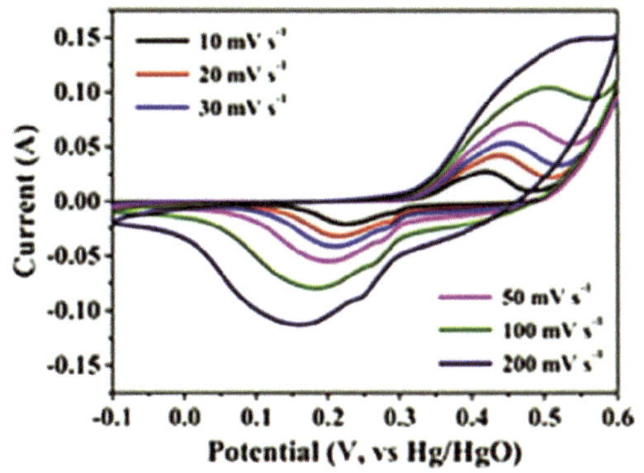

FIGURE 6.4 CV curve of PyrOxin POP at different scan rates.

Reprinted with permission from Elsevier, 2018.

During the electrochemical process, pyridinic nitrogen takes hydroxide ions from KOH to form a hydroxyl group-containing pyridinyl radical. Subsequently, phenol groups get converted to carbonyl quinone under basic conditions and both the pyridinic nitrogen and quinone type groups serve as redox-active sites and function as pseudocapacitors.

6.4.3 Heterogeneous Catalysis

Over years, many homogeneous and heterogeneous POP catalysts have been developed. In contrast with homogeneous catalysts, heterogeneous catalysts are more appropriate for catalyzing the reactions due to their chemical stability, easy separation, recovery, and recycling. The catalytic process is activated by the availability of porosity, high surface area, and functional groups in the structure of the support matrix. To date, various heterogeneous catalysts have been prepared for various types of reactions.

6.4.3.1 Cu(II) Complex on Pyridine-based POP for Nitroarene Reduction

4-nitrophenol (4NP) is broadly used in the agricultural industry but is harmful and its detoxification should be done. But due to its high chemical stability, it is not easy to remove it by traditional water treatment techniques. Therefore, it is reduced to 4-aminophenol (4AP) which is widely used in the plastic and pharmaceutical industry. Earlier this reduction was carried out by using precious metals like Au, Pt, Pd, Ru, etc., which are very costly and difficult to recycle. To achieve a cost-effective methodology for the reduction of 4NP to 4AP use of Cu metal is suggested.

By considering the importance of lone pair of nitrogen and amide functional group, an organic porous polymer has been developed that influences the catalytic activity of loaded copper metal and is used for the reduction of nitroarenes. Khatioda et al.[69] synthesized a pyridine-based 2D POP that is connected by carboxamide functionality (POP-Am1). This is then loaded with Cu(II) and is named Cu@POP-Am1 (Scheme 6.23).

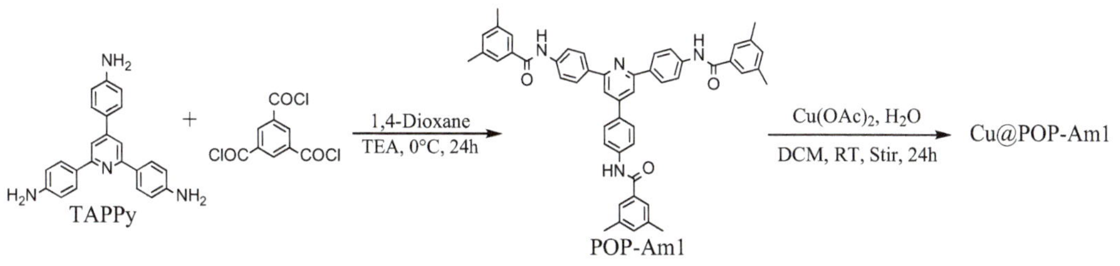

SCHEME 6.23 Synthesis of Cu(II) loaded POP-Am1 i.e. Cu@POP-Am1.

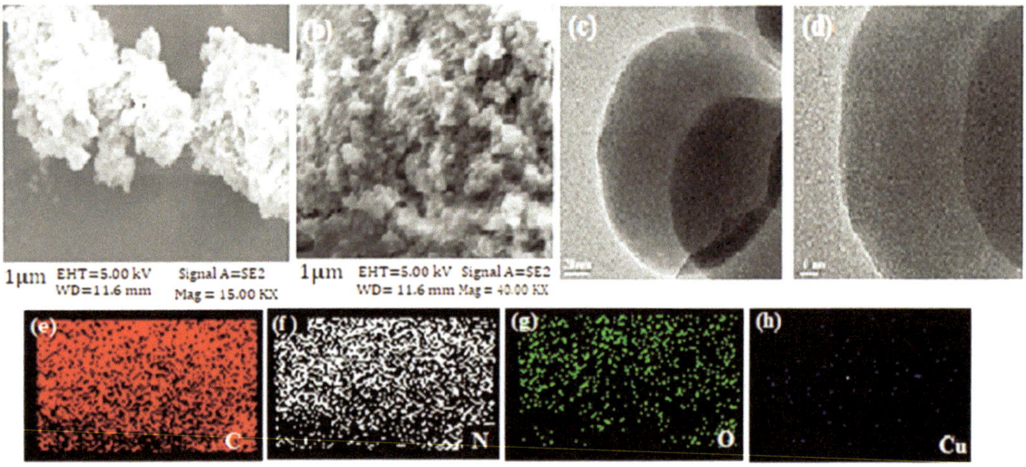

FIGURE 6.5 (a,b) SEM images, (c,d) TEM images and (e,f,g,h) Elemental mapping of Cu(II) loaded POP-Am1 representing dispersion of carbon, oxygen, nitrogen, and copper respectively.

Reprinted with permission from ChemPubSoc Europe, 2018.

POP-Am1 was found to have a BET surface area of 27 m^2 g^{-1} and an average pore diameter of 29.24 Å. SEM images of POP-Am1 suggest that the polymer contains spherical-shaped particles of 1 μm diameter and TEM images verify the layered structure of the polymer (Figure 6.5).

After loading POP-Am1 with Cu(II), the uniform dispersion of Cu ions onto the pores of the polymer was analyzed by its elemental mapping.

Cu@POP-Am1 catalyzes the reduction of 4NP to 4AP using $NaBH_4$ as the reducing agent. Here, $NaBH_4$ is not only a hydrogenating agent but also reduces Cu(II) to its unstable state in situ. $NaBH_4$ then provides a proton to 4-nitrophenolate ion and converts it to 4AP with the release of a water molecule (Scheme 6.24).

6.4.3.2 POP-supported Rhodium for Hydroformylation of Olefins

Hydroformylation is a process for the synthesis of aldehydes and was developed in the 1930s by Roelen.[70] In industries, hydroformylation was done by using homogeneous ligand-modified rhodium catalyst. However, heterogeneous catalysts have many advantages over a homogeneous catalyst such as easy recyclability, separation, etc. Therefore, Jia et al.[71] synthesized a porous organic ligand (POL) loaded with Rh for the hydroformylation of internal and terminal olefins. They took divinyl-functionalized bisphosphoramidite **(vBPa)** and tris(4-vinylphenyl)phosphine **(3vPPh$_3$)** as the comonomers for the synthesis of the POL-BPa&PPh$_3$. **vBPa** was prepared in five steps. Initially, 2,2′-biphenol was reacted with bromine to give a brominated product which was reacted with acetyl chloride to give a diacetylated product. Then, the diacetylated product was reacted with potassium ethenyltrifluoroborate via a Pd-catalyzed Suzuki–Miyaura coupling reaction to give a product which was further hydrolyzed in a basic medium to give divinyl-functionalized biphenol. The resulting biphenol reacted with chlorodipyrolyphosphine in Et_3N to give the comonomer **vBPa**. The other comonomer **3vPPh$_3$** was synthesized by the reaction of (4-vinylphenyl)-magnesium bromide with PCl_3. Copolymerization of **vBPa** and **3vPPh$_3$** initiated by AIBN in THF resulted in the formation of POL-BPa&PPh$_3$ which was reacted with Rh(acac)$(CO)_2$ in THF to produce the desired Rh-loaded porous organic ligand (Rh/POL-BPa&PPh$_3$) (Scheme 6.25).

The catalytic activity of Rh/POP-BPa&PPh$_3$ was tested for the hydroformylation of 1-hexene. On increasing the temperature from 80 °C to 100 °C, the conversion of 1-hexene increased but with

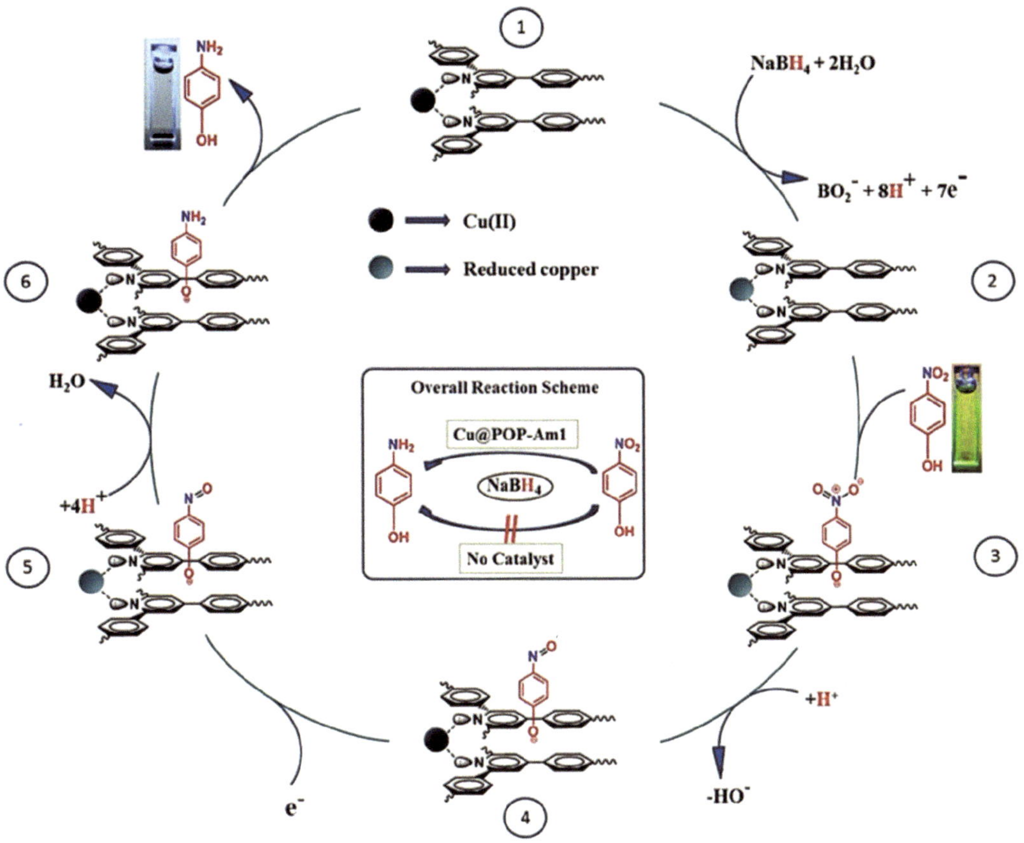

SCHEME 6.24 Mechanistic pathway of reduction of 4-NP to 4-AP by Cu@POP-Am1.

Reprinted with permission from ChemPubSoc Europe, 2018.

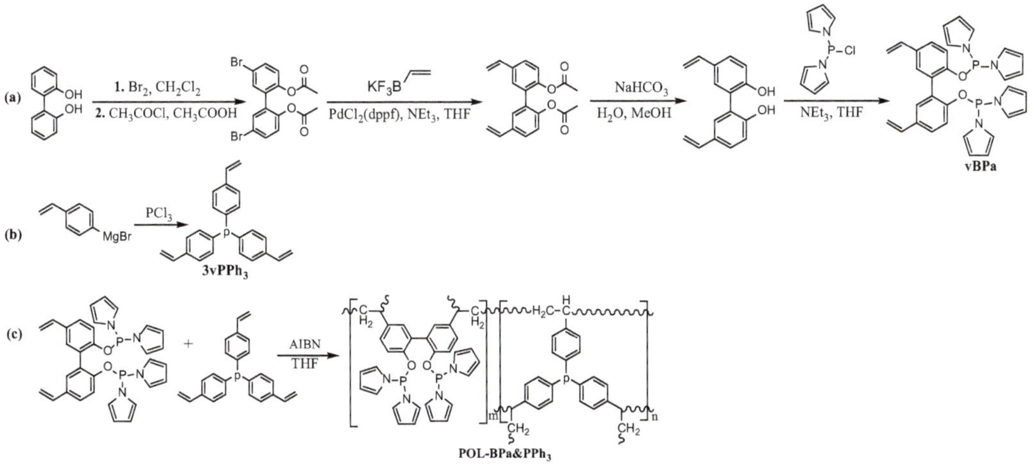

SCHEME 6.25 Synthesis of POL-BPa&PPh$_3$.

a decrease in regioselectivity toward the linear aldehyde. Also, the conversion was observed to be the maximum when H_2/CO pressure was 10/10 bar. The increase in the reaction time also helped in improving the conversion of 1-hexene (Scheme 6.26).

High π-acidity of the N-pyrrolylphosphorous unit, PPh_3 group, and high ligand density resulted in the excellent regioselectivity, activity, and recyclability of the prepared heterogeneous catalyst.

6.4.3.3 Ni(II)-metallated POP for Suzuki–Miyaura Crosscoupling Reaction

Conventionally, the Suzuki–Miyaura crosscoupling reaction for the C–C bond formation is catalyzed by very expensive Pd. As compared to Pd, the first-row transition metal Ni is cheap, more reactive, and abundant, and therefore, Ni-based catalysts are a sustainable alternative to Pd for C–C bond formation. Also, the α-diimine-based metal catalyst has attracted chemists toward themselves as they are easy to synthesize, stable, and highly active as a catalyst. Combining the advantage of both, Dong et al.[72] synthesized a Ni(II)-α-diimine-POP via Sonogashira coupling of Ni(II) α-diimine and 1,3,5-triethynylbenzene in solvothermal conditions. Initially, a Schiff-base condensation reaction was performed between 4-iodo-2,6-diisopropylbenzenamine and butane-2,3-dione, followed by metalation with DME($NiBr_2$), then coupled with 1,3,5-triethynylbenzene to form Ni(II)-α-diimine POP (Scheme 6.27).

SEM images show that C, N, Ni, and Br are uniformly dispersed in the polymer matrix.

Ni(II)-α-diimine-POP was used for the coupling of iodobenzene and phenylboronic acid using toluene as solvent and $K_3PO_4.3H_2O$ as a base at 80 °C (Scheme 6.28).

The catalyst was found to be effective up to five catalytic cycles, after that leaching of Ni occurred. Also, the catalyst was able to catalyze the coupling reaction of substrates having various functional groups such as $-OCH_3$, $-CO_2Me$, $-CN$, etc. Aryl iodides were more reactive than aryl bromides in the coupling reaction.

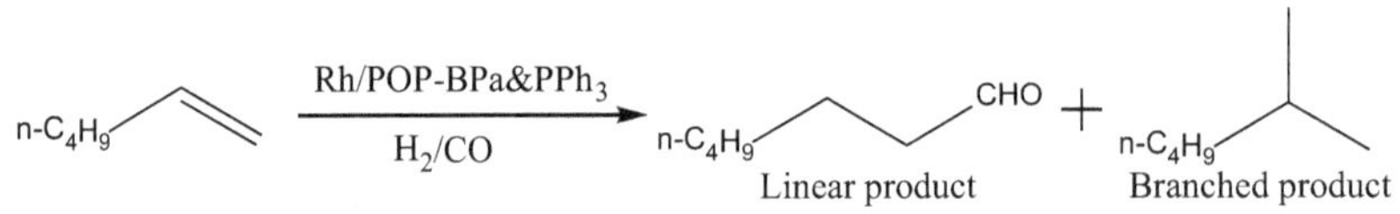

SCHEME 6.26 Hydroformylation of 1-hexene by Rh/POP-BPa&PPh_3.

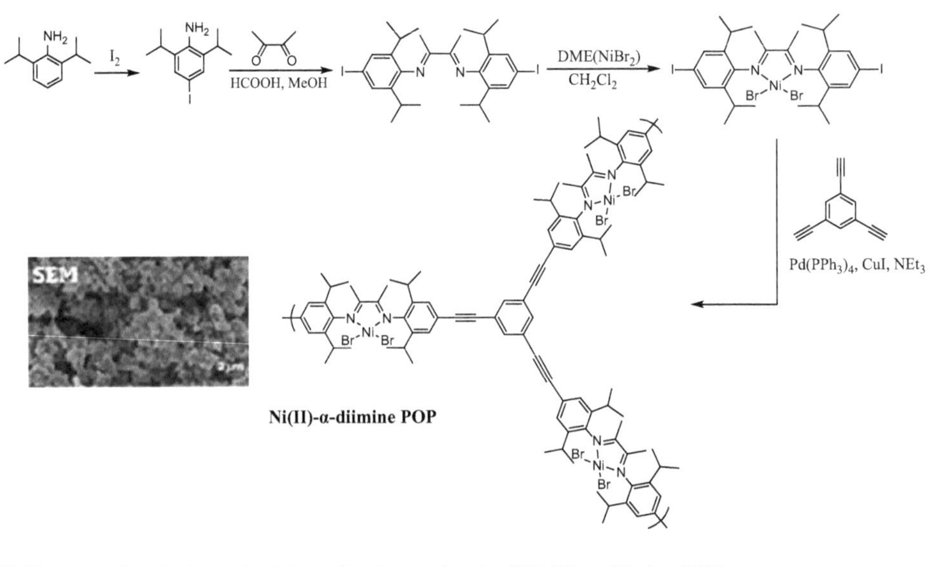

SCHEME 6.27 Synthetic methodology for the synthesis of Ni(II)-α-diimine POP.

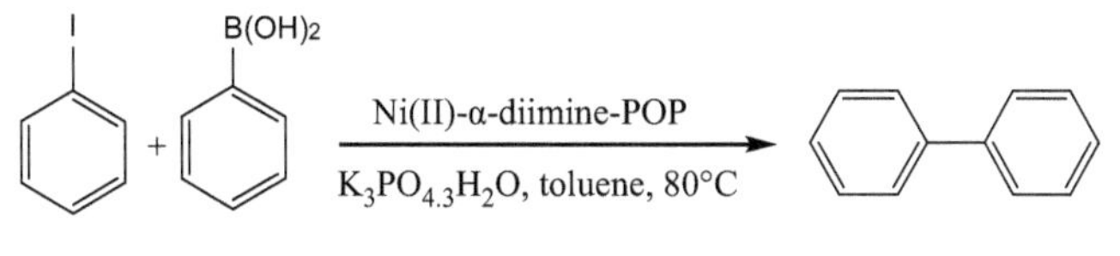

SCHEME 6.28 Coupling reaction of iodobenzene and phenylboronic acid by Ni(II)-α-diimine-POP.

6.4.3.4 Ru-loaded POP for Decomposition of Formic Acid to H_2

Formic acid (FA) has several benefits like easy transportation and storage, nontoxicity, and high volumetric hydrogen density which make it suitable for hydrogen storage and production.[73,74] Zhang et al.[75] synthesized two POPs, POPs-TPP, and POPs-PPh_3 from organic ligands having nitrogen and phosphorous donors respectively. POP-PPh_3 was synthesized from polymerization of tris(4-vinylphenyl)phosphine (3v-PPh_3) in AIBN at 100 °C in an autoclave which was further loaded with Ru(acac)$(CO)_2$ to give Ru@POP-PPh_3 as the catalyst.[76] POP-TPP was prepared by the polymerization of tetra(4-vinylphenyl)porphyrin (TPP) and loaded with Ru(acac)$(CO)_2$ to give Ru@POP-TPP.[77] TEM of unused and used Ru@POP-PPh_3 and Ru@POP-TPP showed that Ru nanoparticles were not detectable on fresh polymer catalysts. Even after the reaction, Ru nanoparticles were detectable only in Ru@POP-TPP which shows the remarkable stability of Ru@POP-PPh_3 (Figure 6.6).

Ru@POP-PPh_3 was found to be more effective for the decomposition of FA. It has a TOF of 7284 h^{-1} with high H_2 selectivity with approx. 0.1% CO, whereas Ru@POP-TPP has a TOF of 2308 h^{-1} with a comparatively high CO content of 2.221%. These results show that the properties of the ligands of the polymer framework could greatly affect the selectivity and activity of Ru@POPs catalysts. The catalyst can be effectively recovered and reused up to five times without any loss of activity.

6.4.3.5 Porphyrin-based POP to Support Mn Heterogeneous Catalysts for Selective Oxidation of Alcohols

Selective oxidation of alcohols is done to yield aldehydes and ketones which are used extensively in the pharmaceutical and plastic industry. These reactions are generally catalyzed by Au, Ru, or Pd catalysts due to their high activity, selectivity, and reusability but their limited availability and high cost have restricted their applications. Therefore, developing low-cost and environmentally benign transition metal-based catalysts is highly desirable. Porphyrin is a conjugated heterocycle with four pyrrolidine units interconnected by methine bridges. The four N atoms of porphyrin coordinate with various metal ions to give metalloporphyrins which are widely used as catalysts. These N atoms can strongly bind with Mn species leading to the formation of stable heterogeneous Mn-based catalysts.

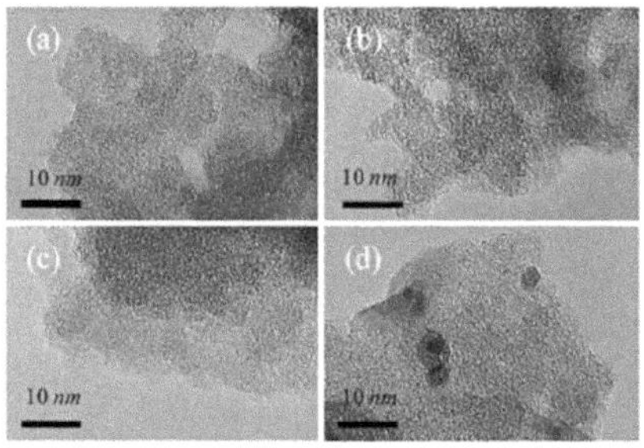

FIGURE 6.6 TEM images of (a) Fresh Ru@POP-PPh_3, (b) Used Ru@POP-PPh_3, (c) Fresh Ru@POP-TPP, and (d) Used Ru@POP-TPP.

Reprinted with permission from Elsevier, 2019.

TABLE 6.1
Comparison of Porosity of the POPs

Sample	BET Surface Area (m^2/g)	Pore Volume (m^3/g)	Average Pore Size (nm)
TFP-DPM	17.2	0.0142	6.6
TFP-DPM-2	30.8	0.0913	23.7
TFP-DPM-3	284.4	0.5782	16.3
TFP-DPM-4	178.6	0.8170	39.4

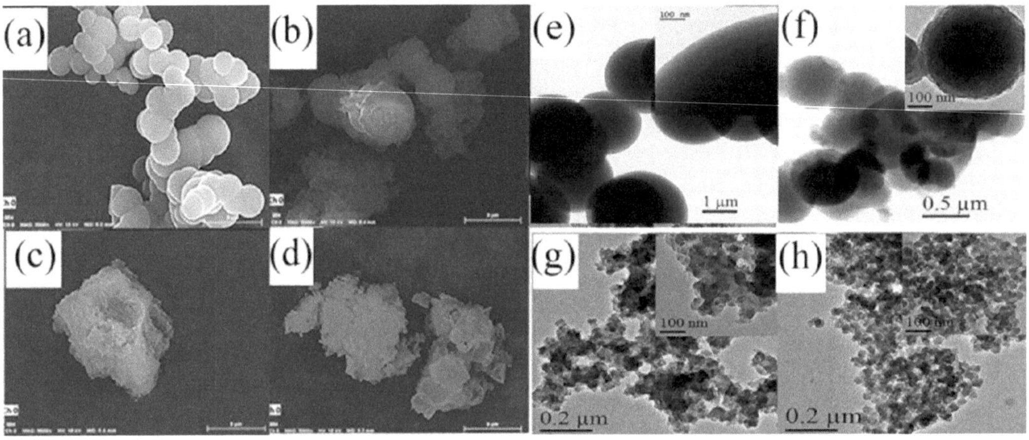

FIGURE 6.7 (a–d) SEM images of TFP-DPM, TFP-DPM-2, TFP-DPM-3, and TFP-DPM-4; (e–h) TEM images of TFP-DPM, TFP-DPM-2, TFP-DPM-3, and TFP-DPM-4.

Reprinted with permission from Wiley, 2019.

Additionally, the high conjugation helps in the flow of electrons to and from the valence ions resulting in an easy reduction of metals and enhanced reactivity of these catalysts in alcohol oxidation. Chen et al.[78] synthesized various porphyrin-based POP by taking 1,3,5-Triformylphloroglucinol (TFP) and 2,2′-dipyrrolylmethane (DPM) as the precursor. The POPs were further reacted with $Mn(OAc)_2.4H_2O$ to give various Mn/TFP-DPMs.

The comparison of the BET surface area, average pore size, and pore volume of various TFP-DPMs are given in Table 6.1.

The larger the pore size and BET surface area, the larger is the catalytic efficiency of the polymer. Also, SEM images of TFP-DPMs confirm their microscale spherical morphology and TEM images show that TFP-DPM and TFP-DPM-2 have compact surface and TFP-DPM-3 and TFP-DPM-4 have randomly distributed pores over the polymer surface (Figure 6.7).

These catalysts were used to carry out the oxidation of 1-phenylethanol where $PhI(OAc)_2$ acts as an optimum oxidant and KBr, a perfect additive. The maximum yield of the oxidized product was obtained when 1,4-dioxane was used as the co-solvent.

6.4.3.5.1 Mechanism of the Oxidation of Alcohols by TFP-DPMs

The mechanism of catalysis can be explained based on Corey's mechanism (Scheme 6.29).

Initially, PIDA and KBr react to form Br_2 in an aqueous solution.

$$PhI(OAc)_2 + 2KBr \rightarrow PhI + 2KOAc + Br_2$$

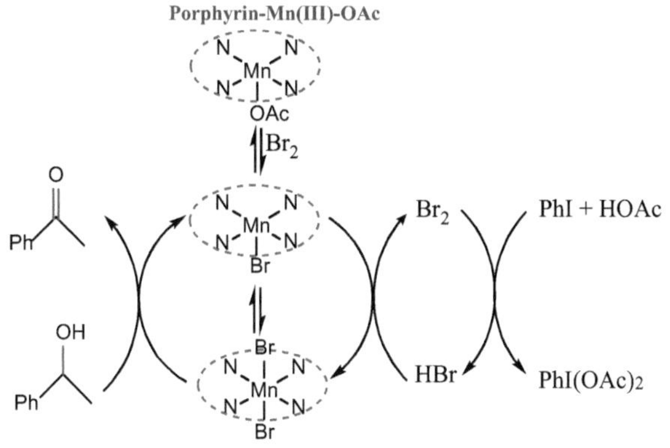

SCHEME 6.29 Mechanistic pathway for oxidation of 1-phenylethanol by Mn/TFP-DPMs.

Side by side HOBr species is produced in the reaction medium which exists in equilibrium with Br_2 in the presence of H_2O.[79]

$$PhI(OAc)_2 + KBr + H_2O \rightarrow PhI + KOAc + HOAc + HOBr$$

Br_2 brominates porphyrin-Mn(III)-OAc to porphyrin-Mn(V)-Br and [porphyrin-Mn(V)-Br_2]$^+$ which further reacts with the alcohol to give corresponding oxidized products.[79, 80]

6.4.4 Photocatalysis

Photocatalysis is the catalysis of the reaction by light. Various homogeneous photocatalyst-bearing transition metals are known but they are costly, toxic, or unstable. Thus, the development of stable heterogeneous catalysts is crucial and essential. Various heterogeneous photocatalysts have been developed, some of which are explained below:

6.4.4.1 Conjugated Porous Polymer Based on Phenanthrene Units

Conjugated Porous Polymers (CPPs) are a great alternative to conventional molecular photocatalysts as they possess high chemical and thermal stability, non-toxic, and easy to synthesize. Monterde et al.[81] designed a strategy for the preparation of nine CPPs via Sonogashira-Hagihara crosscoupling reaction. These CPPs contain various alkynyl comonomers. The introduction of ethynyl groups helps in rotation about phenyl-ethynyl bonds which lead to increased porosity and extended conjugation. Reaction with **1** and **2** gives **CPPs-So1** and **CPPs-So2** respectively having extended conjugation and reaction with **3** giving rise to polymer **CPPs-So3** that don't have extended conjugation. Further, various CPPs have been prepared by carrying out different reactions (Scheme 6.30). The dioxolane groups protect the keto group which makes Sonogashira coupling more effective and simultaneously hinders concatenation during polymerization.

CPP-So-acetals are hydrolyzed by trifluoromethanesulfonic acid under microwave radiation to form **CPP-So-OOH** which is further reacted with 2,4-difluoro-6-hydroxybenzaldehyde and ammonium acetate to give various **CPP-So-PhIm-2F** (phenantromidazole derivatives) (Scheme 6.31). The hydrolyzed CPPs possess a higher BET surface area than the ones obtained directly from 2,7-d iiodophenanthrene-9,10-dione.

CPP-So2-PhIm-2F represents high porosity, stability, and excellent photoredox properties and can be recycled six times without significant loss of activity and selectivity. CPP-So1-PhIm-2F and CPP-So3-PhIm-2F have a specific surface area of 568 and 456 m^2 g^{-1}, whereas CPP-So2-PhIm-2F has a high surface area of 684 m^2 g^{-1}. Lastly, the scanning electron microscopy (SEM) analysis clearly shows that the CPP-So-PhIm-2F polymer series have regular rearrangement (Figure 6.8).

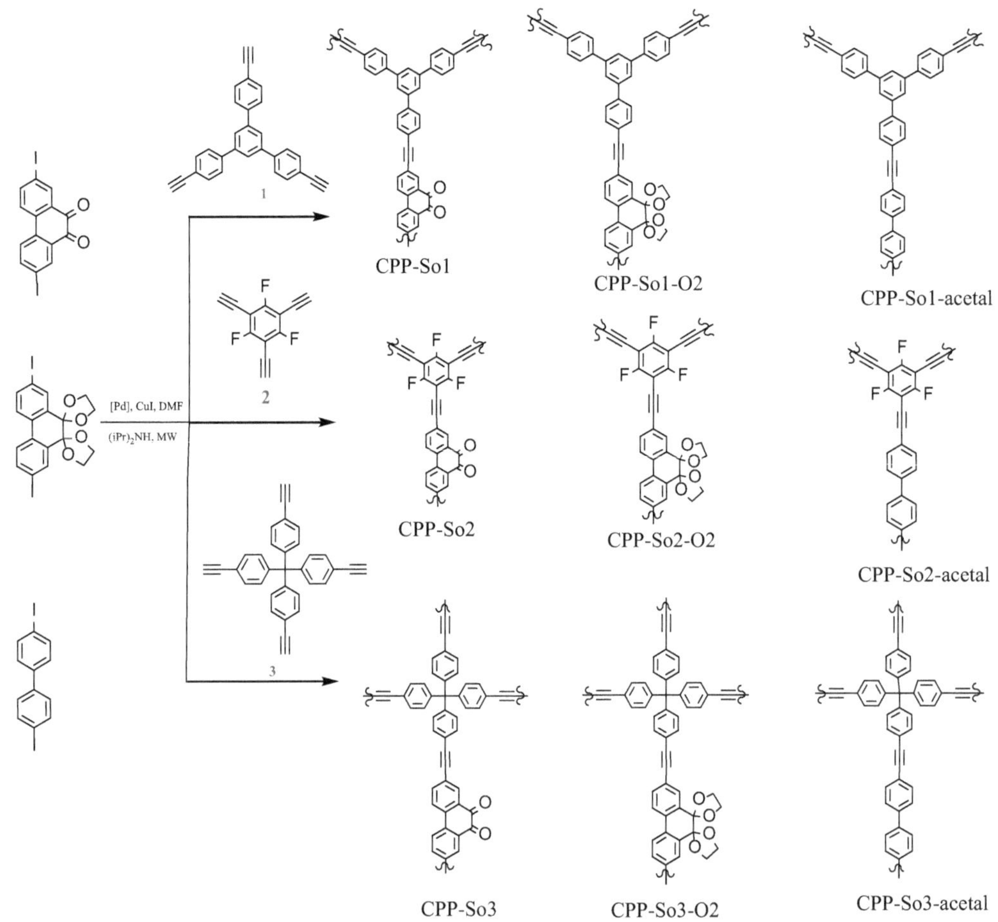

SCHEME 6.30 Synthesis of CPP-Sonogashira frameworks.

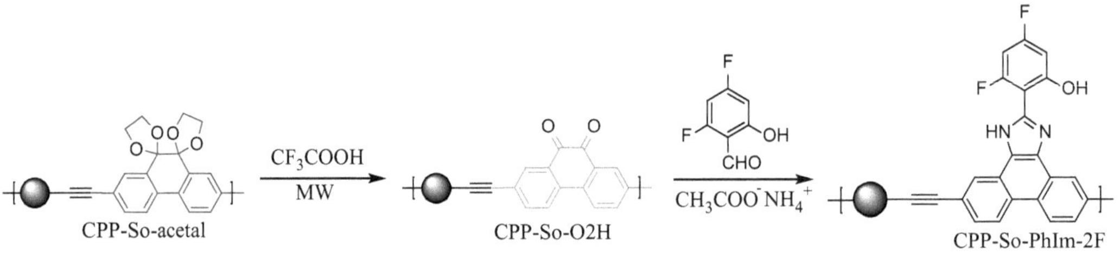

SCHEME 6.31 Hydrolysis of dioxolane polymer to diketo derivative, followed by preparation of phenantroimidazole derivatives.

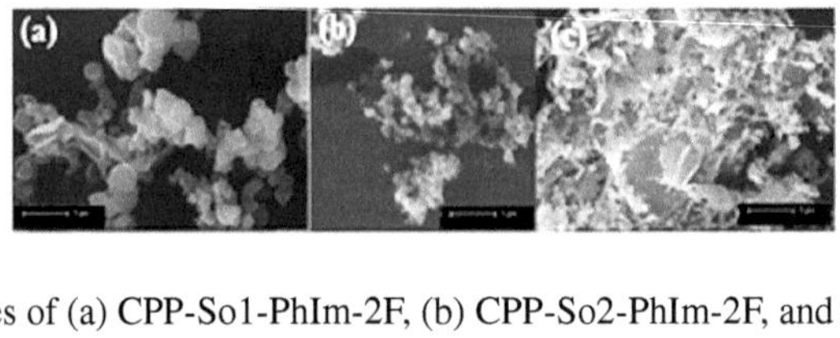

FIGURE 6.8 SEM images of (a) CPP-So1-PhIm-2F, (b) CPP-So2-PhIm-2F, and (c) CPP-So3-PhIm-2F.

Reprinted with permission from ACS, 2020.

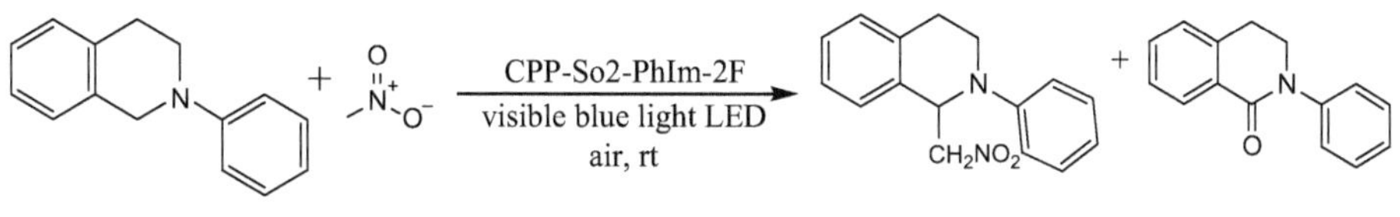

SCHEME 6.32 Photocatalyzed Aza-Henry reaction of N-phenyltetrahydroisoquinoline and nitromethane.

The activity of phenantromidazole derivatives was tested in carrying out the Aza-Henry reaction of N-phenyltetrahydroisoquinoline and nitromethane induced by the visible light (Scheme 6.32).

6.4.4.2 (dipyrrin)(bipyridine)ruthenium(II) Visible Light Photocatalyst

Generally, the heterogeneous visible light photocatalyst comprises metal oxides, dyes, plasmonic nanoparticles supported on metal oxides or polymeric graphitic carbon nitride.[82] But these approaches are neither sustainable nor greener. Therefore, having a ruthenium-based photocatalyst is a better approach for carrying out the organic transformations. Dipyrrin ligands can form complexes with metal ions such as ruthenium. But there is a disadvantage of dipyrrin that it cannot be manipulated easily in the lab and therefore to overcome this limitation BODIPY dyes are used that use BF_2 group as protecting group of dipyrrin in its structure.

Liras et al.[83] used BODIPY as organic support for the synthesis of a conjugated microporous polymer (dipyrrin)(bipyridine)ruthenium(II) (CMPBDP-Ru) via Sonogashira-Hagihara coupling reaction. BODIPY moiety is used as it provides the required orthogonality to synthesize CMPs with appropriate porosity.[84] CMPBDP was synthesized using 1,3,5,7,8-pentamethyl-2,6-diiodo-4,4-difluoro-4-bora-3a,4a-diaza-s-indacene[85] and 4,4″-diethynyl-5′-(4-ethynyl phenyl)-1,1′:3′,1″-terphenyl.[86] Then, dipyrrin moiety of CMPBDP was deprotected by reaction with BBr_3 and hydrolysis by water/acetone mixture.[87] Then, the free dipyrrin moieties of CMPBDP are complexed with $Ru(bpy)_2Cl_2$ to give (dipyrrin)(bipyridine)ruthenium(II) complex (Scheme 6.33).

PXRD pattern shows that the synthesized polymer is amorphous in nature and SEM analysis of the polymer shows that the materials possess granular particles. The BET surface area of CMPBDP and CMPBDP-Ru was found to be 518 and 306 $m^2\ g^{-1}$ respectively. The photocatalytic activity of CMPBDP-Ru was investigated in the Aza-Henry reaction. When the sp^3 C-H bond that is adjacent to the nitrogen is oxidized catalytically, it yields reactive iminium ions that can easily react with nucleophiles in the reaction of functionalization of amine or dehydrogenation to imines to give several products as represented in Scheme 6.34.

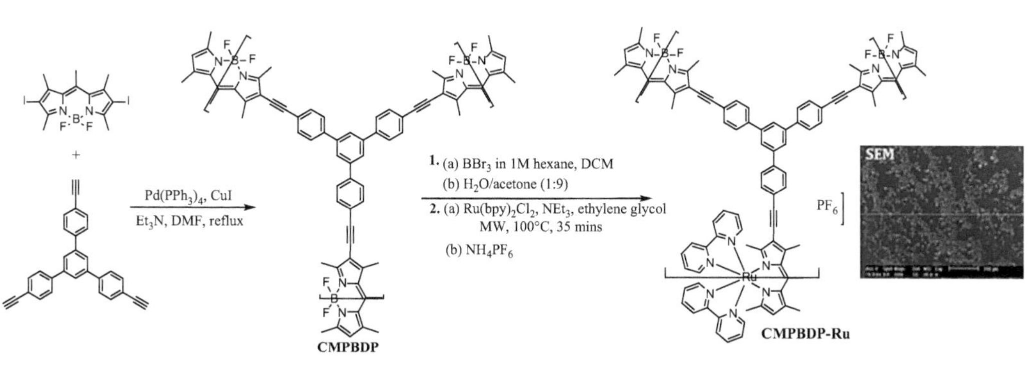

SCHEME 6.33 Synthesis of CMPBDP and its functionalized form, CMPBDP-Ru.

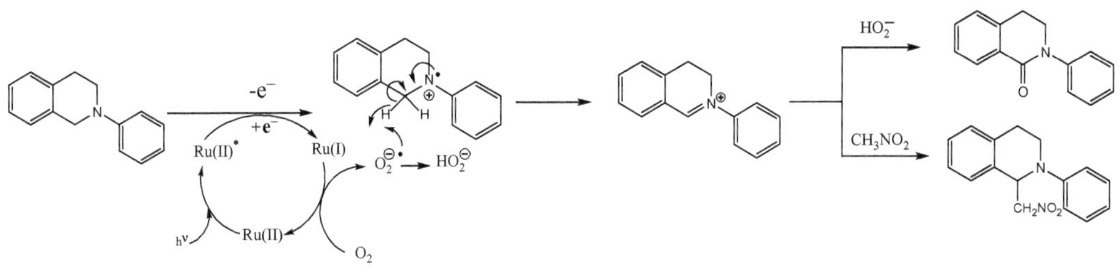

SCHEME 6.34 Plausible mechanism of Aza-Henry reaction of N-Phenyltetrahydroisoquinoline. Here, CMPBDP-Ru was used for the photocatalytic reaction of tetrahydroisoquinoline and nitroalkanes at room temperature and was found to have good activity and selectivity. Studies show that no product was formed when the catalyst was used in dark or when no catalyst was used in light. CMPBDP-Ru is highly thermally stable, photochemically stable, and can be recovered and reused seven times without any significant loss in activity and selectivity.

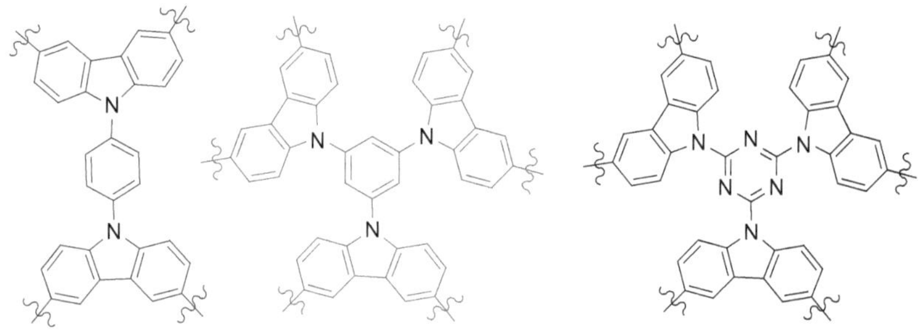

FIGURE 6.9 Structures of CMP-CSU5, CMP-CSU6, and CMP-CSU7.

6.4.4.3 Carbazole-based CMPs for C-3 Functionalization of Indoles

3-substituted indoles are a potential intermediate for medical applications. C-3 functionalization of indoles is generally performed by formylation and thiocyanate reaction. Due to the rigidity, non-planar structure, high light-absorbing capacity, and electron-donating ability, carbazoles have gained much attention in the synthesis of optoelectronic materials such as perovskite solar cells,[88] dye-sensitized solar cells,[89] and organic field-effect transistor.[90, 91] Based on this concept, Zhang et al.[92] have synthesized various carbazole-based CMPs for photolytically catalyzing the C-3 functionalization of indoles (Figure 6.9).

The prepared CMP-CSUs were used for the C-3 formylation and thiocyanate of indoles (Scheme 6.35).

Among all the prepared CMPs, CMP-CSU6 has the highest oxidation potential. The catalytic performance of CMP-CSU6 was found to be better than the pre-existing photocatalysts such as rose Bengal and $Ru(bpy)_3Cl_2$. Also, it doesn't require any base, acid, strong oxidant, or high temperature for its activity.

The BET surface area of CMP-CSU5, CMP-CSU6, and CMP-CSU7 was found to be 978, 1281, and 811 $m^2\ g^{-1}$ respectively and CMP-CSU6 possess the hierarchical micro-mesoporous structure of which mesopores contribute approx. 43% which is effective for the reaction of large substrates. SEM analyses of CMP-CSU5, CMP-CSU6, and CMP-CSU7 show different morphological structures – graphic flake-like, fiber-like, and plate-like as shown in Figure 6.9. CMP-CSU6 can be recovered and reused five times without any significant loss in their activity.

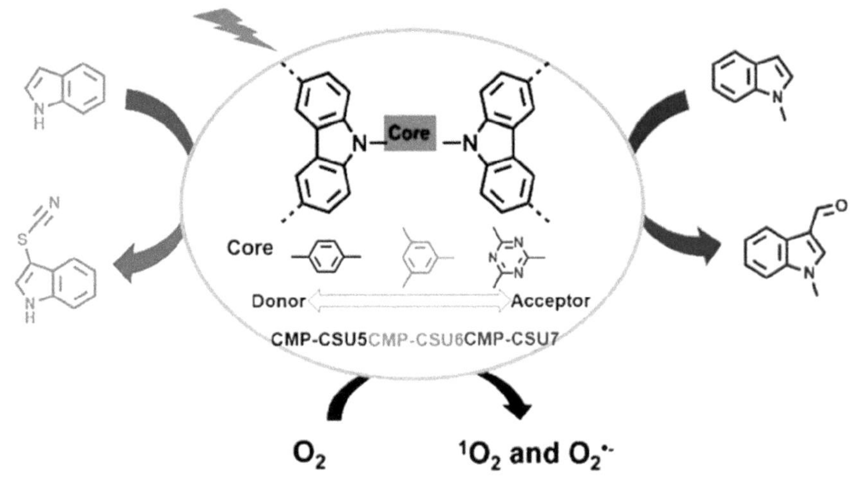

SCHEME 6.35 C-3 functionalization of indoles by CMP-CSUs via photocatalysis.

Reprinted with permission from RSC, 2019.

6.4.4.3.1 Mechanism of C-3 Formylation of N-methylindole by CMP-CSU6 Polymer Catalyst

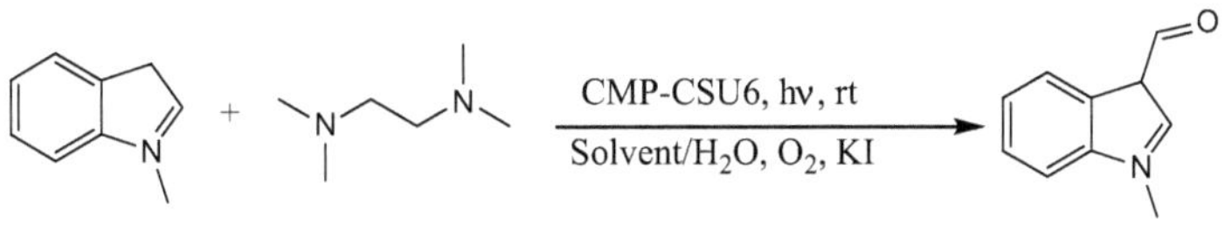

SCHEME 6.36 C-3 formylation of N-methylindole.

In C-3 formylation, initially, the excitation of CMP-CSU6 takes place to form CMP-CSU6* which is quenched by O_2 to give superoxide radical anion (O_2•–) and CMP-CSU6$^+$ by a single electron transfer (SET). The latter oxidize TMEDA to give radical cations **1a** and to reproduce CMP in a photoredox cycle. In due course, a hydrogen atom of **1a** goes to O_2•– to form an H_2O_2 anion and an iminium ion. Then, the electrophilic addition of iminium ion on N-methylindole takes place to form an indole iminium ion which releases a proton to give an intermediate **2b** and H_2O_2. In another light photoredox cycle (**2b→2c**), an iminium ion is again formed which undergoes hydrolysis to give the final product, 3-formyl-N-methylindole (**2c→2d**) (Scheme 6.37).

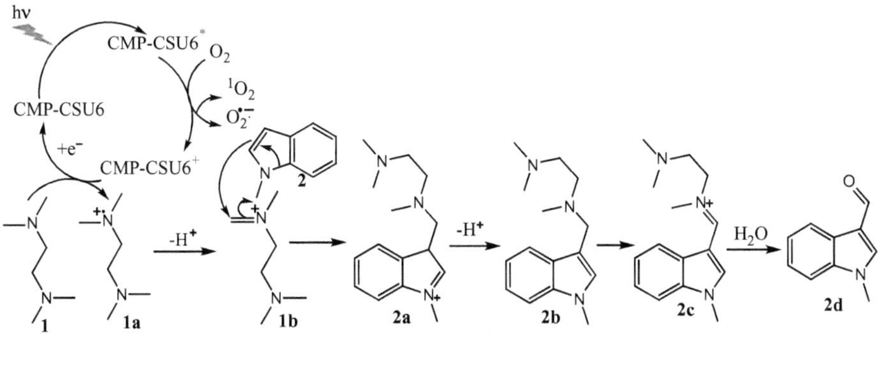

SCHEME 6.37 Mechanism for the C-3 formylation of N-methylindole.

6.4.4.3.2 The Mechanism for C-3 Thiocyanation of 1H-indole

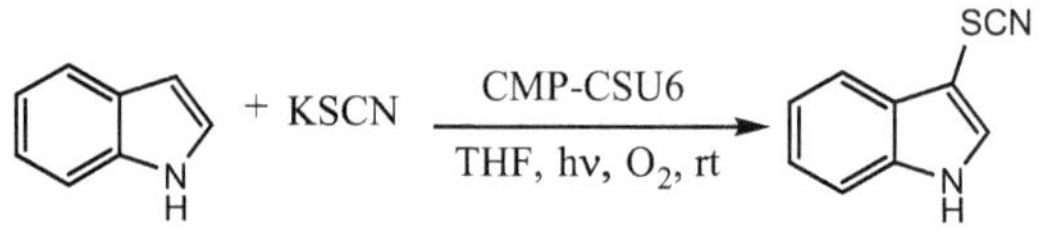

SCHEME 6.38 C-3 thiocyanate of 1H-indole.

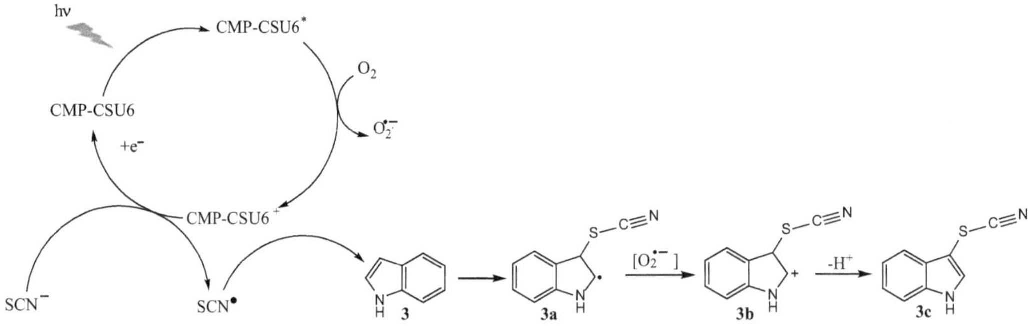

SCHEME 6.39 Mechanism for C-3 thiocyanate of 1H-indole.

Afterward, the activity of CMP-CSU6 for the C-3 thiocyanate of 1H-indoles was investigated. Ninety-five percent of the thiocyanate product was obtained when NH_4SCN and THF was used. Further experiments suggested that catalyst, light, and oxygen play a vital role in transformation reactions. When radiations are applied, excitation of CMP-CSU6 takes place to form CMP-CSU6* which is quenched by O_2 to generate H_2O_2 and CMPCSU6$^+$ via SET. The photoredox cycle ended by an electron transfer between CMP-CSU6$^+$ and –SCN to give CMP-CSU6 and •SCN. The latter then attacks 1H-indole to yield the intermediate **3a**. **3a** was then attacked by O_2•– and oxidized to give **3b**. **3b** loses a proton and undergoes aromatization to give the final product, 3-thiocyanoindole **3c** (Scheme 6.39).

6.4.5 Electrocatalysis

Electrocatalysis is the heterogeneous catalysis of electrochemical reactions. These reactions occur at the interface of the electrode and electrolyte and where the electrode plays the role of electron donor/acceptor as well as the catalyst.

6.4.5.1 Redox-active N-containing CPP for Oxygen Reduction Reaction (ORR)

Electrochemical energy is a form of renewable energy whose demand is increasing due to the need for sustainable development. Among all the known electrochemical devices, fuel cells are most common and for hydrogen fuel cells the main reaction is oxygen reduction (ORR). Conventionally, Pt is used as the catalyst for these reactions but its high cost, limited availability, potential carbon monoxide poisoning, etc., have restricted their application. For a few decades, the main focus is on developing an active catalyst for ORR that is cheap, abundant, stable, and possesses efficiency comparable with Pt catalyst. Due to the lightweight, covalent bonds, π-conjugation, and high chemical and thermal stability, conjugated microporous polymers are suitable to be used as a catalyst.

Bandyopadhyay et al.[92] designed an N-doped CPP by taking a monomer having both pyridine and pyrazine moiety via Sonogashira-Hagihara crosscoupling reaction (Scheme 6.40).

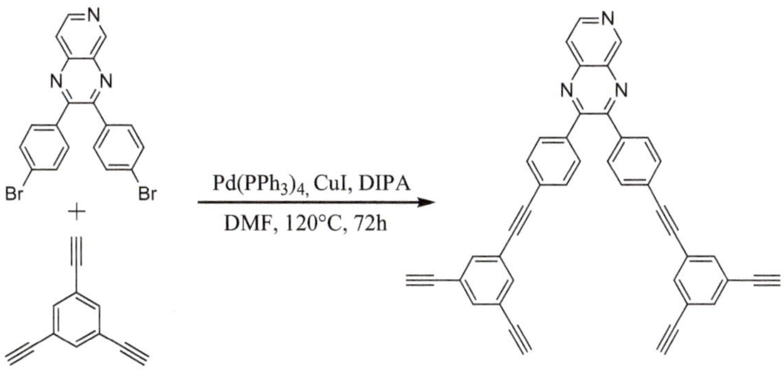

SCHEME 6.40 Synthesis of conjugated porous polymer (CPP–P1) from 2,3-bis(4-bromophenyl)-pyrido [4-b] pyrazine and 1,3,5-triethyl benzene.

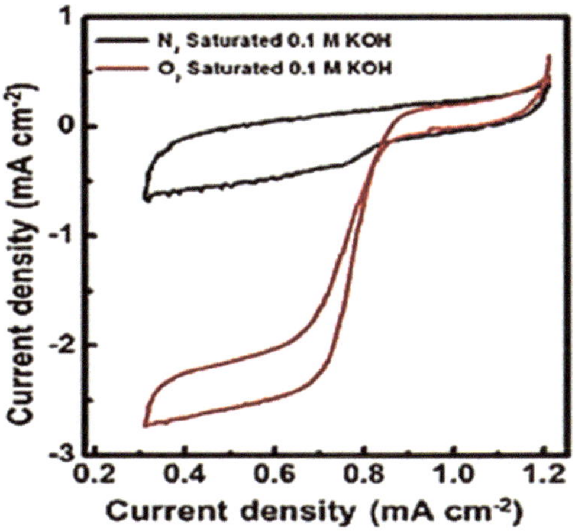

FIGURE 6.10 CV curves for CPP under O_2-saturated and N_2-saturated 0.1M KOH electrolyte.

The synthesized polymer has a redox-active pyrazine group which is responsible for carrying out the ORR reaction. The reaction takes place through a 4-electron ORR pathway. To evaluate the electrocatalysis property of synthesized CPP, cyclic voltammetry (CV) measurements were carried out in a 0.1 M KOH (aq) solution. CPP was used as a working electrode, a graphite rod as a counter electrode, and Ag/AgCl (3.0 M KCl) was taken as a reference electrode. In 0.1 M aqueous KOH solution saturated with molecular oxygen, a clear peak was observed at 0.78 V (vs RHE) that shows oxygen reduction. An experiment by taking N_2-saturated 0.1 M aqueous KOH solution as the electrolyte was also carried out and no peak was observed in that case (Figure 6.10). Hence, CPP is an effective catalyst for oxygen reduction.

The BET surface area of CPP was 109 $m^2\ g^{-1}$. Atomic force microscopy (AFM) and field-emission scanning electron microscopy (FESEM) were done for determining the morphology of the synthesized polymer and the FESEM image confirms the fused sphere-like structure having a dimension around 1 μm and low surface area.[93]

From DFT calculations, it was found that the traditional route for the ORR in alkaline media (formation of *–OOH and *–O groups) is highly energetically unfavorable for this system. Thus, Bandyopadhyay and co-workers suggested a suitable mechanism of the ORR according to which both pyrazine and pyridine moieties are involved in the reaction.

REFERENCES

1. Gates, B. C. *Catalytic Chemistry*; Wiley: New York, 1992.
2. Somorjai, G. A. *Nature* 2004, 430, 730.
3. Wan, K. T., Davis, M. E. *Nature* 1994, 370, 449–450.
4. Su, D. S., Perathoner, S., Centi, G. *Chem. Rev.* 2013, 113, 5782–5816.
5. Li, C.-J., Trost, B. M. *PNAS* 2008, 105, 13197–13202.
6. Dioumaev, V. K., Bullock, R. M. *Nature* 2000, 424, 530–532.
7. De Vos, D. E., Dams, M., Sels, B. F., Jacobs, P. *Chem. Rev.* 2002, 102, 3615–3640.
8. Cole-Hamilton, D. J. *Science* 2003, 299, 1702–1706.
9. (a) Davis, M. E. *Nature*, 2002, 417, 813; (b) Díaz, U., Brunel, D., Corma, A., *Chem. Soc. Rev.*, 2013, 42, 4083.
10. (a) Corma, A. *Chem. Rev.*, 1997, 97, 2373; (b) Meng, X., F.-S. Xiao, *Chem. Rev.*, 2014, 114, 1521.
11. Côté, A. P., Benin, A. I., Ockwig, N. W., O'Keeffe, M., Matzger, A. J., Yaghi, O. M. *Science* 2005, 310, 1166–1170.
12. El-Kaderi, H. M., Hunt, J. R., Mendoza-Cortés, J. L., Côté, A. P., Taylor, R. E., O'Keeffe, M., Yaghi, O. M. *Science* 2007, 316, 268–272.
13. Ding, S.-Y., Wang, W. *Chem. Soc. Rev.* 2013, 42, 548–568.
14. Fang, Q., Zhuang, Z., Gu, S., Kaspar, R. B., Zheng, J., Wang, J., Qiu, S., Yan, Y. *Nat. Commun.* 2014, 5, 4503–4511.
15. Kaur, P., Hupp, J. T., Nguyen, S. T. *ACS Catal.* 2011, 1, 819–835.
16. Zhang, Y., Riduan, S. N. *Chem. Soc. Rev.* 2012, 41, 2083–2094.
17. Zhang, Y., Ying, J. Y. *ACS Catal.* 2015, 5, 2681–2691.
18. Dawson, R., Cooper, A. I., Adams, D. J. *Prog. Polym. Sci.* 2012, 37, 530–563.
19. Rose, M. *ChemCatChem* 2014, 6, 1166–1182.
20. Li, B., Zhang, Y., Krishna, R., Yao, K., Han, Y., Wu, Z., Ma, D., Shi, Z., Pham, T., Space, B., Liu, J., Thallapally, P. K., Liu, J., Chrzanowski, M., Ma, S. *J. Am. Chem. Soc.* 2014, 136, 8654–8660.
21. Xu, Y., Jin, S., Xu, H., Nagai, A., Jiang, D. *Chem. Soc. Rev.* 2013, 42, 8012–8031.
22. Gopalakrishnan, D., Dichtel, W. R. *J. Am. Chem. Soc.* 2013, 135, 8357–8362.
23. Jiang, J.-X., Su, F., Trewin, A., Wood, C. D., Campbell, N. L., Niu, H., Dickinson, C., Ganin, A. Y., Rosseinsky, M. J., Khimyak, Y. Z., Cooper, A. I. *Angew. Chem. Int. Ed.* 2007, 46, 8574–8578.
24. Chen, L., Yang, Y., Jiang, D. *J. Am. Chem. Soc.* 2012, 132, 9138–9143.
25. Yue, Y., Mayes, R. T., Kim, J., Fulvio, P. F., Sun, X.-G., Tsouris, C., Chen, J., Brown, S., Dai, S. *Angew. Chem. Int. Ed.* 2013, 52, 13458–13462.
26. Zhang, Y., Wei, S., Liu, F., Du, Y., Liu, S., Ji, Y., Yokoi, T., Tatsumi, T., Xiao, F.-S. *Nano Today* 2009, 4, 135–142.
27. McKeown, N. B., Gahnem, B., Msayib, K. J., Budd, P. M., Tattershall, C. E., Mahmood, K., Tan, S., Book, D., Langmi, H. W., Walton, A. *Angew. Chem. Int. Ed.* 2006, 45, 1804–1807.
28. Kandambeth, S., Mallick, A., Lukose, B., Mane, M. V., Heine, T., Banerjee, R. *J. Am. Chem. Soc.* 2012, 134, 19524–19527.
29. Chandra, S., Kandambeth, S., Biswal, B. P., Lukose, B., Kunjir, S. M., Chaudhary, M., Babarao, R., Heine, T., Banerjee, R. *J. Am. Chem. Soc.* 2013, 135, 17853–17861.
30. Hunt, J. R., Doonan, C. J., LeVangie, J. D., Côté, A. P., Yaghi, O. M. *J. Am. Chem. Soc.* 2008, 130, 11872–11873.
31. Dong, K., Sun, Q., Meng, X., Xiao, F.S., *Catal. Sci. Technol.* 2017, 7, 1028–1039.
32. Liu, F., Meng, X., Zhang, Y., Ren, L., Nawaz, F., Xiao, F.-S. *J. Catal.* 2010, 271, 52–58.
33. Wang, L., Zhang, J., Sun, J., Zhu, L., Zhang, H., Liu, F., Zheng, D., Meng, X., Shi, X., Xiao, F.-S. *ChemCatChem* 2013, 5, 1606–1613.
34. Bunck, D. N., Dichtel, W. R. *Angew. Chem. Int. Ed.* 2012, 51, 1885–1889.
35. Nagai, A., Guo, Z., Feng, X., Jin, S., Chen, X., Ding, X., Jiang, D. *Nat. Commun.* 2011, 2, 536–543.
36. Okuhara, T. *Chem. Rev.* 2002, 102, 3641–3666.
37. Liu, F., Kong, W., Qi, C., Zhu, L., Xiao, F.-S. *ACS Catal.* 2012, 2, 565–572.
38. Román-Leshkov, Y., Chheda, J. N., Dumesic, J. A. *Science* 2006, 312, 1933–1937.
39. Wang, L., Wang, H., Liu, F., Zheng, A., Zhang, J., Sun, Q., Lewis, J. P., Zhu, L., Meng, X., Xiao, F.-S. *ChemSusChem*, 2014, 7, 402.

40. (a) Kim, D. W., Chi, D. Y. *Angew. Chem. Int. Ed.*, 2004, 43, 483; (b) Leng, Y., Wang, J., Zhu, D. R., Ren, X. Q., Ge, H. Q., Shen, L. *Angew. Chem. Int. Ed.*, 2009, 48, 168; (c) Parvulescu, V. I., Hardacre, C. *Chem. Rev.* 2007, 107, 2615.
41. Liu, F., Wang, L., Sun, Q., Zhu, L., Meng, X., F.-S. Xiao, *J. Am. Chem. Soc.* 2012, 134, 16948.
42. Okuhara, T., *Chem. Rev.* 2002, 102, 3641.
43. Sun, Q., Lv, Z., Du, Y., Wu, Q., Wang, L., Zhu, L., Meng, X., Chen, W., Xiao, S., *Chem. Asian. J.* 2013, 8, 2822.
44. (a) Noyori, R., *Angew. Chem. Int. Ed.*, 2002, 41, 2008; (b) Hong, L., Koelsch, G., Lin, X., Wu, S., Terzyan, S., Ghosh, A. K., Zhang, X. C., Tang, J., *Science* 2000, 290, 150; (c) Akiyama, T., *Chem. Rev.* 2007, 107, 5744.
45. (a) Magano, J., Dunetz, J. R., *Chem. Rev.* 2011, 111, 2177; (b) Canali, L., Sherrington, D. C., *Chem. Soc. Rev.* 1999, 28, 85.
46. Sun, Q., Meng, X., Liu, X., Zhang, X., Yang, Y., Yang, Q., F.-S. Xiao, *Chem. Commun.*, 2012, 48, 10505.
47. Wang, X. S., Chrzanowski, M., Yuan, D., Sweeting, B. S., Ma, S., *Chem. Mater.* 2014, 26, 1639.
48. Chen, L., Yang, Y., Jiang, D., *J. Am. Chem. Soc.* 2012, 132, 9138.
49. Sun, Q., Jiang, M., Shen, Z., Jin, Y., Pan, S., Wang, L., Meng, X., Chen, W., Ding, Y., Li, J., F.-S. Xiao, *Chem. Commun.* 2014, 50, 11844.
50. Sun, Q., Dai, Z., Liu, X., Sheng, N., Deng, F., Meng, X., F.-S. Xiao, *J. Am. Chem. Soc.* 2015, 137, 5204.
51. Huangfu, Y., Sun, Q., Pan, S., Meng, X., F.-S. Xiao, *ACS Catal.* 2015, 5, 1556–1559.
52. Sun, Q., Ma, S., Dai, Z., Meng, X., F.-S. Xiao, *J. Mater. Chem. A* 2015, 3, 23871.
53. Ma, L., Wanderley, M. M., Lin, W., *ACS Catal.* 2011, 1, 691. 8, 202.
54. (a) DeCoste, J. B., Peterson, G. W., *Chem. Rev.* 2014, 114, 5695. (b) Didas, S. A., Choi, S., Chaikittisilp, W., Jones, C. W., *Acc. Chem. Res.* 2015, 48, 2680.
55. Dai, Z., Sun, Q., Liu, X., Bian, C., Wu, Q., Pan, S., Wang, L., Meng, X., Deng, F., F.-S. Xiao, *J. Catal.* 2016, 338, 202.
56. Wang, J., Sng, W., Yi, G., Zhang, Y., Imidazolium salt-modified porous hypercrosslinked polymers for synergistic CO_2 capture and conversion. *Chem Commun* 2015, 51, 12076–12079.
57. Wang, W., Li, C., Yan, L., Wang, Y., Jiang, M., Ding, Y., Ionic liquid/Zn-PPh3 integrated porous organic polymers featuring multifunctional sites: highly active heterogeneous catalyst for cooperative conversion of CO_2 to cyclic carbonates. *ACS Catal.* 2016, 6(9), 6091–6100.
58. Wu, S., Teng, C., Cai, S., Jiang, B., Wang, Y., Meng, H., Tao, H. Triphenylphosphine-based functional porous polymer as an efficient heterogeneous catalyst for the synthesis of cyclic carbonates from CO_2. *Nanoscale Res. Lett.* 2017, 12(1), 609–617.
59. Zhang, Z., Fan, F., Xing, H., Yang, Q., Bao, Z., Ren, Q. Efficient synthesis of cyclic carbonates from atmospheric CO_2 using a positive charge delocalized ionic liquid catalyst. *ACS Sustain. Chem. Eng.* 2017, 5, 2841–2846.
60. Xue, Z., Jiang, J., Ma, M., Li, M., Mu, T., Gadolinium-based metal–organic framework as an efficient and heterogeneous catalyst to activate epoxides for cycloaddition of CO_2 and alcoholysis. *ACS Sustain. Chem. Eng.* 2017, 5, 2623–2631.
61. Xu, B., Wang, J., Sun, J., Huang, Y., Zhang, J., Zhang, X., Zhang, S., Fixation of CO_2 into cyclic carbonates catalyzed by ionic liquids: a multi-scale approach. *Green Chem.* 2015, 17, 108–122.
62. Dai, W., Jin, B., Luo, S., Luo, X., Tu, X., Au, C., Polymers anchored with carboxyl-functionalized di-cation ionic liquids as efficient catalysts for the fixation of CO_2 into cyclic carbonates. *Catal. Sci. Technol.* 2014, 4, 556–562.
63. Kim, K., Kim, S., Talapaneni, S. N., Buyukcakir, O., Almutawa, A. M. I., Polychronopoulou, K., Coskun, A. Transition metal complex directed synthesis of porous cationic polymers for efficient CO_2 capture and conversion. *Polymer* 2017, 126, 296–302.
64. Liu, S., Zhao, Y., Zhang, B., Xia, H., Zhou, J., Xie, W., Li, H., *J. Power Sources* 2018, 381, 116.
65. Kaverlavni, S. K., Moosavifard, S. E., Bakouei, A., *J. Mater. Chem. A*, 2017, 5, 14301.
66. Acharya, U., Bober, P., M. Trchová, Zhigunov, A., Stejskal, J., Pfleger, J., *Polymer*, 2018, 150, 130.
67. Chen, J., Du, C., Zhang, Y., Wei, W., Wan, L., Xie, M., Tian, Z., Constructing porous organic polymer with hydroxyquinoline as electrochemical-active unit for high-performance supercapacitor. *Polymer* 2019, 162, 43–49.
68. Wu, D., Nese, A., Pietrasik, J., Liang, Y., He, H., Kruk, M., Huang, L., Kowalewski, T., Matyjaszewski, K., *ACS Nano* 2012, 6, 6208.

69. Khatioda, R., Pathak, D., Sarma, B., Cu(II) complex onto a pyridine-based porous organic polymer as a heterogeneous catalyst for nitroarene reduction. *ChemistrySelect* 2018, 3(23), 6309–6320.
70. (a) Roelen, O. German Patent DE 849548, 1938/1952; U.S. Patent 2327066, 1943. *Chem. Abstr.* 1944, 38, 363. (b) van Leeuwen, P. W. N. M., Claver, C. *Rhodium Catalyzed Hydroformylation*; Kluwer Academic Publishers: Dordrecht, 2000.
71. Jia, X., Liang, Z., Chen, J., Lv, J., Zhang, K., Gao, M., Zong, L., Xie, C. *Org. Lett.* 2019, 21(7), 2147–2150.
72. Dong, Y., Jv, J.-J., Li, Y., Li, W.-H., Chen, Y.-Q., Sun, Q., Ma, J.-P., Dong, Y.-B. Nickel-metalated porous organic polymer for Suzuki–Miyaura cross-coupling reaction. *RSC Adv.* 2019, 9(35), 20266–20272.
73. Liu, X., Li, S. S., Liu, Y. M., Cao, Y., *Chin. J. Catal.* 2015, 36, 1461–1475.
74. Yang, X. C., Xu, Q., *Chin. J. Catal.* 2016, 37, 1594–1599.
75. Zhang, Y., Lyu, Y., Wang, Y., Li, C., Jiang, M., Ding, Y. Highly active and stable porous polymer heterogenous catalysts for decomposition of formic acid to produce H2. *Chinese J. Catal.* 2019, 40(2), 147–151.
76. Sun, Q., Jiang, M., Shen, Z., Jin, Y., Pan, S., Wang, L., Meng, X., Chen, W., Ding, Y., Li, J., Xiao, F. S., *Chem. Commun.* 2014, 50, 11844–11847.
77. Dai, Z. F., Sun, Q., Liu, X. L., Bian, C. Q., Wu, Q. M., Pan, S. X., Wang, L., Meng, X. J., Deng, F., Xiao, F. S., *J. Catal.* 2016, 338, 202–209.
78. Chen, J., Zhang, Y., Zhu, D., Li, T., Selective oxidation of alcohols by porphyrin-based porous polymer-supported manganese heterogeneous catalysts. *Appl. Organomet. Chem.* 2019, e5259.
79. Brown, M. K., Blewett, M. M., Colombe, J. R., Corey, E. J., *J. Am. Chem. Soc.* 2010, 132, 11165.
80. Wu, Z., Chen, L., Liu, J., Parvez, K., Liang, H., Shu, J., Sachdev, H., Graf, R., Feng, X., Müllen, K., *Adv. Mater.* 2013, 26, 1450.
81. Monterde, C., Pintado-Sierra, M., Navarro, R., Sánchez, F., Iglesias, M., An effective approach towards conjugated porous organic frameworks based on phenanthrene building blocks. Metal-free heterogeneous photocatalysts. *ACS Appl. Mater. Interfaces* 2020, 12 (13), 15108–15114.
82. Lang, X., Chen, X., Zhao, J., *Chem. Soc. Rev.* 2014, 43, 473.
83. Liras, M., Pintado-Sierra, M., Iglesias, M., Sánchez, F., A deprotection strategy of a BODIPY conjugated porous polymer to obtain a heterogeneous (dipyrrin)(bipyridine)ruthenium(ii) visible light photocatalyst. *J. Mater. Chem. A* 2016, 4(44), 17274–17278.
84. Liras, M., Iglesias, M., Sánchez, F., *Macromolecules*, 2016, 49, 1666.
85. Bonardi, L., Ulrich, G., Ziessel, R., *Org. Lett.*, 2008, 10, 2183.
86. Weber, E., *J. Chem. Soc. Perkin Trans.* 1988, 2, 1251.
87. Lundrigan, T., Thompson, A., *J. Org. Chem.*, 2013, 78, 757.
88. Xu, B., Sheibani, E., Liu, P., Zhang, J., Tian, H., Vlachopoulos, N., Boschloo, G., Kloo, L., Hagfeldt, A., Sun, L. Carbazole based hole transport materials for efficient solid-state dye sensitized solar cells and Perovskite solar cells. *Adv. Mater.* 2014, 26, 6629–6634.
89. Wang, Y., Chen, B., Wu, W., Li, X., Zhu, W., Tian, H., Xie, Y. Efficient solar cells sensitized by porphyrins with an extended conjugation framework and a carbazole donor: from molecular design to cosensitization. *Angew. Chem. Int. Ed.* 2014, 53, 10779–10783.
90. Li, Z., Liu, Y., Yu, G., Wen, Y., Guo, Y., Ji, L., Qin, J., Li, Z. A new carbazole-constructed hyperbranched polymer: convenient one-pot synthesis, hole-transporting ability, and field-effect transistor properties. *Adv. Funct. Mater.* 2009, 19, 2677–2683.
91. Zhang, W., Tang, J., Yu, W., Huang, Q., Fu, Y., Kuang, G., ... Yu, G., Visible light-driven C-3 functionalization of indoles over conjugated microporous polymers. *ACS Catal.* 2018, 8084–8091.
92. Bandyopadhyay, S., Boukhvalov, D. W., Nayak, A. K., Ha, S. R., Shin, H. J., Kwon, J., Song, T., Choi, H., Redox active nitrogen-containing conjugated porous polymer: An organic heterogeneous electrocatalysts for oxygen reduction reaction. *Dyes Pigm.* 2019, 120, 107557.
93. Bandyopadhyay, S., Anil, A. G., James, A., Patra, A., Multifunctional porous organic polymers: tuning of porosity, CO_2, and H_2 storage and visible-light-driven photocatalysis. *ACS Appl. Mater. Interf.* 2016, 8, 27669–27678.

7 Triazine Porous Frameworks

Faiza Wahad, Zeeshan Abid, Sughra Gulzar, and Syed Arfan Haider
Government College University Lahore, Lahore, Pakistan

Munazza Shahid
University of Education, Lahore, Pakistan

Muhammad Altaf and Raja Shahid Ashraf
Government College University Lahore, Lahore, Pakistan

CONTENTS

7.1 INTRODUCTION

Porous organic frameworks possessing a heteroatom-rich core have led to a paradigm shift in catalysis, molecular separation, and energy applications. The heteroatom effect (HAE) induced by the introduction of heteroatoms dramatically improves the performance of these materials, as evident by the superior performance of nitrogen-enriched porous frameworks in contrast with their carbon-based analogs. Reports suggest that doping of nitrogen atoms into the materials influences the distribution of electrons in the carbon material enhancing the electroactive surface area (Hall et al. 2010, Hulicova-Jurcakova et al. 2009). Likewise, nitrogen-doping is used as a technique to improve applications of molecular separation and gas absorption as well as sensing and selectivity in

DOI: 10.1201/9781003169604-7

electrocatalysis (Liebl and Senker 2013, Dey, Bhunia, Breitzke, et al. 2017, Bügel, Spieß, and Janiak 2021, Wang, Chen, et al. 2018, Tao et al. 2016).

Triazines are a class of nitrogen-containing heterocycles that have been under the spotlight as nitrogen-enriching materials. In the late 2000s, the synthesis of polymers based on triazine linkage led to the concept of covalent triazine framework (CTF), which later on became a new class of promising porous organic polymers (POPs) with high-performance for useful applications (Kuhn, Antonietti, and Thomas 2008, Diercks and Yaghi 2017, Liu, Zan, et al. 2017, Liu, Huang, et al. 2018, Yu, Mahmood, et al. 2018). As a type of covalent organic framework (COF), CTFs involve covalent linkages resulting from triazine ring formation reaction between organic building blocks thus leading to a porous framework. The choices for nitrogen-rich materials are vast for applications where high nitrogen content or HAE is required (Park et al. 2017, Gunasekar and Yoon 2021). However, the triazine-linkage contribution to the CTF nitrogen content is more advantageous in comparison to the other materials such as polymers of intrinsic microporosity (PIM), hyper-crosslinked polymers (HCPs) and other COFs. Consequently, a wide range of different porous materials containing triazine rings is reported in the literature (Wang, Huang, et al. 2016, Li, Liu, et al. 2020, Liu et al. 2014, Hug et al. 2014, Gu et al. 2015, Mohamed, EL-Mahdy, et al. 2020, Li, Zhou, and Wang 2020, Bhunia, Vasylyeva, and Janiak 2013, Yuan et al. 2018, Wen et al. 2021, Liu et al. 2020, Bhunia et al. 2013).

CTFs feature a whole range of unique characteristics such as aromatic linkage of carbon and nitrogen (C=N) and lack of weak bonds despite aromatic substitutions. Many of the CTFs reported are amorphous or semicrystalline owing to the presence of ultra-strong bonds that, under the influence of high dynamic condensation, hampers the formation of an ordered structure. Hence, CTFs were usually listed as conjugated microporous polymers (CMPs), a subclass of amorphous porous polymer, because of their property of combining the π-conjugated skeleton of CTFs with permanent nanopores, in comparison to various other available porous materials which are conventional nonporous polymers or polymers lacking conjugation (Xu et al. 2013). The distinguishing features of CTFs provided them with high nitrogen content, greater chemical stability, HAE and endowed them with many practical applications, including carbon dioxide (CO_2) capture (Dong, Wang, Wang, et al. 2020, Mukhtar et al. 2021), separation and storage of materials (Zhu et al. 2012, Liebl and Senker 2013, Zhao et al. 2013, Hug et al. 2015, Yu, Gu, et al. 2018) energy production and storage (Hao, Ning, et al. 2015, Xu et al. 2017, Liu, Huang, et al. 2019, Talapaneni et al. 2016, Liu, Lyu, et al. 2018, Kaleeswaran et al. 2017, Wu, Maier, and Yu 2020, Guo, Wu, et al. 2018, Dey et al. 2019, Krishnaraj et al. 2020), heterogeneous catalysis (Puthiaraj, Lee, Zhang, et al. 2016, Zhao et al. 2018), and photocatalysis (Bai et al. 2015, Bi et al. 2015, Guo, Wu, et al. 2018, Xu et al. 2021, Wang et al. 2020).

In the last decade, several synthetic strategies were developed for CTFs, including ionothermal trimerization (Kuhn, Antonietti, and Thomas 2008, Mohamed, EL-Mahdy, et al. 2020), amide-based polycondensation (Liu, Huang, et al. 2018, Wang, Yang, et al. 2017), phosphorus pentoxide (P_2O_5) catalyzed reaction (Yu, Mahmood, et al. 2018), Friedel-Crafts reaction (Puthiaraj et al. 2015, Dey et al. 2016, Puthiaraj, Kim, and Ahn 2016), and superacid catalyzed method (Ren et al. 2012). With the development of these synthesizing methods, researchers discovered that ordered structure materials could be achieved with ultra-strong bonds of CTFs under certain conditions (Liu, Zan, et al. 2017, Liu, Huang, et al. 2018, Tan et al. 2021, Bojdys et al. 2010, Zhang, Cheng, et al. 2020). Hence formation of a more ordered structure makes CTFs appealing for its increased demand in the future. In accordance with the ongoing research, two types of CTFs are used commonly: amorphous and crystalline.

In this chapter, we have discussed different synthetic procedures and applications of CTFs which have triazine rings as a connecting unit and with no weak linkage bonds present. This brief coverage of CTF synthetic strategies and applications will help to comprehend ongoing research and future perspectives of CTFs and challenges associated with it.

7.2 SYNTHETIC PROCEDURES OF CTFs AND THEIR STRUCTURAL DESIGNS

7.2.1 Ionothermal Trimerization Strategy

The strategy was first introduced by Thomas et al. who used high-temperature reversible reaction for addressing the problem arising from crystallization of CTFs with triazine linkage, formed by the ultra-strong covalent bond of aromatic C=N (Ding and Wang 2013). This high-temperature reversible reaction procedure was thus called an ionothermal trimerization reaction. This strategy uses ionothermal conditions to carry out dynamic trimerization reaction, where self-optimizing and reversible polymerization occurred in order to form crystalline CTF. At high temperatures with the use of molten zinc chloride, good solubility was shown by nitriles due to the Lewis acid-base strong interaction, resulting in a clear solution of all the monomers. Here, zinc chloride enables the reversible reaction for trimerization and acted as a good catalyst for the reaction (Sun, Wang, and Xu 2020).

This strategy was used for synthesizing the first crystalline hexagonal packing triazine-based organic framework (CTF-1). The surface area (SA) of CTF-1 calculated by Brunauer-Emmett-Teller (BET) was found to be 7991 m^2 g^{-1}. It was found that other monomers such as DCP, DCT, DCBP and TCT were not able to form crystalline CTFs. The reason behind this was the structural arrangement of the nitrile group. TCT and DCBP lack planner geometry but have contorted structure, whereas DCP and DCT do not show linearly arranged nitrile group. Later on, highly crystalline CTF-2 material was generated by 2,6-dicyanonaphthaline planner monomer (Bojdys et al. 2010), but it has a very low surface area of 90 m^2 g^{-1} due to material layer arrangement in a partially staggered orientation. Then, by 1,3,5-tricyanobenzene trimerization reaction in molten $ZnCl_2$, a covalent triazine framework (CTF-0) was synthesized, having a surface area of 500 m^2 g^{-1} (Algara-Siller et al. 2014, Katekomol et al. 2013, Mahmood et al. 2015). Figure 7.1 shows the amorphous or semi-crystalline CTFs from other monomers.

Different strategies are used for improving structural design such as the introduction of heteroatoms (Guo, Niu, et al. 2018), specific block building and pore size adjustment (Zhang, Hu, et al. 2021). Adjustment of pore size is usually done by improving the monomer molecule's structural features and size. For instance, CTF-1 has a larger pore size than CTF-0, because CTF-0 has a 1,3,5-tricyanobenzene group having three functional groups in it. Similarly, CTF-2 has a bigger pore size of 1.6 nm than CTF-1 which has a pore size of 1.2 nm, because of the longer monomer of CTF-2 (Mukherjee et al. 2019).

Moreover, the performance of the CTF structure is improved by the addition of heteroatoms into the structure of triazine for some applications. In order to increase the performance of CO_2 capture in CTFs, Zhao et al. gave designing and synthesizing strategy of perfluorinated covalent triazine-based frameworks (FCTF-1) having enhanced CO_2 absorption ability owing to the C–F bond having hydrophobic nature characteristics (Zhao et al. 2013). Hence, the introduction of heteroatoms such as sulfur, oxygen, nitrogen and fluorine were adopted as a common strategy for various applications into the structure of CTFs, resulting in increased performance of structures under certain conditions (Wang, Chen, et al. 2018, Artz and Palkovits 2015, Xu et al. 2018, Wen et al. 2021, Jena et al. 2018). For, instance, the introduction of the carbazole group into the CTF structures had surprisingly improved gas capture and photocatalysis (Wu et al. 2014, Guo, Niu, et al. 2018, Chen et al. 2021, Tan et al. 2021). The introduction of acetylacetone units improved the performance of the CTF structure to heterogeneous catalysis hence stabilizing the catalyst (Jena et al. 2018). Therefore, possible designing strategies, such as pore designing strategies, heteroatom introduction, and importing specific units, could be adopted as needed to increase the performance of CTF structures and for various applications.

On the other hand, synthesis strategies such as $ZnCl_2$ ratio adjustment and temperature controlling strategy play an important part in order to prepare high-performance CTF structures (Osadchii et al. 2017). Unnecessary decomposition of CTF structures is avoided by optimizing synthetic conditions

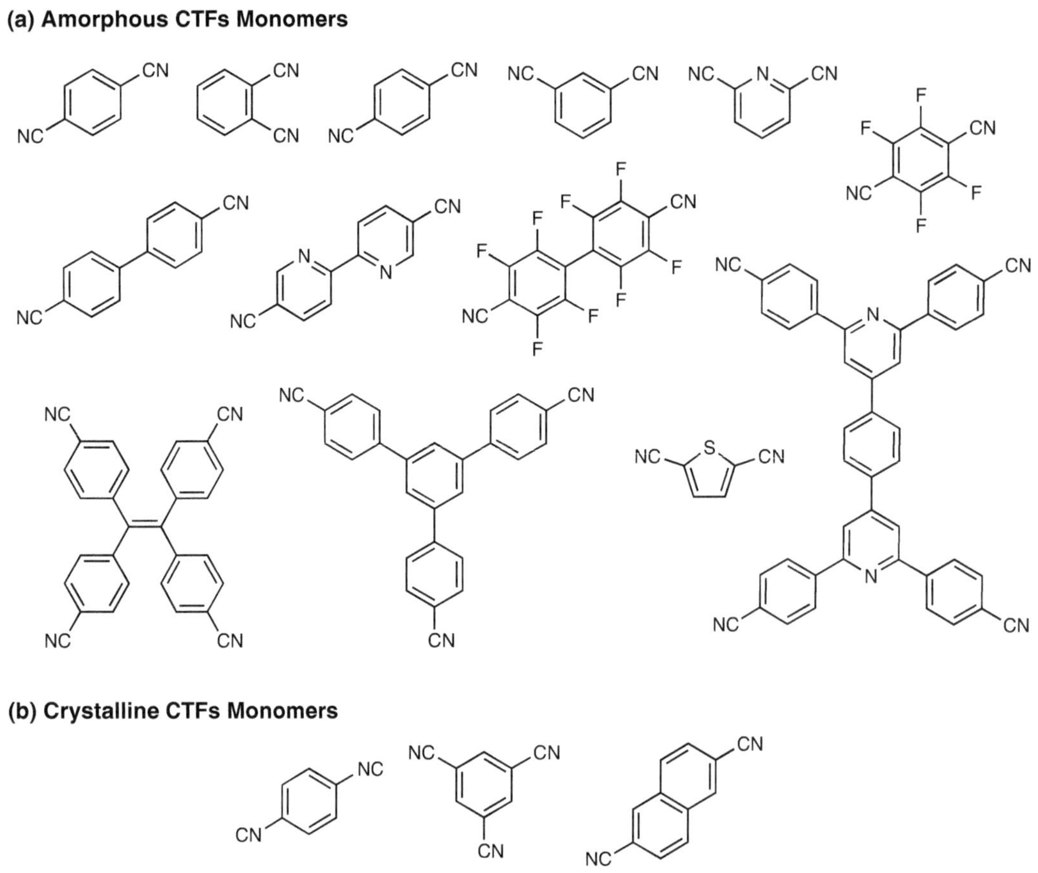

FIGURE 7.1 Monomers for ionothermal synthesis of Amorphous and Crystalline CTFs.

such as synthesizing CTF structures at 400 °C. It is observed that CTF samples usually have black color, and this is due to the partial carbonization and structural variation because of high operating temperature. However, increasing the temperature increases the BET surface area of the structure. Wang et al. recently reported a synthetic scheme of a series of perfluorinated CTFs i.e., FN-CTF-X where X is for synthesis temperature in which molten $ZnCl_2$ ionothermal assistance was given by the use of a building block of tetrafluoroter-ephthalonitrile (Wang, Chen, et al. 2018). The BET surface area of the structure shows that increasing the synthesis temperature increases the pore numbers as shown by FN-CTF-550 (1131 $m^2\ g^{-1}$), FN-CTF-400 (817 $m^2\ g^{-1}$), FN-CTF-900 (2162 $m^2\ g^{-1}$) and CTF-700 (2040 $m^2\ g^{-1}$).

Hence, it is obvious that the crystalline structure of CTF is highly influenced by temperature conditions. Amorphous CTF-2 was obtained at a high temperature of 450 °C, whereas crystalline CTF-2 at 400 °C for 40 h (Bojdys et al. 2010). For a better understanding of this behavior, detailed information has been summarized in Table 7.1, including optimum temperature parameter, crystalline state, pore size properties and optimum $ZnCl_2$ ratio. The drawback of the method is the removal of zinc chloride completely from the CTF structure (Liu, Guo, et al. 2019).

7.2.2 High Temperature Phosphorus Pentoxide (P_2O_5)-Catalyzed Method

Baek was the first scientist who gave the novel strategy for synthesizing CTF structures by the use of phosphorous pentoxide (P_2O_5) as a catalyst (Liu, Guo, et al. 2019). A direct condensation approach was used for condensing the aromatic primary amide group of the ring into s-triazine ring by using

TABLE 7.1
CTFs Polymerization Conditions, Structure and Porosity through Ionothermal Method

CTFs	Monomer	Temperature T [°C]	Time t [h]	Crystalline State	Monomer/ $ZnCl_2$ (mol eq)	BET Surface Area S_{BET} [m^2g^{-1}]	Total Pore Volume V_{total} [cm^3g^{-1}]	Ref.
CTF-1	dicyanobenzene	400	10	Crystalline	1.0	791	0.40	(Kuhn, Antonietti, and Thomas 2008)
CTF-1	dicyanobenzene	400	40	Amorphous	0.1	1123	0.58	
Pt-CTF	2,6-dicyanopyridine	400	4	Amorphous	0.1	730	0.934	(Palkovits, Antonietti, Kuhn, Thomas, and Schüth 2009)
CTF-2	2,6-dicyanonaphthalene	400	40	Crystalline	1.0	90	1.51	(Bojdys et al. 2010)
CTF-1	1,4-dicyanobenzene	600	-	Amorphous	1.0	2814	1.79	(Chan-Thaw et al. 2010)
bipy-CTF	2,2-bipyridine	400	48	Amorphous	0.2	671	0.37	(Hug et al. 2012)
		600	48	Amorphous	0.2	2393	1.19	
		700	48	Amorphous	0.2	3219	1.66	
PCTF-1	tetra(4-cyanophenyl)ethylene	400	20	Amorphous	0.1	2235	1.56	(Bhunia, Vasylyeva, and Janiak 2013)
		400	20	Amorphous	0.2	784	0.76	
PCTF-4	1,3,5-tris(4-cyanophenyl)adamantane	400	48	Amorphous	0.2	1090	0.75	(Bhunia et al. 2013)
PCTF-5	1,3,5,7-tetrakis(4-cyanophenyl) adamantane	400	20	Amorphous	0.1	1183	0.70	
fl-CTF400	9*H*-fluorene-2,7-dicarbonitrile	400	1	Amorphous	0.2	2862	0.78	(Hug et al. 2014)
CTF-a	1,3-dicyanobenzene	600	10	Amorphous	0.2	2439	1.96	(Artz and Palkovits 2015)
CTF-b	2,6-pyridinedicarbonitrile	600	10	Amorphous	0.2	1179	0.64	
CTF-c	1,4-dicyanobenzene	600	10	Amorphous	0.2	2071	1.36	
CTF-d	4,4′-biphenyldicarbonitrile	600	10	Amorphous	0.2	1683	2.63	
CTF-CSU1	3,6-dicyanocarbazole	400	80	Amorphous	1.0	685	0.36	(Yu, Gu, et al. 2018)
CTF-PO71	Pigment Orange-71	400	40	-	0.42	1401	0.84	(Lu et al. 2018)
F_{12}CTF-3	Fluorinated trinitrile monomer	600	10	-	0.1	2085	1.32	(Yang et al. 2019)
2,5-DCP-CTF-0	2,5-dicyanopyridine	400	40	-	0.1	544	0.33	(Li et al. 2019)
2,5-DCP-CTF-1		600	40	-	0.1	1332	0.68	
2,5-DCP-CTF-2		400/600	20/20	-	0.12	1768	0.90	
2,5-DCP-CTF-3		400/600	20/20	-	0.1	1707	089	
2,5-DCP-CTF-4		400/600	20/60	-	0.12	1447	0.94	

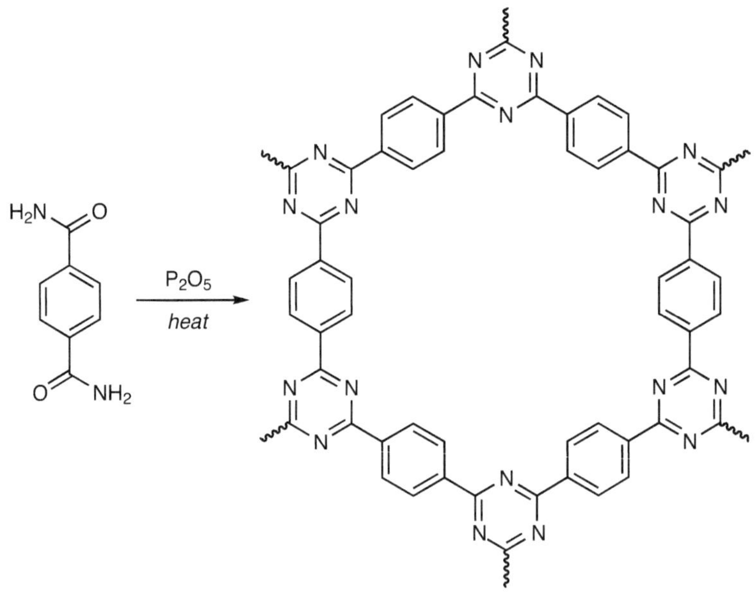

FIGURE 7.2 Scheme of P_2O_5 catalyzed direct condensation of terephthalamide to CTFs.

P_2O_5 as a catalyst (Figure 7.2). This procedure involves two steps to form triazine rings. At first, aromatic primary amide group C(=O)-NH_2 underwent catalyzation into nitrile (C≡N) followed by consequent condensation into s-triazine rings. This resulted in the availability of both nitrile and aromatic amide groups for synthesizing CTFs under the presence of P_2O_5 catalyst. PCTF-1 obtained by this approach showed a higher surface area of 2034 $m^2\ g^{-1}$ and good crystallinity than that of lower surface area CTF-1 (791 $m^2\ g^{-1}$). This method gains substantial importance because of the eco-friendly synthetic approach as well as the most important property to clean the catalyst in comparison to the classical CTF synthetic approach by the use of $ZnCl_2$ catalyst. Moreover, the operating temperature for this approach is also set at 400 °C, which ultimately results in the carbonization of CTF structure (Yu, Mahmood, et al. 2018). Figure 7.2 shows a general scheme for this reaction.

7.2.3 Amidine-based Polycondensation Methods

A mild synthesis strategy was adopted by Tan et al. to prepare CTFs by the use of various amidine units as a monomer and made them react with monomers of aldehydes (Guo, Wu, et al. 2018, Wang, Yang, et al. 2017). The synthetic procedure involves the condensation reaction of amidine dihydrochloride with aldehyde including the formation of Schiff base followed by Michael addition. Under certain conditions, the optimum reaction conditions were achieved by using DMSO solvent and cesium carbonate as a base, making this strategy different from other CTF synthesis procedures. Moreover, a series of CTFs were synthesized by the use of four different monomers of aldehydes, for example, CTF-HUST-1 by using 1,4-benzene-dialdehyde, CTF-HUST-2 by using 4,4'-biphenyl-dialdehyde, CTF-HUST-3 by using tris(4-formylphenyl) amine and CTF-HUST-4 by using 4-formylbiphenyl)-amine. They reacted with each other under a mild temperature condition of 120 °C and in the presence of terephthalamidine dihydrochloride. These CTF structures shared a common feature of yellow color sample, which means they did not undergo carbonization, as the operating temperature was set low. By this method, different structural diversity and design were obtained enriching the toolbox of building blocks, but the CTFs thus obtained are amorphous mainly (Wang, Yang, et al. 2017, Tahir et al. 2019).

In order to synthesize crystalline-ordered CTFs, in-situ oxidation strategy was adopted (Liu, Huang, et al. 2018). Oxidation of alcohols in the presence of DMSO and air resulted in aldehyde formation, specifically in the presence of a suitable base (Liao et al. 2020, Tiwari, Akash, and Bhanage 2015, You et al. 2015). Using this strategy of in-situ oxidation, aldehyde monomers are formed as intermediate by oxidation of alcohol monomers, resulting in a controlled nucleation rate. By this approach, a series of crystalline CTFs were achieved, having good photocatalytic performance and large BET surface area, because of high crystallization. The resulted crystalline CTFs have a synthesis temperature of 180 °C and could be synthesized in an open system, leading to their higher potential to be used for industrial-scale production. The given approach gives hope as well as challenges for better defining CTFs (Tahir et al. 2019).

7.2.4 Superacid Catalyzed Method

Cooper et al. developed a new strategy to cope with the high-temperature issue for synthesizing CTF materials, by using a superacid catalyzed strategy and synthesized these materials using trifluoromethanesulfonic acid (TFMS) at microwave-assisted and room temperature conditions (Ren et al. 2012). TFMS was used as a catalyst in this strategy. By using TFMS-catalyzed reaction, CTF-type polymers were synthesized such as P1M-P6M and P1-P6 at microwave-assisted heating and room temperature, respectively. Room temperature-synthesized CTFs were amorphous by nature. Whereas microwave-assisted synthesized CTFs polymers such as P1M, P2M, and P4M showed less evidence of crystallinity and preferable orientation. Later on, the interfacial polymerization strategy and TFMS vapor methods were modified for synthesizing uncarbonized CTFs for, different applications (Liu, Zan, et al. 2017, Huang et al. 2021, Bi et al. 2015, Huang et al. 2017, Huang et al. 2016, Meier et al. 2017). It can be observed that products obtained using the TFMS-catalyzed method are fluorescent powders, free-flowing, having photoluminescence and absorption spectra depending on monomer choice, in comparison to products obtained from ionothermal method having black colored appearance. CTF-type polymers P1-P6 and P1M-P6M, thus synthesized have surface area varying from 2 to 1152 $m^2 g^{-1}$. Hence, it was also observed that the lower temperature plays a vital role in preparing membranes. Dai et al. used aromatic nitrile trimerization reaction to synthesize fluorescent, porous, triazine-based membranes having a great porosity ratio (Zhu et al. 2012). They reported that cross-linking reaction can be easily catalyzed by using superacids at low temperatures, which allows the facile derivation of porous polymer membrane architecture. A viscous red solution of 4,4'-biphenyldicarbonitrile and CF_3SO_3H was allowed to stir for 1.5 h and allowed to spread on a clean glass dish forming a thin layer. At a heating temperature of 100 °C for approximately 20 mins, a film having a soft thin coating with a surface area of 738 $m^2 g^{-1}$ was achieved successfully. Hence the low temperature assisted the TFMS-catalyzed reaction, thus avoiding the weaker chemical bond cleavage, undesired decomposition and carbonization (Zhu et al. 2012). Unfortunately, the TFMS-catalyzed method gave CTF-type polymer a limited long-range order degree in the polymer materials (Huang et al. 2021).

Lately, the TFMS-catalyzed method was modified by Xu et al., using interfacial polymerization thus successfully synthesizing CTF-1 with high crystallinity having AB stacking structure (Liu, Zan, et al. 2017, Liu, Lyu, et al. 2018, Yang et al. 2020). This method made use of a solvent having good solubility to dissolve monomers such as CH_2Cl_2, and for nitrile cyclotrimerization, CF_3SO_3H was used as a catalyst. The immiscibility of the catalyst and solvent under powerful stirring still resulted in CH_2Cl_2/CF_3SO_3H interfaces, playing a role in achieving crystalline CTF structure. Due to the interface dynamics, the Van der walls epitaxial effect and reaction reversibility were easy to control, which helped in the crystal growth of CTF materials. This reaction was achieved in fewer minutes (tens of mins or less), in comparison to the other reaction requiring hours or even several days. Unfortunately, the low equipment utilization rate and bulk solvent consumption make it difficult for this method to be used on an industrial scale.

7.2.5 Friedel–Crafts Reaction Method

Friedel-Crafts reaction approach was used for synthesizing porous CTF materials, by reaction of aromatic building blocks with cyanuric chloride (Puthiaraj et al. 2015, Dey et al. 2016, Geng et al. 2018, Fu et al. 2018). The catalyst used for this strategy is a strong proton acid such as anhydrous aluminum chloride and methane sulfonic acid, plating a substantial role in polymerization reaction (Xiong et al. 2014, Luo et al. 2017, An, Zheng, et al. 2020). The characterization techniques used for understanding structural and chemical composition, morphology, stability and photoluminescence properties of these materials are solid-state NMR, FE-SEM, HR-TEM, FT-IR, TGA, PXRD, EA, UV-vis and photoluminescence analysis (Puthiaraj, Lee, Zhang, et al. 2016). The method has great advantages including being non-toxic, inexpensive, and more atom efficient and simpler in comparison to other methodologies. High chemical and thermal stability have also been exhibited by these materials, in the range of aqueous and organic solvents. Despite all these advantages, the drawback of this synthesis approach is polymers thus formed are irreversible, as in the dynamic trimerization strategy. It has also provided a variety of synthesized monomers that do not need a cyano functionalized group for further activity. These methods of Friedel-Crafts reactions are found to be neither eco-friendly nor economical, hence do not found application on an industrial scale.

A new mechanochemical method under Friedel-Crafts alkylation was developed by Borchardt et al. for synthesizing CTFs materials (Troschke et al. 2017). In this method, for the triazine unit, cyanuric chloride was used, which acted as a nucleophilic reagent having electron-rich density, $ZnCl_2$ as a bulking agent and $AlCl_3$ was used in stoichiometric ratio as an activating agent, respectively. The energy impact on reactants is determined by density, in mechanochemical reactions. So, to achieve high density, tungsten carbide balls were used which played a part in the polymerization reaction. Using the given route produces a series of different CTF products having different surface area ranges such as CTF-AC, CTF-BZ, CTF-CBZ, CTF-TPM and CTF-NT having a surface area of 470 $m^2\ g^{-1}$, 170 $m^2\ g^{-1}$, 570 $m^2\ g^{-1}$, 590 $m^2\ g^{-1}$ and 110 $m^2\ g^{-1}$, respectively. Furthermore, mechanochemistry gained importance because of its time-efficient production, solvent-free property and scalability in comparison to other common synthesis routes for CTFs. The method paved the way for synthesizing new generation CTF materials (Troschke et al. 2017).

7.3 APPLICATIONS OF CTFs

The CTF materials being considered as a type of porous organic polymer (POP), possesses good thermal and chemical stability, high BET surface area, having various applications such as gas separation and absorption, energy storage, photocatalysis and heterogeneous catalysis, etc. CTF materials' properties and performance are highly dependent on diversified building block and synthetic methodologies used. Until now no efforts have been made in optimizing the properties of CTFs (Krishnaraj et al. 2020, Zhang and Jin 2019, Wang et al. 2019, Park et al. 2017).

7.3.1 Adsorption and Separation

CTFs are promising materials having high porosity and BET surface area. They are used for carrying out various processes such as absorption and separation of metal ions, organic pollutants and various gases. It is noteworthy that the presence of triazine ring in CTF structure offers various advantages including wastewater treatment and CO_2 capture and separation (CSS).

7.3.1.1 CO_2 Capture and Separation

The performance of CO_2 absorption and separation is determined by CTFs networks' physical and chemical environment. CO_2 pore size is 0.33 nm; hence CO_2 storage is prominent in porous materials with microporous and ultra-microporous structures with a diameter of <1 nm. Therefore, the ability to capture CO_2 in CTFs is largely enhanced by these high degrees of microporous bodies and ultra-microporous bodies (Wang, Huang, et al. 2016). CTFs achieved permanent porosity during reaction procedures; hence, the porosity of the material would be highly affected by the reaction factors and synthetic pathways (Wang, Huang, et al. 2016, Kuhn, Antonietti, and Thomas 2008). Janiak et al. reported the difference in CTF material's pore size when prepared with different strategies. CTFs prepared using the TFMS method show reduced pore size than CTFs synthesized by the ionothermal method (Bhunia et al. 2013). It is noteworthy that changing the reaction conditions such as temperature and catalyst amount resulted in CTFs with a difference in pore size distribution, though using the same monomers for starting the reaction, resulted in varying CO_2 uptake capacity of the materials. These results are summarized in Table 7.2 (Hug et al. 2014, Bhunia, Vasylyeva, and Janiak 2013, Wang, Leus, et al. 2017, Jena et al. 2018).

Hence, it is found that CO_2 capture can be increased by increasing the micropore size ratio and BET surface area of the CTF materials, leading to enhanced building blocks length. Later on, a series of CTFs were synthesized by Yang et al., having triphenylamine core and varying the length of branched arm as a building block, resulting in decreased CO_2 capture performance with increasing arm length (Gu et al. 2015). Another approach used to improve the pore size of POPs was the non-coplanar and twisted topology of the materials. Jian et al. carried out phthalazinone-based CTF material synthesis and found out that these materials have 1845 $m^2\ g^{-1}$ surface area and CO_2 uptake capacity of 17 wt%. CTFs networks could be further modified increasing cross-linking density or function group. Hence, CO_2 capture and porosity of CTFs can be tuned by changing the ratio of different monomers (Dey et al. 2019, Dey, Bhunia, Breitzke, et al. 2017, Dey, Bhunia, Boldog, and Janiak 2017).

Another decisive factor of porous CTFs is their chemical environment, which affects the performance of CO_2 uptake. Few experiments were carried out for developing the understanding of functional elements in CTFs. A group of scientists reported CTFs having three nitrogens based on phenyl, pyrimidine, lutidine and bipyridine units. Results have shown that bipyridine-CTFs have CO_2 uptake capacity of 5.58 mmol g^{-1} at 1 bar and 273 K temperature, being the second-highest POPs reported value. Whereas, highest among POPs is pyridine-based CTFs achieving high CO_2 selectivity toward its N_2, according to ideal absorbed solution theory (Hug et al. 2015). Furthermore, CO_2 uptake activity is also enhanced by a fluorine atom, as micropore fraction is increased by the release of fluorous at high temperature (Dey, Bhunia, Boldog, Janiak, et al. 2017). Han et al. reported, for one thing, polar C–F bond enhances the CO_2 absorption thermodynamically; for another, the ultra-micropores amount is enhanced by the release of fluorine, offering increased kinetic selectivity as well as potential to absorb gas for CO_2-N_2 separation. CO_2 uptake target of 26.3% (273K, 1 bar) was achieved by Voort et al. by synthesizing CTFs containing fluorine atoms (Xu et al. 2018). Later researches showed that metal complexes of Pd and Zn also have the ability to increase CO_2 uptake for CTFs (Kaleeswaran et al. 2017). Scientists have reported the Zn-porphyrin ability to uptake CO_2 is 13.9 wt% at 273 K, 1 bar, as metal complex has provided CTFs with more active sites (Liu, Li, et al. 2013).

Another smart strategy for increasing CO_2 uptake is employing building blocks having charge or high polarity (Xu et al. 2018). Coskun et al. were the first to report that high CO_2 uptake affinity was achieved by charged CTFs, which was 26.12 wt% at conditions of 1 bar, 273 K and Q_{st} (isosteric heat of absorption) at zero loadings was found to be more than 40 Kj mol^{-1} (Buyukcakir et al. 2017). Likewise, for increasing CO_2 and CTFs interaction, researches were also conducted. Coskun et al. activated the CTFs by using KOH and results exhibited an increase of 26.4 wt% in modified CTFs for CO_2 capture, which is a higher value than 11.9 wt% for pristine-CTF at same conditions (Lee, Talapaneni, and Coskun 2017). Yu and his co-workers reported the use of acetohydrazine and carboxyl groups using pre-synthesis strategy and using post-synthesis strategy into CTF structures. Better performance was exhibited by the carboxyl group coupled post-modified strategy achieving wt% of 5.7% and 15.95% (Fu et al. 2017, Dong, Wang, and Wang 2020).

TABLE 7.2
The CTFs Capacity for CO_2 Uptake

Monomer	CTFs	Synthesis condition: Method	Synthesis condition: Temperature (°C)	Synthesis condition: Monomer:$ZnCl_2$	CO_2 capture (mmol/g): 273 K	CO_2 capture (mmol/g): 298 K	BET surface area (m^2 g^{-1}) S_{BET}	Ref.
1,3,6,8-cyanopyrene	Pyrene-CTF-10	Ionothermal method	500	1:10	5.10	2.82	819	(Mohamed, El-Mahdy, et al. 2020)
	Pyrene-CTF-20			1:20	3.43	2.54	1019	
7,7,8,8-tetracyanoquinodimethane	TCNQ-CTF-900	Amidine Poly-condensation trimerization	400	1:15	5.99	3.41	4000	(Deng et al. 2020)
tetra(4-cyanophenyl)ethylene	PCTF-1	Ionothermal method	400	1:10	14.4	8.8	2235	(Bhunia, Vasylyeva, and Janiak 2013)
	PCTF-2			1:5	8.2	4.8	784	
4,4′,4″,4″′-methanetetrayltetrabenzonitrile	P6	TFMS catalyzed method	Room temp.	-	14.8 wt%	-	1152	(Ren et al. 2012)
	P6M		Microwave	-	18.4 wt%	-	947	
Phthalazinone and triphenylamine	PHCTF-8	Ionothermal method	400	1:15	17.8 wt%	11.2 wt%		(Yuan et al. 2018)
1,4-Phenylenediamine	CTF	Poly-condensation method	400	-	-	3.995	-	(Mukhtar et al. 2021)
	CTF-NH			-	-	7.780	-	
9H-fluorene-2,7-dicarbonitrile	Fl-CTF-300	Ionothermal method	300	2:1	5.46	3.38	15	(Hug et al. 2014)
	Fl-CTF-350		350	2:1	18.4 wt%	10.1 wt%	1235	
2,2′,3,3′,5,5′,6,6′-octafluoro-4,4′-biphenyldicarbonitrile	F-DCBP-CTF-1	Ionothermal method	400	1:5	26.31 wt%	16.81 wt%	1574	(Xu et al. 2018)

1,3-Bis(4-benzocynide)imidazoline	CTF-Cl-1	Ionothermal method	400	1:5	40.0 wt%	26.3 wt%	516	(Xu et al. 2019)
	CTF-Cl-2			1:10	43.4 wt%	28.1 wt%	599	
	CTF-PF-3			1:5	41.1 wt%	27.0 wt%	590	
	CTF-PF-4			1:10	44.7 wt%	28.5 wt%	889	
4,4′-malonyldibenzonitrile	acac-CTFs-5-400	Ionothermal method	400	1:5	2.87	1.89	1131	(Jena et al. 2018)
	acac-CTFs-10-400		400	1:10	3.06	1.96	1150	
	acac-CTFs-5-500		500	1:5	3.30	1.97	1556	
	acac-CTFs-10-500		500	1:10	3.16	1.91	1626	
1,4-benzene-dialdehye / Benzene-1,4-dicarboximidamide	CTF-HUST-1	Amidine based polycondensation	120	-	12.56 wt%	-	663	
Terephthalamide	pCTF-1	Phosphorus pentoxide (P_2O_5)-catalyzed	High temp.	-	22 wt%	-	2034	(Xu et al. 2018)
1,1′-Bis(cyanomethyl)-[4,4′-bipyridine]-1,1′- diium Chloride	ICTF-Cl	Ionothermal method	400	1:1	5.7	3.4	751	(Zhu et al. 2020)
Triphenylamine core	PCTF-4	Ionothermal method	400	1:10	20.5 wt%	12.6 wt%	1404	(Gu et al. 2015)
1,4-dicyanobenzene	CaCTF-1-700	Ionothermal method			26.4 wt%	15.6 wt%	2367	(Lee, Talapaneni, and Coskun 2017)

7.3.1.2 The Removal of Pollutants

A promising technique called solid-phase extraction (SPE) is based on easy recovery for adsorbents, high separation efficiency and simple operation (Heydari et al. 2021). A great potential is offered to CTFs by having high porosity and BET surface area for enhancing absorption ability. π-conjugated systems help in increasing CTFs interaction with aromatic compounds, whereas triazine ring present in them enhances their bond with metals, enhancing CTFs absorption capacity and efficiency as absorbent (Khan, Ahmad, and Ahmad 2020). Li and coworkers designed CTFs with different elements for absorption of uranium (U), and the results obtained exhibited a strong affinity of U with both designs of CTFs. The selectivity of U with twelve different types of cations can go to 54% in water, whereas in large conjugated systems, different elements causing different effects were weakened (Bai et al. 2015). According to the reported literature, from the aqueous solution the sodium dodecyl sulfate and polyglycol ether can reach an absorption capacity of up to 4 g g^{-1}, a value 20 times more than carbon black (Bhunia et al. 2015). Grossman and his coworkers fabricated a 2D membrane of CTF, for desalinated seawater, and the results thus achieved showed two order higher magnitude performance of these permeable membranes of CTFs than polyamide membranes. Furthermore, CTF-0 2D membranes were also used for purification of H2 and helium separation (Wang, Li, et al. 2016). Moreover, plastic packaging materials made of phthalate esters, pharmaceutical antibiotics and other organic dyes are efficiently absorbed by CTF (Liu, Zhou, et al. 2017, Liu, Chen, et al. 2013, An, Liu, et al. 2020). Further findings demonstrated that first reported POPs i.e., CTFs have the ability to separate acetylene and ethylene (Lu et al. 2018).

7.3.2 Heterogeneous Catalysis

Out of all other applications, CTFs are a promising class for catalyst support and catalysis due to high porosity and surface area, which are helpful in catalytic processes for the transport of mass (Puthiaraj, Lee, Zhang, et al. 2016, Chang et al. 2015, He et al. 2017). Under harsh conditions, high chemical and thermal stability make CTFs' performance a stable cycle. Nitrogen-rich content helps in stabilizing loading metal nanoparticles, resulting in durability and good catalytic efficiency of CTF materials (Qian, Wang, and Zhang 2021, Chan-Thaw et al. 2010, 2011, Kamiya 2020).

The porosity of CTFs impacts the catalytic performance, therefore reaction conditions and building blocks were undergone detailed investigation (Chan-Thaw et al. 2012, Zhao et al. 2018, Kaczmarek et al. 2020, Qian, Wang, and Zhang 2021). Gascon et al. were able to synthesize CTFs using monomers of 4,4′-biphenyldicarbonitrile and 2,6-pyridinedicarbonitrile in 2:1 ratio, which were then coordinated as Ir^{III} CP^{*} with a turnover frequency (TOF) of 27000 h^{-1}, in order to make hydrogen from formic acid (Bhanja et al. 2017).

In addition, the catalytic performance of CTFs synthesized from other porous materials was also investigated. Gascon and coworkers conducted catalytic experiments by preparing thin-coated CTFs on monoliths support, with the loading of metals like Ir^{III} or Pt (Bavykina et al. 2017). Monolith-supported CTFs showed good performance than powdered form CTFs due to accelerated mass transport through thin films and monolith channels. Furthermore, CTFs can also act as a template. For example, y-alumina which is a CTF-based template was fabricated exhibiting microporous crystallinity and a surface area of 349 m^2 g^{-1} (Zhang et al. 2017). In catalytic processes, recycling of the catalyst should also be considered. Hence in order to recycle the catalyst, Gascon et al. used polyimide Matrimid as a binding agent and synthesized spherical bipyridine-CTFs having a large size (Bavykina et al. 2016). These materials were combined with Ir^{III} CP^{*} and then used in CO_2 hydrogenation to formic acid resulting in good recyclability and high activity.

An important and more convenient reaction is the transfer hydrogenation reaction in the presence of a non-H_2 source between the starting materials (Wang and Astruc 2015). Chen and coworkers used 4 wt% Pd as catalyst loaded on CTF-1, to carry out the hydrogenation process (He et al. 2015). The results thus obtained showed 3.6 times higher catalytic efficiency in comparison to Pd/activated carbon. These intensified results of hydrogenation and heterolytic dihydrogen splitting were obtained because of electron-rich nitrogen with basic CTF support. Another type of TH reaction is also conducted on CTF support called Hydroformylation. The catalytic performance was determined after several efforts being made such as CO_2 hydrogenation to formates in presence of Ir/CTFs, (Park et al. 2015) 1-octene catalysis with Rh/CTFs (Liu, Guo, et al. 2019, Pilaski et al. 2016) and formation of alcohol by carbonyl compound in presence of Ir/CTFs (Sudakar et al. 2016).

CTFs offer obvious advantages owing to their high affinity to CO_2 for conversion and fixation of CO_2 (Buyukcakir et al. 2017). Carbene-based N-heterocyclic CTFs in coordination with Ir were synthesized by Yoon and co-workers. Catalyst-catalyzed CO_2 conversion into formate gave the highest reported values of TON which is 24300 h^{-1} and TOF of 16000 h^{-1} (Gunasekar et al. 2017). Additionally, the chemical fixation of CO_2 to epoxides was presented by pristine CTFs. Obvious differences were seen by loading different metals on different CTFs, with the same starting materials (Rajendiran, Park, et al. 2017, Rajendiran, Natarajan, and Yoon 2017).

One of the big challenges is the oxidation of methane to form methanol, because to make CO_2 the methane undergoes overoxidation readily by breakage the CH_3-H bond which required 435 KJ mol^{-1} of energy. Thus, such reactions require selective and active catalysts for addressing this problem (ten Brink, Arends, and Sheldon 2000, Muehlhofer, Strassner, and Herrmann 2002). This problem was solved by using Pt/pyridine-CTFs (Palkovits, Antonietti, Kuhn, Thomas, and Schüth 2009). Metal-loaded CTFs are important in chemical reactions and also find their application in C-C cross-coupling reactions (Yin and Liebscher 2007, Lin et al. 2015). Using Pd as an oxidation catalyst and loading it on CTFs, a high conversion of 100% is achieved for CO to CO_2, when used in Suzuki-miyaura coupling and Heck-type reactions, resulting in good selectivity and activity (Mullangi et al. 2015, Baba and Miyaji 2020).

7.3.3 Applications for Energy Storage and Conversion

7.3.3.1 Metal-Ion Batteries

For addressing the challenges in poor cycle performance and energy density, the development of metal-ion batteries such as lithium-ion battery (LIB) and sodium-ion battery (SIB) is of crucial importance. Based on rich nitrogen content and porosity, CTFs also exhibited great performance in ion batteries (Wu et al. 2021, Wang et al. 2021, Li et al. 2016).

CTF-loaded sulfur worked as a sulfur cathode in nanoporous frameworks, to improve cyclability and solve the soluble lithium polysulfide loss for lithium-sulfur batteries. The results showed that in CTF-1/S at 155 °C, the sulfur content can reach 34%, whereas it reaches 97% in the second cycle due to charge-discharge coulombic efficiency (Liao et al. 2014). Furthermore, nucleophilic aromatic substitution (S_NAr) reaction can enhance the sulfur content. This reaction was carried out between sulfur and perfluoroaryl units and was supposed to increase the S-content by 86% in SF-CTF (Je et al. 2017).

CTFs also find their application in Li batteries as the anode material. The low conductivity of silicon (Si) anodes limited their use in LIBs as a promising anode (Zhu et al. 2019). The Si-coated on carbonized CTF was presented as core-shell structures to be used in lithium-ion batteries as an anode, showing a capacity of 1390 mAh g^{-1} for conductive anode at 0.5 A g^{-1}, whereas a capacity retention of 107% was also observed after 200 cycles (Zhu, Yang, et al. 2017). Graphene-coupled CTFs undergo pyrolysis to synthesize sandwich-like 2D porous carbon (G-PCs) materials for their application in energy storage. These G-PCs exhibited high performance because of abundant nitrogen content and high surface area, in both LIBs and SIBs giving 235 mAh g^{-1} capacitance at 5 A g^{-1} for 3000 cycles and 138 mAh g^{-1} at 1 A g^{-1} for 500 cycles, respectively (Zhu, Zhuang, et al. 2017, He et al. 2019).

7.3.3.2 Supercapacitors

In supercapacitors, energy is stored at the electrode-electrolyte interface during the process of adsorption-desorption reversible reaction of electrolyte ions offering the potential to CTFs owing to rich nitrogen content and porosity (Kötz and Carlen 2000, Sakaushi et al. 2012, Wu et al. 2021, Bhanja et al. 2017). 2D CTFs were synthesized through ionothermal method by Hao et al. using monomers of 1,3,5 tricyanobenzene, and they provide an understanding of micropore and nitrogen effect on ionic-liquid based supercapacitors (Hao, Ning, et al. 2015). The results put forward the idea of nitrogen impact on electrolytes materials relative permittivity and improving supercapacitors performance (Troschke et al. 2020).

Furthermore, trimerization of tetracyanoquinodimethane was carried out using ionothermal method by (Zhang and Jin 2019). After carrying out the reaction at different temperatures, results suggested that CTFs have rich nitrogen content and high surface area, and increasing heating temperature increases the conductivity of CTFs. Out of all designed molecules, TCNQ-CTF-800 showed a capacitance of 383 F g^{-1} at a current density of 0.1 A g^{-1}showing great performance of this CTF under alkaline conditions (Zhang and Jin 2019).

7.3.4 Electrocatalysis

To design the materials, the use of porous CTFs as electrocatalysts is another approach for converting the starting material into a work-full designing material. But, the poor conductivity of CTFs inhibited their use to carry out electrocatalysis (Guo et al. 2013). First-ever reported reaction that catalyzed CTFs as electrocatalysts was the oxygen reduction reaction (ORR) (Kamiya et al. 2014). Conductive carbon nanoparticles (CP) coexisted with CTFs enhance their conductivity through in-situ polymerization, then Pt coupled with hybrid material was used as catalyst, exhibiting good ORR activity showing high methanol tolerance. The Pt anode catalyst works for the ORR as well as the hydrogen oxidation reaction (HOR), for the fuel cell system of the polymer electrolyte (Takeguchi et al. 2012, Saida et al. 2012). By varying the Pt-loaded content on CTF/CP, the results obtained show that the ORR activity became very low by increasing the Pt contents from 0.29 wt% to 2.8 wt%, thus inhibiting the corrosive degradation of the cathode (Kamai et al. 2016).

Furthermore, Pt-free catalysts having low cost were studied for ORR applications. Kamiya et al. used Cu-CTF/CP in the ORR and achieved good results with an onset potential of 810 mV and high stability, in comparison to other Cu-based catalysts used in electrocatalysis (Iwase et al. 2015). Li et al. coupled iron (Fe) with CTFs and synthesized electrocatalyst Fe-Nx-C heated at 700 °C (Tong et al. 2017). FB7 electrocatalyst out of other Fe-based catalysts exhibited excellent performance with an onset potential of 14 mV in $HClO_4$ solution of 0.1 M, slightly lower potential than Pt/C available commercially. Later on, Co_3O_4 supported on CTFs was also prepared through the hydrothermal method showing higher methanol tolerance and better stability in ORR, in comparison to Pt/C available commercially (Chen et al. 2018).

On other hand, the conductivity of CTFs was improved by using the carbonation approach. In this approach, carbonized CTFs were used as electrocatalysts without using metal support. The bottom-up approach was used to synthesize thermalized triazine-based frameworks (TTFs) having 3D orientation (Hao, Zhang, et al. 2015). The quaternary nitrogen provided high ORR activity, whereas the electrocatalytic activity of TFF was increased by providing a high temperature. In this work, a heteroatom doping approach was also used and results showed that fluorine gave high ORR activity and bromine improved the selectivity (Xing et al. 2020).

7.3.5 Photocatalysis

CTFs are promising candidates for photocatalysis due to their robust conjugated structure, nitrogen-rich content and absorption of light in the visible range. CTFs are used to carry out various photocatalytic reactions such as the photocatalytic conversion of chemical species, photodegradation of organic pollutants, photoreduction of CO_2 and photocatalytic water splitting, etc. (Chao et al. 2021, Zhu et al. 2018, Niu et al. 2014, Yang et al. 2018). Table 7.3 summarizes CTFs photocatalysts reported in the literature for hydrogen evolution by water splitting. By using an organic catalyst such as g-C_3N_4 for photocatalytic hydrogen production through water splitting has become a hotspot application for hydrogen evolution under visible light (Wang, Zhang, et al. 2018, Wang, Maeda,

TABLE 7.3
Summary of Hydrogen Evolution of CTFs Photocatalyst by Water Splitting

		Synthesis Condition		Photocatalytic Condition					
Building Block	**CTFs**	**Method Used**	**Temperature (°C)**	**wt% Pt**	**Visible light illumination (nm)**	**Photocatalyst used (mg)**	**Vol% TEOA**	**HER (mol h^{-1} g^{-1})**	**Ref.**
Terephthalonitrile	CTF-1	TFMS catalyzed	Room temp.	1	>420	80	14	250	(Bi et al. 2015)
1,4- dicyanobenzene	PTO-300-15	Iono-theromal	300	2.2	>420	10	10	1076	(Schwinghammer et al. 2015)
4,4'-biphenyldicarbonitrile	CTF-2	TFMS catalyzed	Room temp.	3	>420	25	10	296	(Meier et al. 2017)
1,4- dicyanobenzene	CTF-1_10 min	TFMS catalyzed Iono-thermal	40 400	3	>420	25	10	1072	(Kuecken et al. 2017)
tetra(4-cyanophenyl) ethylene	CTFs	TFMS catalyzed	0	3	-	10	10	118	(Bhunia et al. 2016)
CTF-1	CTF-BP-Pt	Photodeposition	Room temp.	0.05	>420	10	-	614.6	(Zheng et al. 2020)
1,4- dicyanobenzene	CTF-S_{10}	TFMS catalyzed	Room temp.	1	>420	20	10	2000	(Tao et al. 2016)
1,4-benzene-dialdehye / Benzene-1,4-dicarboximidamide	CTF-HUST-1	Amidine-based polycondensation	120	3	>420	50	10	1460	(Wang, Yang, et al. 2017)
1,4-Benzenedimethanol / Benzene-1,4-dicarboximidamide	CTF-C1	Amidine-based polycondensation	180	3	>420	50	10	5100	(Liu, Huang, et al. 2018)
p-xylylenediamine / Benzene-1,4-dicarboximidamide	CTF-HUST-A1	Amidine-based polycondensation	400	3	>420	-	-	9200	(Zhang, Cheng, et al. 2020)
1,4-dicyanobenzene	MoS2/CTF-T1	TFMS catalyzed	Room temp.	1	>420	20	10	560	(Jiang et al. 2018)
1,4-dicyanobenzene	CTF-1-100 w	TFMS catalyzed (microwave)	-	3	>420	50	10	5500	(Xie et al. 2018)
4,4′-biphenyldicarbonitrile	CTP-2	TFMS catalyzed	Room temp.	3	>420	50	10	500	(Lan et al. 2019)

Thomas, et al. 2009, Wang, Maeda, Chen, et al. 2009). CTF has a suitable bandgap and it is indicated by theoretical results that it can be used for producing hydrogen under visible light (Tan et al. 2021, Bi et al. 2015). However, low photocatalytic efficiency has resulted from the destruction of semiconductor characteristics and conjugated structure of CTFs by ionothermal method with carbonizing CTFs. Hence photocatalytic performance can be enhanced by decreasing the carbonization of CTFs and using strategies like ionothermal method and TFMS catalytic method (Meier et al. 2017, Schwinghammer et al. 2015, Kuecken et al. 2017, Guan et al. 2021, Bhunia et al. 2016, Jiang et al. 2018). Amidine polycondensation approach was also used and the conjugated structure of CTF was retained and high hydrogen evolution rate (HER) results were achieved (Wang, Yang, et al. 2017, Zhang, Wang, et al. 2020, Gong et al. 2021).

Heteroatom doping changes the distribution of charges and provides active sites to the system. As reported in the literature, a high HER performance of 2000 $\mu mol\ h^{-1}$ was exhibited by sulfur-doped CTFs with Pt as cocatalyst under visible light irradiation (Tao et al. 2016). High CTF-S_x catalytic performance is supposed to increase fast electrons and holes separation, low recombination of charges and increased absorption ability. Similarly, ordered structure also promotes increased photocatalytic performance and charge transfer (Tan et al. 2021, Liu, Huang, et al. 2018).

Building blocks also play a crucial part in changing photocatalytic performance in conjugated CTFs structures. For enhancing charge separation and electron transfer, the donor-acceptor strategy is effective and different donors gave different photocatalytic performances (Guo, Niu, et al. 2018). The donor ability toward photocatalysis was tuned by substituting 9-position with different heteroatoms (N, S, O) using the same triazine acceptor rings. The resulting CTF-N gave 10760 $\mu mol\ h^{-1}$ HER activity which is the highest till then and also showed stability under visible light irradiation (Guo, Niu, et al. 2018).

Another half-reaction much difficult than hydrogen evolution reaction is photocatalytic oxygen evolution used for water splitting (Jena et al. 2020). Oxidation of H_2O into O_2 with a higher HOMO value of 1.23 eV could be dynamically possible. Wu and coworkers reported the use of RuO_2 as a cocatalyst for CTFs adopted to water oxides (Liu et al. 2014). Great potential for water splitting was demonstrated by using Co as cocatalyst loaded on CTFs by oxygen evolution reaction conducted by Wang et al. Similarly, Tang et al. conducted oxygen evolution reaction on CTFs through microwave chemistry, exhibiting OER of 140 $\mu mol\ h^{-1}$, a value much higher than g-C_3N_4. Photocatalysts such as CTF-rhenium hybrid could also be used to carry out photoreduction of CO_2 in a solid-gas system into CO, another useful application of CTFs (Mukhtar et al. 2021, Xu et al. 2018, Qian, Wang, and Zhang 2021).

Mesoporous silica such as SBA15 was also used as a template for understanding the photocatalytic behavior of CTFs. Zhang et al. were able to synthesize CTF-Th@SBA15 under TfOH vapor using 2,5-dicyanothiophene monomer, the resulting CTF material thus formed was used as a catalyst and oxidized benzyl alcohol to benzaldehyde giving approximately 100% selectivity and efficiency. Similarly, asymmetric CTFs were also fabricated using the same procedure and benzophosphole oxides were produced with these asy-CTFs giving higher photocatalytic performance. This was because the coexistence of different energy levels makes a cascade for efficient electron transfer (Wu et al. 2020, Huang et al. 2018).

7.3.6 Other Applications of CTFs

CTFs having inherent porous network property enable them to be used for other applications too. They can be used as a drug carrier in which the release of the drug is controlled by tunning chemical environment or pore size of the CTFs. The literature reported for this exhibit that quercetin and ibuprofen have great drug delivery performance (Vyas et al. 2016, Luo et al. 2017). Furthermore, CTFs can also be used as sensors owing to their fluorescence property (Wang, Zhang, et al. 2016). The fluorescence intensity of CTFs is quenched by Hg^{2+} owing to their nitroaromatic analyte adjunction (Luo et al. 2017, Geng et al. 2018, Baig et al. 2020, Mohamed, Zhang, et al. 2020, Zhang, Xue, et al. 2021, Bhunia et al. 2016).

REFERENCES

Algara-Siller, G., N. Severin, S. Y. Chong, T. Bjorkman, R. G. Palgrave, A. Laybourn, M. Antonietti, Y. Z. Khimyak, A. V. Krasheninnikov, J. P. Rabe, U. Kaiser, A. I. Cooper, A. Thomas, and M. J. Bojdys. 2014. Triazine-based graphitic carbon nitride: a two-dimensional semiconductor. *Angewandte Chemie International Edition* 53 (29):7450–5. doi: 10.1002/anie.201402191.

An, Fengxia, Jingliang Liu, Zhaoyi Xu, and Shourong Zheng. 2020. Efficient removal of three dyes using porous covalent triazine frameworks: adsorption mechanism and role of pore distribution. *Water Science Technology* 82 (12):3023–3031.

An, Wan-Kai, Shi-Jia Zheng, Ya-Nan Du, San-Yuan Ding, Zhi-Jun Li, Song Jiang, Yuchen Qin, Xiaobiao Liu, Pi-Feng Wei, and Zhan-Qi Cao. 2020. Thiophene-embedded conjugated microporous polymers for photocatalysis. *Catalysis Science Technology* 10 (15):5171–5180.

Artz, Jens, and Regina Palkovits. 2015. Base-free aqueous-phase oxidation of 5-hydroxymethylfurfural over ruthenium catalysts supported on covalent triazine frameworks. *ChemSusChem* 8 (22):3832–3838.

Baba, Toshihide, and Akimitsu Miyaji. 2020. "Heterogeneous and Homogeneous Catalytic Partial Oxidations of Methane to Methanol and Its Derivatives." In *Catalysis and the Mechanism of Methane Conversion to Chemicals*, 53–71. Springer.

Bai, Chiyao, Meicheng Zhang, Bo Li, Yin Tian, Shuang Zhang, Xiaosheng Zhao, Yang Li, Lei Wang, Lijian Ma, and Shoujian Li. 2015. Three novel triazine-based materials with different O/S/N set of donor atoms: one-step preparation and comparison of their capability in selective separation of uranium. *Journal of Hazardous Materials* 300:368–377.

Baig, Noorullah, Suchetha Shetty, Saleh Al-Mousawi, and Bassam Alameddine. 2020. Synthesis of conjugated polymers via cyclopentannulation reaction: promising materials for iodine adsorption. *Polymer Chemistry* 11 (17):3066–3074.

Bavykina, Anastasiya V, Alma I Olivos-Suarez, Dmitrii Osadchii, Rahul Valecha, Robert Franz, Michiel Makkee, Freek Kapteijn, and Jorge Gascon. 2017. Facile method for the preparation of covalent triazine framework coated monoliths as catalyst support: applications in c1 catalysis. *ACS Applied Materials Interfaces* 9 (31):26060–26065.

Bavykina, Anastasiya V, Elena Rozhko, Maarten G Goesten, Tim Wezendonk, Beatriz Seoane, Freek Kapteijn, Michiel Makkee, and Jorge Gascon. 2016. Shaping covalent triazine frameworks for the hydrogenation of carbon dioxide to formic acid. *ChemCatChem* 8 (13):2217–2221.

Bhanja, Piyali, Kousik Bhunia, Sabuj K Das, Debabrata Pradhan, Ryuto Kimura, Yuh Hijikata, Stephan Irle, and Asim Bhaumik. 2017. A new triazine-based covalent organic framework for high-performance capacitive energy storage. *ChemSusChem* 10 (5):921–929.

Bhunia, A., S. Dey, M. Bous, C. Zhang, W. von Rybinski, and C. Janiak. 2015. High adsorptive properties of covalent triazine-based frameworks (CTFs) for surfactants from aqueous solution. *Chem Commun (Camb)* 51 (3):484–6. doi: 10.1039/c4cc06393g.

Bhunia, A., V. Vasylyeva, and C. Janiak. 2013. From a supramolecular tetranitrile to a porous covalent triazine-based framework with high gas uptake capacities. *Chem Commun (Camb)* 49 (38):3961–3. doi: 10.1039/c3cc41382a.

Bhunia, Asamanjoy, Ishtvan Boldog, Andreas Möller, and Christoph Janiak. 2013. Highly stable nanoporous covalent triazine-based frameworks with an adamantane core for carbon dioxide sorption and separation. *Journal of Materials Chemistry A* 1 (47):14990–14999.

Bhunia, Asamanjoy, Dolores Esquivel, Subarna Dey, Ricardo Fernández-Terán, Yasutomo Goto, Shinji Inagaki, Pascal Van Der Voort, and Christoph Janiak. 2016. A photoluminescent covalent triazine framework: CO_2 adsorption, light-driven hydrogen evolution and sensing of nitroaromatics. *Journal of Materials Chemistry A* 4 (35):13450–13457.

Bi, Jinhong, Wei Fang, Liuyi Li, Jinyun Wang, Shijing Liang, Yunhui He, Minghua Liu, and Ling Wu. 2015. Covalent Triazine-Based Frameworks as Visible Light Photocatalysts for the Splitting of Water. *Macromolecular Rapid Communications* 36 (20):1799–1805.

Bojdys, Michael J, Jekaterina Jeromenok, Arne Thomas, and Markus Antonietti. 2010. Rational extension of the family of layered, covalent, triazine-based frameworks with regular porosity. *Advanced Materials* 22 (19):2202–2205.

Bügel, Stefanie, Alex Spieß, and Christoph Janiak. 2021. Covalent triazine framework CTF-fluorene as porous filler material in mixed matrix membranes for CO2/CH4 separation. *Microporous Mesoporous Materials* 316:110941.

Buyukcakir, Onur, Sang Hyun Je, Siddulu Naidu Talapaneni, Daeok Kim, Ali Coskun, and interfaces. 2017. Charged covalent triazine frameworks for CO2 capture and conversion. *ACS Applied Materials Interfaces* 9 (8):7209–7216.

Chan-Thaw, C. E., A. Villa, G. M. Veith, K. Kailasam, L. A. Adamczyk, R. R. Unocic, L. Prati, and A. Thomas. 2012. Influence of periodic nitrogen functionality on the selective oxidation of alcohols. *Chemistry: An Asian Journal* 7 (2):387–93. doi: 10.1002/asia.201100565.

Chan-Thaw, Carine E, Alberto Villa, Phisan Katekomol, Dangsheng Su, Arne Thomas, and Laura Prati. 2010. Covalent triazine framework as catalytic support for liquid phase reaction. *Nano Letters* 10 (2):537–541.

Chan-Thaw, Carine E, Alberto Villa, Laura Prati, and Arne Thomas. 2011. Triazine-based polymers as nanostructured supports for the liquid-phase oxidation of alcohols. *Chemistry–A European Journal* 17 (3):1052–1057.

Chang, Fei, Jianping Guo, Guotao Wu, Lin Liu, Miao Zhang, Teng He, Peikun Wang, Pei Yu, and Ping Chen. 2015. Covalent triazine-based framework as an efficient catalyst support for ammonia decomposition. *RSC Advances* 5 (5):3605–3610.

Chao, Xiaofang, Yaqian Xu, Hui Chen, Diejing Feng, Jinxing Hu, and Yan Yu. 2021. TiO_2-Based photocatalyst modified with a covalent triazine-based framework organocatalyst for carbamazepine photodegradation. *RSC Advances* 11 (12):6943–6951.

Chen, Bo, Lang Chen, Zijun Yan, Jinyang Kang, Shan-Yong Chen, Yongdong Jin, Lijian Ma, Hongjian Yan, and Chuanqin Xia. 2021. Conjugated microporous polymers as visible light responsible platform for photoredox conversion of biomass derived chemicals. *Green Chemistry* 23 (10):3607–3611.

Chen, Sicong, Yuanzhi Zhu, Danyun Xu, Wenchao Peng, Yang Li, Guoliang Zhang, Fengbao Zhang, and Xiaobin Fan. 2018. Covalent triazine framework anchored with Co_3O_4 nanoparticles for efficient oxygen reduction. *ChemElectroChem* 5 (4):717–721.

Deng, Q-W, G-Q Ren, Y-J Li, L Yang, S-L Zhai, T Yu, L Sun, W-Q Deng, A Li, and Y-H Zhou. 2020. Hydrogen and CO_2 storage in high surface area covalent triazine–based frameworks. *Materials Today Energy* 18:100506.

Dey, Subarna, Asamanjoy Bhunia, Ishtvan Boldog, and Christoph Janiak. 2017. A mixed-linker approach towards improving covalent triazine-based frameworks for CO_2 capture and separation. *Microporous Mesoporous Materials* 241:303–315.

Dey, Subarna, Asamanjoy Bhunia, Hergen Breitzke, Pedro B Groszewicz, Gerd Buntkowsky, and Christoph Janiak. 2017. Two linkers are better than one: enhancing CO_2 capture and separation with porous covalent triazine-based frameworks from mixed nitrile linkers. *Journal of Materials Chemistry A* 5 (7):3609–3620.

Dey, Subarna, Asamanjoy Bhunia, Dolores Esquivel, and Christoph Janiak. 2016. Covalent triazine-based frameworks (CTFs) from triptycene and fluorene motifs for CO_2 adsorption. *Journal of Materials Chemistry A* 4 (17):6259–6263.

Dey, Subarna, Stefanie Bügel, Sara Sorribas, Alexander Nuhnen, Asamanjoy Bhunia, Joaquín Coronas, and Christoph Janiak. 2019. Synthesis and characterization of covalent triazine framework CTF-1@ polysulfone mixed matrix membranes and their gas separation studies. *Frontiers in Chemistry* 7:693.

Diercks, CS, and OM Yaghi. 2017. The atom, the molecule, and the covalent organic framework. *Science* 355 (6328). doi: 10.1126/science.aal1585.

Ding, San-Yuan, and Wei Wang. 2013. Covalent organic frameworks (COFs): from design to applications. *Chemical Society Reviews* 42 (2):548–568.

Dong, Bin, Dong-Yue Wang, and Wen-Jing Wang. 2020. Post-functionalization of hydroxyl-appended covalent triazine framework via borrowing hydrogen strategy for effective CO_2 capture. *Microporous Mesoporous Materials* 292:109765.

Dong, Bin, Dong-Yue Wang, Wen-Jing Wang, Xi-Long Tian, and Gang Ren. 2020. Post synthesis of a glycine-functionalized covalent triazine framework with excellent CO_2 capture performance. *Microporous Mesoporous Materials* 306:110475.

Fu, Yu, Zhiqiang Wang, Xianbiao Fu, Jun Yan, Cheng Liu, Chunyue Pan, and Guipeng Yu. 2017. Acid/hydrazide-appended covalent triazine frameworks for low-pressure CO_2 capture: pre-designable or post-synthesis modification. *Journal of Materials Chemistry A* 5 (40):21266–21274.

Fu, Yu, Wenguang Yu, Weijie Zhang, Qiao Huang, Jun Yan, Chunyue Pan, and Guipeng Yu. 2018. Sulfur-rich covalent triazine polymer nanospheres for environmental mercury removal and detection. *Polymer Chemistry* 9 (30):4125–4131.

Geng, Tongmou, Weiyong Zhang, Zongming Zhu, Guofeng Chen, Lanzhen Ma, Sainan Ye, and Qingyuan Niu. 2018. A covalent triazine-based framework from tetraphenylthiophene and 2, 4, 6-trichloro-1, 3, 5-triazine motifs for sensing o-nitrophenol and effective I 2 uptake. *Polymer Chemistry* 9 (6):777–784.

Gong, Lele, Xiaowei Wang, Tao Zheng, Jerry Liu, Jie Wang, Yu-Chia Yang, Jing Zhang, Xiao Han, Lipeng Zhang, and Zhenhai Xia. 2021. Catalytic mechanism and design principle of coordinately unsaturated single metal atom-doped covalent triazine frameworks with high activity and selectivity for CO_2 electroreduction. *Journal of Materials Chemistry A* 9 (6):3555–3566.

Gu, Chunyang, Deyu Liu, Wei Huang, Jie Liu, and Renqiang Yang. 2015. Synthesis of covalent triazine-based frameworks with high CO_2 adsorption and selectivity. *Polymer Chemistry* 6 (42):7410–7417.

Guan, L., G. Cheng, B. Tan, and S. Jin. 2021. Covalent triazine frameworks constructed via benzyl halide monomers showing high photocatalytic activity in biomass reforming. *Chem Commun (Camb)*. doi: 10.1039/d1cc01102b.

Gunasekar, Gunniya Hariyanandam, Kwangho Park, Vinothkumar Ganesan, Kwangyeol Lee, Nak-Kyoon Kim, Kwang-Deog Jung, and Sungho Yoon. 2017. A covalent triazine framework, functionalized with ir/n-heterocyclic carbene sites, for the efficient hydrogenation of CO_2 to formate. *Chemistry of Materials* 29 (16):6740–6748.

Gunasekar, Gunniya Hariyanandam, and Sungho Yoon. 2021. "New aspects of covalent triazine frameworks in heterogeneous catalysis." In *Emerging Carbon Materials for Catalysis*, 1–32. Elsevier.

Guo, Bixuan, Chuanguang Wu, Qing Su, Ziqian Liu, Xiaodong Li, Guanghua Li, and Qiaolin Wu. 2018. A Zn-salen based covalent triazine framework as a promising candidate for CO2 capture. *Materials Letters* 221:236–239.

Guo, J, Y Xu, S Jin, L Chen, T Kaji, Y Honsho, MA Addicoat, J Kim, A Saeki, H Ihee, S Seki, S Irle, M Hiramoto, J Gao, and D Jiang. 2013. Conjugated organic framework with three-dimensionally ordered stable structure and delocalized pi clouds. *Nature Communications* 4 (1):2736. doi: 10.1038/ncomms3736.

Guo, Liping, Yingli Niu, Haitao Xu, Qingwei Li, Shumaila Razzaque, Qi Huang, Shangbin Jin, and Bien Tan. 2018. Engineering heteroatoms with atomic precision in donor–acceptor covalent triazine frameworks to boost photocatalytic hydrogen production. *Journal of Materials Chemistry A* 6 (40):19775–19781.

Hall, Peter J, Mojtaba Mirzaeian, S Isobel Fletcher, Fiona B Sillars, Anthony JR Rennie, Gbolahan O Shitta-Bey, Grant Wilson, Andrew Cruden, and Rebecca Carter. 2010. Energy storage in electrochemical capacitors: designing functional materials to improve performance. *Energy Environmental Science* 3 (9):1238–1251.

Hao, L, J Ning, B Luo, B Wang, Y Zhang, Z Tang, J Yang, A Thomas, and L Zhi. 2015. Structural evolution of 2D microporous covalent triazine-based framework toward the study of high-performance supercapacitors. *Journal of the American Chemical Society* 137 (1):219–25. doi: 10.1021/ja508693y.

Hao, Long, Shuangshuang Zhang, Rongji Liu, Jing Ning, Guangjin Zhang, and Linjie Zhi. 2015. Bottom-up construction of triazine-based frameworks as metal-free electrocatalysts for oxygen reduction reaction. *Advanced Materials* 27 (20):3190–3195.

He, Juan, Fujian Xu, Jing Hu, Shanling Wang, Xiandeng Hou, and Zhou Long. 2017. Covalent triazine framework-1: a novel oxidase and peroxidase mimic. *Microchemical Journal* 135:91–99.

He, Teng, Lin Liu, Guotao Wu, and Ping Chen. 2015. Covalent triazine framework-supported palladium nanoparticles for catalytic hydrogenation of N-heterocycles. *Journal of Materials Chemistry A* 3 (31):16235–16241.

He, Yafei, Xiaodong Zhuang, Chaojun Lei, Lecheng Lei, Yang Hou, Yiyong Mai, and Xinliang Feng. 2019. Porous carbon nanosheets: synthetic strategies and electrochemical energy related applications. *Nano Today* 24:103–119.

Heydari, Maryam, Mohammad T Jafari, Mohammad Saraji, Roozbeh Soltani, and Mohammad Dinari. 2021. Covalent triazine-based framework-grafted functionalized fibrous silica sphere as a solid-phase microextraction coating for simultaneous determination of fenthion and chlorpyrifos by ion mobility spectrometry. *Microchimica Acta* 188 (1):1–11.

Huang, W, J Byun, I Rorich, C Ramanan, PWM Blom, H Lu, D Wang, L Caire da Silva, R Li, L Wang, K Landfester, and KAI Zhang. 2018. Asymmetric covalent triazine framework for enhanced visible-light photoredox catalysis via energy transfer cascade. *Angewandte Chemie International Edition* 57 (27):8316–8320. doi: 10.1002/anie.201801112.

Huang, W, B Li, Y Wu, Y Zhang, W Zhang, S Chen, Y Fu, T Yan, and H Ma. 2021. In situ-doped superacid in the covalent triazine framework membrane for anhydrous proton conduction in a wide temperature range from subzero to elevated temperature. *ACS Applied Materials & Interfaces* 13 (11):13604–13612. doi: 10.1021/acsami.1c01134.

Huang, Wei, Beatriz Chiyin Ma, Hao Lu, Run Li, Lei Wang, Katharina Landfester, and Kai AI Zhang. 2017. Visible-light-promoted selective oxidation of alcohols using a covalent triazine framework. *ACS Catalysis* 7 (8):5438–5442.

Huang, Wei, Zi Jun Wang, Beatriz Chiyin Ma, Saman Ghasimi, Dominik Gehrig, Frédéric Laquai, Katharina Landfester, and Kai AI Zhang. 2016. Hollow nanoporous covalent triazine frameworks via acid vapor-assisted solid phase synthesis for enhanced visible light photoactivity. *Journal of Materials Chemistry A* 4 (20):7555–7559.

Hug, Stephan, Maria B Mesch, Hyunchul Oh, Nadine Popp, Michael Hirscher, Jürgen Senker, and Bettina V Lotsch. 2014. A fluorene based covalent triazine framework with high CO_2 and H_2 capture and storage capacities. *Journal of Materials Chemistry A* 2 (16):5928–5936.

Hug, Stephan, Linus Stegbauer, Hyunchul Oh, Michael Hirscher, and Bettina V Lotsch. 2015. Nitrogen-rich covalent triazine frameworks as high-performance platforms for selective carbon capture and storage. *Chemistry of Materials* 27 (23):8001–8010.

Hug, Stephan, Michael E Tauchert, Shen Li, Ursula E Pachmayr, and Bettina V Lotsch. 2012. A functional triazine framework based on N-heterocyclic building blocks. *Journal of Materials Chemistry A* 22 (28):13956–13964.

Hulicova-Jurcakova, Denisa, Masaya Kodama, Soshi Shiraishi, Hiroaki Hatori, Zong Hua Zhu, and Gao Qing Lu. 2009. Nitrogen-enriched nonporous carbon electrodes with extraordinary supercapacitance. *Advanced Functional Materials* 19 (11):1800–1809.

Iwase, Kazuyuki, Tatsuro Yoshioka, Shuji Nakanishi, Kazuhito Hashimoto, and Kazuhide Kamiya. 2015. Copper-modified covalent triazine frameworks as non-noble-metal electrocatalysts for oxygen reduction. *Angewandte Chemie International Edition* 54 (38):11068–11072.

Je, Sang Hyun, Hyeon Jin Kim, Jiheon Kim, Jang Wook Choi, and Ali Coskun. 2017. Perfluoroaryl-elemental sulfur SNAr chemistry in covalent triazine frameworks with high sulfur contents for Lithium–Sulfur batteries. *Advanced Functional Materials* 27 (47):1703947.

Jena, H. S., C. Krishnaraj, S. Parwaiz, F. Lecoeuvre, J. Schmidt, D. Pradhan, and P. Van Der Voort. 2020. Illustrating the role of quaternary-N of BINOL covalent triazine-based frameworks in oxygen reduction and hydrogen evolution reactions. *ACS Applied Materials & Interfaces* 12 (40):44689–44699. doi: 10.1021/acsami.0c11381.

Jena, Himanshu Sekhar, Chidharth Krishnaraj, Guangbo Wang, Karen Leus, Johannes Schmidt, Nicolas Chaoui, and Pascal Van Der Voort. 2018. Acetylacetone covalent triazine framework: an efficient carbon capture and storage material and a highly stable heterogeneous catalyst. *Chemistry of Materials* 30 (12):4102–4111.

Jiang, Qianqian, Long Sun, Jinhong Bi, Shijing Liang, Liuyi Li, Yan Yu, and Ling Wu. 2018. MoS2 Quantum Dots-Modified Covalent Triazine-Based Frameworks for Enhanced Photocatalytic Hydrogen Evolution. *ChemSusChem* 11 (6):1108–1113. doi: 10.1002/cssc.201702220.

Kaczmarek, Anna M, Sara Abednatanzi, Dolores Esquivel, Chidharth Krishnaraj, Himanshu Sekhar Jena, Guangbo Wang, Karen Leus, Rik Van Deun, Francisco J Romero-Salguero, and Pascal Van Der Voort. 2020. Amine-containing (nano-) Periodic Mesoporous Organosilica and its application in catalysis, sorption and luminescence. *Microporous Mesoporous Materials* 291:109687.

Kaleeswaran, Dhananjayan, Rajendran Antony, Abhishek Sharma, Ateeque Malani, and Ramaswamy Murugavel. 2017. Catalysis and CO_2 Capture by Palladium-Incorporated Covalent Organic Frameworks. *ChemPlusChem* 82 (10):1253–1265. doi: 10.1002/cplu.201700342.

Kamai, R, K Kamiya, K Hashimoto, and S Nakanishi. 2016. Oxygen-Tolerant Electrodes with Platinum-Loaded Covalent Triazine Frameworks for the Hydrogen Oxidation Reaction. *Angewandte Chemie International Edition* 55 (42):13184–13188. doi: 10.1002/anie.201607741.

Kamiya, Kazuhide. 2020. Selective single-atom electrocatalysts: a review with a focus on metal-doped covalent triazine frameworks. *Chemical Science* 11 (32):8339–8349.

Kamiya, Kazuhide, Ryo Kamai, Kazuhito Hashimoto, and Shuji Nakanishi. 2014. Platinum-modified covalent triazine frameworks hybridized with carbon nanoparticles as methanol-tolerant oxygen reduction electrocatalysts. *Nature Communications* 5 (1):1–6.

Katekomol, Phisan, Jérôme Roeser, Michael Bojdys, Jens Weber, and Arne Thomas. 2013. Covalent triazine frameworks prepared from 1, 3, 5-tricyanobenzene. *Chemistry of Materials* 25 (9):1542–1548.

Khan, Adnan Ali, Rashid Ahmad, and Iftikhar Ahmad. 2020. Density functional theory study of emerging pollutants removal from water by covalent triazine based framework. *Journal of Molecular Liquids* 309:113008.

Kötz, Rüdiger, and MJEA Carlen. 2000. Principles and applications of electrochemical capacitors. *Electrochimica Acta* 45 (15–16):2483–2498.

Krishnaraj, Chidharth, Himanshu Sekhar Jena, Karen Leus, and Pascal Van Der Voort. 2020. Covalent triazine frameworks–a sustainable perspective. *Green Chemistry* 22 (4):1038–1071.

Kuecken, Sophie, Amitava Acharjya, Linjie Zhi, Michael Schwarze, Reinhard Schomäcker, and Arne Thomas. 2017. Fast tuning of covalent triazine frameworks for photocatalytic hydrogen evolution. *Chemical Communications* 53 (43):5854–5857.

Kuhn, Pierre, Markus Antonietti, and Arne Thomas. 2008. Porous, covalent triazine-based frameworks prepared by ionothermal synthesis. *Angewandte Chemie International Edition* 47 (18):3450–3453.

Lan, ZA, Y Fang, X Chen, and X Wang. 2019. Thermal annealing-induced structural reorganization in polymeric photocatalysts for enhanced hydrogen evolution. *Chemical Communications (Camb)* 55 (54):7756–7759. doi: 10.1039/c9cc02966d.

Lee, Yoon Jeong, Siddulu Naidu Talapaneni, and Ali Coskun. 2017. Chemically activated covalent triazine frameworks with enhanced textural properties for high capacity gas storage. *ACS Applied Materials Interfaces* 9 (36):30679–30685

Li, Gen, Xue Zhou, and Zhonggang Wang. 2020. Highly nitrogen-rich microporous polyaminals using N, N-dimethylformamide and formamide as the starting monomers for CO_2 adsorption and separation. *The Journal of Physical Chemistry C* 124 (5):3087–3094.

Li, Jian, Peng Liu, Yuanzhe Tang, Hongliang Huang, Hongzhi Cui, Donghai Mei, and Chongli Zhong. 2020. Single-atom $Pt–N_3$ sites on the stable covalent triazine framework nanosheets for photocatalytic N_2 fixation. *ACS Catalysis* 10 (4):2431–2442.

Li, Yi-Meng, Li Yang, Lei Sun, Lei Ma, Wei-Qiao Deng, and Zhen Li. 2019. Chemical fixation of carbon dioxide catalyzed via covalent triazine frameworks as metal free heterogeneous catalysts without a cocatalyst. *Journal of Materials Chemistry A* 7 (45):26071–26076.

Li, Zhongtao, Jingyan Zhou, Rongfei Xu, Shuiping Liu, Yuankun Wang, Peng Li, Wenting Wu, and Mingbo Wu. 2016. Synthesis of three dimensional extended conjugated polyimide and application as sodium-ion battery anode. *Chemical Engineering Journal* 287:516–522.

Liao, Huaping, Huimin Ding, Bijian Li, Xinping Ai, and Cheng Wang. 2014. Covalent-organic frameworks: potential host materials for sulfur impregnation in lithium–sulfur batteries. *Journal of Materials Chemistry A* 2 (23):8854–8858.

Liao, Longfei, Daniel Ditz, Feng Zeng, Marcelo Alves Favaro, Andree Iemhoff, Kavita Gupta, Heinrich Hartmann, Conrad Szczuka, Peter Jakes, and Peter JC Hausoul. 2020. Efficient photocatalytic oxidation of aromatic alcohols over thiophene-based covalent triazine frameworks with a narrow band gap. *ChemistrySelect* 5 (45):14438–14446.

Liebl, Mario R, and Jürgen Senker. 2013. Microporous functionalized triazine-based polyimides with high CO2 capture capacity. *Chemistry of Materials* 25 (6):970–980.

Lin, Sha, Yuxia Hou, Xiao Deng, Haoliang Wang, Shuzhuang Sun, and Xiaomei Zhang. 2015. A triazine-based covalent organic framework/palladium hybrid for one-pot silicon-based cross-coupling of silanes and aryl iodides. *RSC Advances* 5 (51):41017–41024.

Liu, Cheng, Miao Xia, Manxia Zhang, Kuanyu Yuan, Fangyuan Hu, Guipeng Yu, and Xigao Jian. 2020. One-pot synthesis of nitrogen-rich aminal-and triazine-based hierarchical porous organic polymers with highly efficient iodine adsorption. *Polymer Chemistry* 194:122401.

Liu, J, D Zhou, Z Xu, and S Zheng. 2017. Adsorptive removal of pharmaceutical antibiotics from aqueous solution by porous covalent triazine frameworks. *Environmental Pollution* 226:379–384. doi: 10.1016/j.envpol.2017.03.063.

Liu, Jingjing, Pengbo Lyu, Yu Zhang, Petr Nachtigall, and Yuxi Xu. 2018. New layered triazine framework/exfoliated 2D polymer with superior sodium-storage properties. *Advanced Materials* 30 (11):1705401.

Liu, Jingjing, Wu Zan, Ke Li, Yang Yang, Fanxing Bu, and Yuxi Xu. 2017. Solution synthesis of semiconducting two-dimensional polymer via trimerization of carbonitrile. *Journal of the American Chemical Society* 139 (34):11666–11669.

Liu, Jingliang, Huan Chen, Shourong Zheng, and Zhaoyi Xu. 2013. Adsorption of 4, 4′-(Propane-2, 2-diyl) diphenol from aqueous solution by a covalent triazine-based framework. *Journal of Chemical Engineering Data* 58 (12):3557–3562.

Liu, Manying, Liping Guo, Shangbin Jin, and Bien Tan. 2019. Covalent triazine frameworks: synthesis and applications. *Journal of Materials Chemistry A* 7 (10):5153–5172.

Liu, Manying, Qi Huang, Shaolei Wang, Ziyong Li, Buyi Li, Shangbin Jin, and Bien Tan. 2018. Crystalline covalent triazine frameworks by in situ oxidation of alcohols to aldehyde monomers. *Angewandte Chemie International Edition* 57 (37):11968–11972.

Liu, Xiaoming, He Li, Yuwei Zhang, Bo Xu, A Sigen, Hong Xia, and Ying Mu. 2013. Enhanced carbon dioxide uptake by metalloporphyrin-based microporous covalent triazine framework. *Polymer Chemistry* 4 (8):2445–2448.

Liu, Xigui, Danlian Huang, Cui Lai, Guangming Zeng, Lei Qin, Han Wang, Huan Yi, Bisheng Li, Shiyu Liu, and Mingming Zhang. 2019. Recent advances in covalent organic frameworks (COFs) as a smart sensing material. *Chemical Society Reviews* 48 (20):5266–5302.

Liu, Yao, Shaofei Wu, Gang Wang, Guipeng Yu, Jianguo Guan, Chunyue Pan, and Zhonggang Wang. 2014. Control of porosity of novel carbazole-modified polytriazine frameworks for highly selective separation of $CO_2–N_2$. *Journal of Materials Chemistry A* 2 (21):7795–7801.

Lu, Yue, Jia He, Yanli Chen, Heng Wang, Yunfeng Zhao, Yu Han, and Yi Ding. 2018. Effective acetylene/ethylene separation at ambient conditions by a pigment-based Covalent-Triazine framework. *Macromolecular Rapid Communications* 39 (2):1700468.

Luo, Yali, Jun Liu, Yunfei Liu, and Yinong Lyu. 2017. Porphyrin-based covalent triazine frameworks: porosity, adsorption performance, and drug delivery. *Journal of Polymer Science Part A: Polymer Chemistry* 55 (16):2594–2600.

Mahmood, J, EK Lee, M Jung, D Shin, IY Jeon, SM Jung, HJ Choi, JM Seo, SY Bae, SD Sohn, N Park, JH Oh, HJ Shin, and JB Baek. 2015. Nitrogenated holey two-dimensional structures. *Nature Communications* 6 (1):6486. doi: 10.1038/ncomms7486.

Meier, Christian B, Reiner Sebastian Sprick, Adriano Monti, Pierre Guiglion, Jet-Sing M Lee, Martijn A Zwijnenburg, and Andrew I Cooper. 2017. Structure-property relationships for covalent triazine-based frameworks: the effect of spacer length on photocatalytic hydrogen evolution from water. *Polymer* 126:283–290.

Mohamed, Mohamed Gamal, Ahmed FM El-Mahdy, Yasuno Takashi, and Shiao-Wei Kuo. 2020. Ultrastable conductive microporous covalent triazine frameworks based on pyrene moieties provide high-performance CO_2 uptake and supercapacitance. *New Journal of Chemistry* 44 (20):8241–8253.

Mohamed, Mohamed Gamal, Xian Zhang, Tharwat Hassan Mansoure, Ahmed FM El-Mahdy, Chih-Feng Huang, Martin Danko, Zhong Xin, and Shiao-Wei Kuo. 2020. Hypercrosslinked porous organic polymers based on tetraphenylanthraquinone for CO_2 uptake and high-performance supercapacitor. *Polymer* 205:122857.

Muehlhofer, Michael, Thomas Strassner, and Wolfgang A Herrmann. 2002. New catalyst systems for the catalytic conversion of methane into methanol. *Angewandte Chemie International Edition* 41 (10):1745–1747.

Mukherjee, Soumya, Monojit Das, Anupam Manna, Rajamani Krishna, and Sanjib Das. 2019. Newly designed 1, 2, 3-triazole functionalized covalent triazine frameworks with exceptionally high uptake capacity for both CO_2 and H_2. *Journal of Materials Chemistry A* 7 (3):1055–1068.

Mukhtar, Ahmad, Sami Ullah, Abrar Inayat, Sidra Saqib, Nurhayati Binti Mellon, Mohammed Ali Assiri, Abdullah G Al-Sehemi, Muhammad Bilal Khan Niazi, Zaib Jahan, and Mohamad Azmi Bustam. 2021. Synthesis-structure-property relationship of nitrogen-doped porous covalent triazine frameworks for pre-combustion CO_2 capture. *Energy* 216:119230.

Mullangi, Dinesh, Shyamapada Nandi, Sorout Shalini, Sheshadri Sreedhala, Chathakudath P Vinod, and Ramanathan Vaidhyanathan. 2015. Pd loaded amphiphilic COF as catalyst for multi-fold Heck reactions, CC couplings and CO oxidation. *Scientific Reports* 5 (1):1–12.

Niu, Fang, Liming Tao, Yuchao Deng, Hong Gao, Jingang Liu, and Weiguo Song. 2014. A covalent triazine framework as an efficient catalyst for photodegradation of methylene blue under visible light illumination. *New Journal of Chemistry* 38 (12):5695–5699.

Osadchii, DY, AI Olivos-Suarez, AV Bavykina, and J Gascon. 2017. Revisiting nitrogen species in covalent triazine frameworks. *Langmuir* 33 (50):14278–14285. doi: 10.1021/acs.langmuir.7b02929.

Palkovits, Regina, Markus Antonietti, Pierre Kuhn, Arne Thomas, and Ferdi Schüth. 2009. Solid catalysts for the selective low-temperature oxidation of methane to methanol. *Angewandte Chemie International Edition* 48 (37):6909–6912.

Park, K, GH Gunasekar, N Prakash, KD Jung, and S Yoon. 2015. A highly efficient heterogenized iridium complex for the catalytic hydrogenation of carbon dioxide to formate. *ChemSusChem* 8 (20):3410–3. doi: 10.1002/cssc.201500436.

Park, Kwangho, Kwangyeol Lee, Hyunuk Kim, Vinothkumar Ganesan, Kanghee Cho, Soon Kwan Jeong, and Sungho Yoon. 2017. Preparation of covalent triazine frameworks with imidazolium cations embedded in basic sites and their application for CO_2 capture. *Journal of Materials Chemistry A* 5 (18):8576–8582.

Pilaski, Moritz, Jens Artz, Husn-Ubayda Islam, Andrew M Beale, and Regina Palkovits. 2016. N-containing covalent organic frameworks as supports for rhodium as transition-metal catalysts in hydroformylation reactions. *Microporous Mesoporous Materials* 227:219–227.

Puthiaraj, Pillaiyar, Sung-Min Cho, Yu-Ri Lee, and Wha-Seung Ahn. 2015. Microporous covalent triazine polymers: efficient Friedel–Crafts synthesis and adsorption/storage of CO_2 and CH_4. *Journal of Materials Chemistry A* 3 (13):6792–6797.

Puthiaraj, Pillaiyar, Su-Sung Kim, and Wha-Seung Ahn. 2016. Covalent triazine polymers using a cyanuric chloride precursor via Friedel–Crafts reaction for CO_2 adsorption/separation. *Chemical Engineering Journal* 283:184–192.

Puthiaraj, Pillaiyar, Yu-Ri Lee, Siqian Zhang, and Wha-Seungs Ahn. 2016. Triazine-based covalent organic polymers: design, synthesis and applications in heterogeneous catalysis. *Journal of Materials Chemistry A* 4 (42):16288–16311.

Qian, Zhuangfei, Zi Jun Wang, and Kai AI Zhang. 2021. Covalent triazine frameworks as emerging heterogeneous photocatalysts. *Chemistry of Materials* 33 (6):1909–1926.

Rajendiran, Senkuttuvan, Prakash Natarajan, and Sungho Yoon. 2017. A covalent triazine framework-based heterogenized Al–Co bimetallic catalyst for the ring-expansion carbonylation of epoxide to β-lactone. *RSC Advances* 7 (8):4635–4638.

Rajendiran, Senkuttuvan, Kwangho Park, Kwangyeol Lee, and Sungho Yoon. 2017. Ionic-liquid-based heterogeneous covalent triazine framework cobalt catalyst for the direct synthesis of methyl 3-hydroxybutyrate from propylene oxide. *Inorganic Chemistry* 56 (12):7270–7277.

Ren, Shijie, Michael J Bojdys, Robert Dawson, Andrea Laybourn, Yaroslav Z Khimyak, Dave J Adams, and Andrew I Cooper. 2012. Porous, fluorescent, covalent triazine-based frameworks via room-temperature and microwave-assisted synthesis. *Advanced Materials* 24 (17):2357–2361.

Saida, Takahiro, Oki Sekizawa, Nozomu Ishiguro, Masato Hoshino, Kentaro Uesugi, Tomoya Uruga, Shin-ichi Ohkoshi, Toshihiko Yokoyama, and Mizuki Tada. 2012. 4D visualization of a cathode catalyst layer in a polymer electrolyte fuel cell by 3D laminography–XAFS. *Angewandte Chemie International Edition* 51 (41):10311–10314.

Sakaushi, Ken, Georg Nickerl, Florian M Wisser, Daisuke Nishio-Hamane, Eiji Hosono, Haoshen Zhou, Stefan Kaskel, and Jürgen Eckert. 2012. An energy storage principle using bipolar porous polymeric frameworks. *Angewandte Chemie International Edition* 51 (31):7850–7854.

Schwinghammer, Katharina, Stephan Hug, Maria B Mesch, Jürgen Senker, and Bettina V Lotsch. 2015. Phenyl-triazine oligomers for light-driven hydrogen evolution. *Energy Environmental Science* 8 (11):3345–3353.

Sudakar, P, GH Gunasekar, IH Baek, and S Yoon. 2016. Recyclable and efficient heterogenized Rh and Ir catalysts for the transfer hydrogenation of carbonyl compounds in aqueous medium. *Green Chemistry* 18 (24):6456–6461.

Sun, Tian, Congxu Wang, and Yuxi Xu. 2020. Covalent triazine framework nanosheets for efficient energy storage and conversion. *Chemical Research in Chinese Universities*:1–8.

Tahir, Norini, Chidharth Krishnaraj, Karen Leus, and Pascal Van Der Voort. 2019. Development of covalent triazine frameworks as heterogeneous catalytic supports. *Polymers* 11 (8):1326.

Takeguchi, Tatsuya, Toshiro Yamanaka, Kiyotaka Asakura, Ernee Noryana Muhamad, Kohei Uosaki, and Wataru Ueda. 2012. Evidence of nonelectrochemical shift reaction on a CO-tolerant high-entropy state Pt–Ru anode catalyst for reliable and efficient residential fuel cell systems. *Journal of the American Chemical Society* 134 (35):14508–14512.

Talapaneni, Siddulu Naidu, Tae Hoon Hwang, Sang Hyun Je, Onur Buyukcakir, Jang Wook Choi, and Ali Coskun. 2016. Elemental-sulfur-mediated facile synthesis of a covalent triazine framework for high-performance lithium–sulfur batteries. *Angewandte Chemie International Edition* 55 (9):3106–3111.

Tan, Zunkun, Peng Zhang, Qiaoshan Chen, Shengqiong Fang, Guocheng Huang, Jinhong Bi, and Ling Wu. 2021. Visible-light-driven photocatalyst based upon metal-free covalent triazine-based frameworks for enhanced hydrogen production. *Catalysis Science Technology* 11 (5):1874–1880.

Tao, Liming, Fang Niu, Chao Wang, Jingang Liu, Tingmei Wang, and Qihua Wang. 2016. Benzimidazole functionalized covalent triazine frameworks for CO_2 capture. *Journal of Materials Chemistry A* 4 (30):11812–11820.

ten Brink, Gerd-Jan, Isabel WCE Arends, and Roger A Sheldon. 2000. Green, catalytic oxidation of alcohols in water. *Science* 287 (5458):1636–1639.

Tiwari, Abhishek R, T Akash, and Bhalchandra M Bhanage. 2015. NIS-catalyzed oxidative cyclization of alcohols with amidines: a simple and efficient transition-metal free method for the synthesis of 1, 3, 5-triazines. *Organic Biomolecular Chemistry* 13 (45):10973–10976.

Tong, Lei, ZongGui Shao, YaSheng Qian, and Wen Mu Li. 2017. Facile ionothermal synthesis of mesoporous Fe–N x–C composites as efficient catalysts for oxygen reduction in acid media. *Journal of Materials Chemistry A* 5 (8):3832–3838.

Troschke, Erik, Sven Grätz, Tilo Lübken, and Lars Borchardt. 2017. Mechanochemical Friedel–Crafts alkylation—A sustainable pathway towards porous organic polymers. *Angewandte Chemie international Edition* 56 (24):6859–6863.

Troschke, Erik, Desirée Leistenschneider, Tilo Rensch, Sven Grätz, Johannes Maschita, Sebastian Ehrling, Benjamin Klemmed, Bettina V Lotsch, Alexander Eychmüller, and Lars Borchardt. 2020. In Situ Generation of Electrolyte inside Pyridine-Based Covalent Triazine Frameworks for Direct Supercapacitor Integration. *ChemSusChem* 13 (12):3192.

Vyas, Vijay S, Medhavi Vishwakarma, Igor Moudrakovski, Frederik Haase, Gökcen Savasci, Christian Ochsenfeld, Joachim P Spatz, and Bettina V Lotsch. 2016. Exploiting noncovalent interactions in an imine-based covalent organic framework for quercetin delivery. *Advanced Materials* 28 (39):8749–8754.

Wang, Dong, and Didier Astruc. 2015. The golden age of transfer hydrogenation. *Chemical Reviews* 115 (13):6621–6686.

Wang, Guangbo, Karen Leus, Shuna Zhao, and Pascal Van Der Voort. 2017. Newly designed covalent triazine framework based on novel N-heteroaromatic building blocks for efficient CO_2 and H_2 capture and storage. *ACS Applied Materials Interfaces* 10 (1):1244–1249.

Wang, Han, Danni Jiang, Danlian Huang, Guangming Zeng, Piao Xu, Cui Lai, Ming Chen, Min Cheng, Chen Zhang, and Ziwei Wang. 2019. Covalent triazine frameworks for carbon dioxide capture. *Journal of Materials Chemistry A* 7 (40):22848–22870.

Wang, Han, Hui Wang, Ziwei Wang, Lin Tang, Guangming Zeng, Piao Xu, Ming Chen, Ting Xiong, Chengyun Zhou, and Xiyi Li. 2020. Covalent organic framework photocatalysts: structures and applications. *Chemical Society Reviews* 49 (12):4135–4165.

Wang, Keke, Hongliang Huang, Dahuan Liu, Chang Wang, Jinping Li, and Chongli Zhong. 2016. Covalent triazine-based frameworks with ultramicropores and high nitrogen contents for highly selective CO_2 capture. *Environmental Science Technology* 50 (9):4869–4876.

Wang, Kewei, Li-Ming Yang, Xi Wang, Liping Guo, Guang Cheng, Chun Zhang, Shangbin Jin, Bien Tan, and Andrew Cooper. 2017. Covalent triazine frameworks via a low-temperature polycondensation approach. *Angewandte Chemie International Edition* 56 (45):14149–14153.

Wang, Lei, Ying Zhang, Liang Chen, Hangxun Xu, and Yujie Xiong. 2018. 2D polymers as emerging materials for photocatalytic overall water splitting. *Advanced Materials* 30 (48):1801955.

Wang, Shunli, Ying Liang, Tingting Dai, Yalin Liu, Zhuyin Sui, Xinlong Tian, and Qi Chen. 2021. Cationic covalent-organic framework for sulfur storage with high-performance in lithium-sulfur batteries. *Journal of Colloid Interface Science* 591:264–272.

Wang, Xiaoyan, Chong Zhang, Yang Zhao, Shijie Ren, and Jia-Xing Jiang. 2016. Synthetic control and multifunctional properties of fluorescent covalent triazine-based frameworks. *Macromolecular Rapid Communications* 37 (4):323–329.

Wang, Xinchen, Kazuhiko Maeda, Xiufang Chen, Kazuhiro Takanabe, Kazunari Domen, Yidong Hou, Xianzhi Fu, and Markus Antonietti. 2009. Polymer semiconductors for artificial photosynthesis: hydrogen evolution by mesoporous graphitic carbon nitride with visible light. *Journal of the American Chemical Society* 131 (5):1680–1681.

Wang, Xinchen, Kazuhiko Maeda, Arne Thomas, Kazuhiro Takanabe, Gang Xin, Johan M Carlsson, Kazunari Domen, and Markus Antonietti. 2009. A metal-free polymeric photocatalyst for hydrogen production from water under visible light. *Nature Materials* 8 (1):76–80.

Wang, Yong, Jinping Li, Qingyuan Yang, and Chongli Zhong. 2016. Two-dimensional covalent triazine framework membrane for helium separation and hydrogen purification. *ACS Applied Materials Interfaces* 8 (13):8694–8701.

Wang, Yuanshuang, Junxiang Chen, Genxiang Wang, Yan Li, and Zhenhai Wen. 2018. Perfluorinated covalent triazine framework derived hybrids for the highly selective electroconversion of carbon dioxide into methane. *Angewandte Chemie International Edition* 57 (40):13120–13124.

Wen, Jikai, Yijiang Liu, Hongbiao Chen, Mei Yang, and Huaming Li. 2021. Synthesis and characterization of phenanthroimidazole-containing triazine frameworks and their performance in CO_2 capture. *Microporous Mesoporous Materials* 316:110939.

Wu, Chuanguang, Mingjun Hu, Xiaorong Yan, Guangcun Shan, Jinzhang Liu, and Jun Yang. 2021. Azo-linked covalent triazine-based framework as organic cathodes for ultrastable capacitor-type lithium-ion batteries. *Energy Storage Materials* 36:347–354.

Wu, Chuanguang, Hao Zhang, Mingjun Hu, Guangcun Shan, Jiefeng Gao, Jinzhang Liu, Xianglin Zhou, and Jun Yang. 2020. In situ nitrogen-doped covalent triazine-based multiporous cross-linking framework for high-performance energy storage. *Advanced Electronic Materials* 6 (7):2000253.

Wu, Feixiang, Joachim Maier, and Yan Yu. 2020. Guidelines and trends for next-generation rechargeable lithium and lithium-ion batteries. *Chemical Society Reviews* 49 (5):1569–1614.

Wu, Shaofei, Yao Liu, Guipeng Yu, Jianguo Guan, Chunyue Pan, Yong Du, Xiang Xiong, and Zhonggang Wang. 2014. Facile preparation of dibenzoheterocycle-functional nanoporous polymeric networks with high gas uptake capacities. *Macromolecules* 47 (9):2875–2882.

Xie, Jijia, Stephen A Shevlin, Qiushi Ruan, Savio JA Moniz, Yangrong Liu, Xu Liu, Yaomin Li, Chi Ching Lau, Zheng Xiao Guo, and Junwang Tang. 2018. Efficient visible light-driven water oxidation and proton reduction by an ordered covalent triazine-based framework. *Energy Environmental Science Technology* 11 (6):1617–1624.

Xing, Xiaolin, Rongji Liu, Montaha Anjass, Kecheng Cao, Ute Kaiser, Guangjin Zhang, and Carsten Streb. 2020. Bimetallic manganese-vanadium functionalized N, S-doped carbon nanotubes as efficient oxygen evolution and oxygen reduction electrocatalysts. *Applied Catalysis B: Environmental* 277:119195.

Xiong, Shaohui, Xian Fu, Lu Xiang, Guipeng Yu, Jianguo Guan, Zhonggang Wang, Yong Du, Xiang Xiong, and Chunyue Pan. 2014. Liquid acid-catalysed fabrication of nanoporous 1, 3, 5-triazine frameworks with efficient and selective CO_2 uptake. *Polymer Chemistry* 5 (10):3424–3431.

Xu, Fei, Shuhao Yang, Guangshen Jiang, Qian Ye, Bingqing Wei, and Hongqiang Wang. 2017. Fluorinated, sulfur-rich, covalent triazine frameworks for enhanced confinement of polysulfides in lithium–sulfur batteries. *ACS Applied Materials Interfaces* 9 (43):37731–37738.

Xu, Guangjuan, Yiang Zhu, Wei Xie, Shuran Zhang, Chan Yao, and Yanhong Xu. 2019. Porous cationic covalent triazine-based frameworks as platforms for efficient CO_2 and iodine capture. *Chemistry–An Asian Journal* 14 (19):3259–3263.

Xu, Naizhang, Bowei Cai, Qi Li, Yubing Liu, Jie Tang, Kaiqiang Wang, Bolian Xu, and Yining Fan. 2021. The noble-metal-free Ni2P/CTF composites for efficient photocatalytic hydrogen evolution under visible-light irradiation. *Journal of Alloys Compounds*:159565.

Xu, Rui, Xu-Sheng Wang, Hui Zhao, Hua Lin, Yuan-Biao Huang, Rong Cao. 2018. Rhenium-modified porous covalent triazine framework for highly efficient photocatalytic carbon dioxide reduction in a solid–gas system. *Catalysis Science Technology* 8 (8):2224–2230.

Xu, Yanhong, Shangbin Jin, Hong Xu, Atsushi Nagai, and Donglin Jiang. 2013. Conjugated microporous polymers: design, synthesis and application. *Chemical Society Reviews* 42 (20):8012–8031.

Yang, Can, Wei Huang, Lucas Caire da Silva, Kai AI Zhang, and Xinchen Wang. 2018. Functional conjugated polymers for CO_2 reduction using visible light. *Chemistry–A European Journal* 24 (66):17454–17458.

Yang, Zhenzhen, Hao Chen, Song Wang, Wei Guo, Tao Wang, Xian Suo, De-en Jiang, Xiang Zhu, Ilja Popovs, and Sheng Dai. 2020. Transformation strategy for highly crystalline covalent triazine frameworks: from staggered AB to eclipsed AA stacking. *Journal of the American Chemical Society* 142 (15):6856–6860.

Yang, Zhenzhen, Song Wang, Zihao Zhang, Wei Guo, Kecheng Jie, Mohamed I Hashim, Ognjen Š Miljanić, De-en Jiang, Ilja Popovs, and Sheng Dai. 2019. Influence of fluorination on CO_2 adsorption in materials derived from fluorinated covalent triazine framework precursors. *Journal of Materials Chemistry A* 7 (29):17277–17282.

Yin, Lunxiang, and Jürgen Liebscher. 2007. Carbon–carbon coupling reactions catalyzed by heterogeneous palladium catalysts. *Chemical Reviews* 107 (1):133–173.

You, Qing, Fei Wang, Chaoting Wu, Tianchao Shi, Dewen Min, Huajun Chen, and Wu Zhang. 2015. Synthesis of 1, 3, 5-triazines via Cu (OAc) 2-catalyzed aerobic oxidative coupling of alcohols and amidine hydrochlorides. *Organic Biomolecular Chemistry* 13 (24):6723–6727.

Yu, Soo-Young, Javeed Mahmood, Hyuk-Jun Noh, Jeong-Min Seo, Sun-Min Jung, Sun-Hee Shin, Yoon-Kwang Im, In-Yup Jeon, and Jong-Beom Baek. 2018. Direct Synthesis of a Covalent Triazine-Based Framework from Aromatic Amides. *Angewandte Chemie International Edition* 57 (28):8438–8442.

Yu, Wenguang, Shuai Gu, Yu Fu, Shaohui Xiong, Chunyue Pan, Younian Liu, and Guipeng Yu. 2018. Carbazole-decorated covalent triazine frameworks: novel nonmetal catalysts for carbon dioxide fixation and oxygen reduction reaction. *Journal of Catalysis* 362:1–9.

Yuan, Kuanyu, Chengde Liu, Cheng Liu, Shouhai Zhang, Guipeng Yu, Li Yang, Feihu Yang, and Xigao Jian. 2018. Construction of triphenylamine functional phthalazinone-based covalent triazine frameworks for effective CO_2 capture. *Polymer Chemistry* 151:65–74.

Zhang, Miao, Lin Liu, Teng He, Zhao Li, Guotao Wu, and Ping Chen. 2017. Microporous crystalline γ-Al_2O_3 Replicated from Microporous Covalent Triazine Framework and Its Application as Support for Catalytic Hydrolysis of Ammonia Borane. *Chemistry–An Asian Journal* 12 (4):470–475.

Zhang, Siquan, Guang Cheng, Liping Guo, Ning Wang, Bien Tan, and Shangbin Jin. 2020. Strong-base-assisted synthesis of a crystalline covalent triazine framework with high hydrophilicity via benzylamine monomer for photocatalytic water splitting. *Angewandte Chemie International Edition* 132 (15):6063–6070.

Zhang, Siquan, Shengyao Wang, Liping Guo, Hao Chen, Bien Tan, and Shangbin Jin. 2020. An artificial photosynthesis system comprising a covalent triazine framework as an electron relay facilitator for photochemical carbon dioxide reduction. *Journal of Materials Chemistry C* 8 (1):192–200.

Zhang, Tianpeng, Fangyuan Hu, Ce Song, Shengming Li, Wenlong Shao, Siyang Liu, Hao Peng, Shui Hu, and Xigao Jian. 2021. Constructing covalent triazine-based frameworks to explore the effect of heteroatoms and pore structure on electrochemical performance in Li–S batteries. *Chemical Engineering Journal* 407:127141.

Zhang, Ying, and Shangbin Jin. 2019. Recent advancements in the synthesis of covalent triazine frameworks for energy and environmental applications. *Polymers* 11 (1):31.

Zhang, Yuheng, Wen Xue, Jinhong Liu, Shuyu Wan, Ouwen Xu, and Xiashi Zhu. 2021. 2D/2D Covalent triazine-based heterostructure membrane for highly enhanced photoactivity and anti-fouling ability. *Chemical Engineering Journal*:129829.

TABLE 8.1
Comparison of Porous Polymers against Similar Materials Used in Separation Processes

Properties	Porous Polymers	Porous Carbons	Zeolites	Metal-Organic Frameworks
Pore size	Macro/ meso/ micro	Macro/ meso/ micro	Macro/ meso/ micro	Macro/ meso/ micro
Crystallinity	Amorphous	Amorphous	Crystalline	Crystalline
Chemical stability	Excellent	Excellent	Good	Good
Chemical diversity	Very wide	Moderate	Moderate	Very wide
Thermal stability	Excellent	Very good	Very good	Moderate
Structural diversity	Wide range	High	High	Very high
Processing	Low density, simultaneous synthesis and processing	Good conductivity	Low cost, various use in refinery operations	Structural and chemical control

the intrinsic huge spaces within and thus the extremely high surface area. Structural framework, pore size and surface area are some of the major characteristics of porous polymers governing their various applications which are tabulated in Table 8.1.

8.2 ADVANCED SEPARATION APPLICATIONS

Separation process is one of the integrated steps of modern chemical process plants. The conventional separation processes include distillation, adsorption, absorption, etc., for which the technologies are already well-defined and widely applied. But now some advanced separation processes are continuously gaining interest in modern-day Chemical process plants. These include membrane-based techniques, chromatographic separation, supercritical fluid extraction and porous polymers that are highly applicable for these purposes [7].

Separation plays an important role in a variety of industries, including the petrochemical, chemical as well as pharmaceutical industries, due to its low cost, controllable pore size with high porosity, dynamic surface features and processing simplicity. It makes them more appealing prospects for acquiring desired separation ability [8]. To meet the challenge, researchers have worked hard to develop porous polymeric membranes with promising gas transport properties, such as PIMs, polymers with thermal rearrangement, polyimide polymers, polyethylene glycol, etc. Because of their selective super-anti-wetting and super-wetting capabilities against water/oil, superhydrophobic and superoleophilic porous membranes and submerged superoleophobic porous polymeric membranes have been developed for the effective separation of oil- and water-free mixtures and emulsions [9]. Yuan and his coworkers created a new type of membrane with efficient nano-structured electrospun. It has a wide interface/volume area, membrane surface roughness along with a huge number of micrometer-sized pores [10]. Because of the coupling of CO_2-induced controllable oil/water wetting with porous nanostructure, membranes exhibit controlled oil-to-water separation efficiency. Jin and colleagues used an inversion of phase technique, by inducing salt, to create a PAA-g-PVDF fiber [11]. The added salt acts as a nucleator, causing PAA-g-PVDF micelles to form on the membrane surface, resulting in a micro as well as a nanoscale dynamic framework that gives the superhydrophilic and submerged superoleophobic properties to the membrane. Carta et al. developed a polymer molecular sieve with excellent permeability and very good selectivity regarding the gas separations for smaller gas molecules like H_2 and O_2 against larger ones like CH_4 and N_2 [12]. The framework of PIM is introduced with a ring-containing bicycle bridge-like ethanoanthracene. The Tröger's base polymerization is utilized to form staircase polymer with a tenacious shape [13]. The team of Xu has efficiently developed membranes with a nanofiltration system for use in several separation processes using easily controlled interfacial polymerization [14]. Xu and colleagues organized a group of size-dependent porous polymers to work on the separation of nitrogen from various VOCs by molecular sieving of size-dependent separation porous polymers [15]. The materials obtained have excellent separation properties for VOCs over N_2, with rejection rates of above 98% for different organic molecules, e.g. cyclohexane, hexane, heptane, etc. Ichinose et al. successfully created porous nanofibrous structures with huge free surface areas and a pore radius (~2.0 nm) using a micro-phase separation approach from chilled polymer solutions [16]. The products produced can quickly adsorb CO_2 with the value of Henry's constant being almost three times compared to the bulk polymer. Furthermore, similar polymers with nanoporosity can isolate toluene from aqueous solutions containing very few ppm levels concentrated adsorb a considerable amount of tetrahydrofuran from aqueous solution [17]. Sivaniah and his colleagues showed the crosslinking of PIMs by thermal oxidation utilizing tiny quantities of oxygen left over after the process of heat treatment [18]. The sensitivity of the covalently conjugated structures is three orders of magnitude higher than a standard polymer matrix, demonstrating that fine-tuning the geometrical layout of pores and channels inside polymers of microporosity improves molecular sieving capabilities [19]. Livingston and colleagues created polymer membranes with regulated microporosity that are solvent resistant [20]. Interfacial polymerization is used to create deformity and strongly crosslinked polyacrylate nanofilms with a thickness of 20 nm using mangled and noncontorted monomers. Their increased microporosity and intermolecular network void integration result in excellent separation efficiency in H_2/CO_2 and organic solvents. These results could pave the way for interfacial production of the increasing class of nanoporous polymers, which have a lot of potential for molecular separations [21].

With the improvement of surface chemistry, the wetting of an oil-to-water separating membrane may be substituted, e.g. the reaction between CO_2 and amidine may be regarded as a controlled switching because of its reversible swirling nature with Ar or N_2 [22]. Using a frozen polymer solution as a starting point for a microporous separation method, Ichinose & his team promoted a porous, nanofibrous framework with huge unbound surface areas and pore radius around 2 nm. They discovered that nanoporous polymers can extract toluene from aqueous systems with ppm-level concentration along with the adsorption of a considerable amount of tetrahydrofuran from aqueous solutions [23]. Jiang et al. have synthesized hypercrosslinked as well as ionic mesoporous polymer monolith having surface area 212 m^2/g and pore volume 1.08. They applied it in the desulfurization process of model fuel with hydrogen peroxide as the oxidant. The desulfurization efficiency was nearly about 100% [24]. Jin and colleagues used an inversion of phase technique to create a PAA-g-PVDF fiber by using salts [25]. The added NaCl acts as a nucleator, causing PAA-g-PVDF micelles to form on the membrane surface, resulting in a micro/nanoscale hierarchical structure that gives the membrane superhydrophilic and underwater superoleophobic properties. In addition, Kota et al. defined a new membrane that has both superoleophobic and superhydrophilic properties related to submerged water and air [26]. This can continuously distinguish the emulsions of water-in-oil as well as oil-in water with an appreciable efficiency above 99.5%. Ghanem et al. established a class of polyimide membranes based on triptycene with ultra-micro level porosity. They discovered that the properties of these membranes in the transportation of gas can be adjusted by controlling the handstand sites [27]. This method involves adding small isopropyl molecules to the skeletal bridgeheads. Here the chelating action produces bridgehead-replaced products with adjustable pore shape, higher stiffness, and better solubility. As a result, they perform well in industrial gas separations because they are extremely permeable and sensitive.

If we consider the fields where porous polymers are usually useful for separation are gas separation through adsorption and water treatment by separating the contaminant.

8.3 SEPARATION THROUGH ADSORPTION

Gas adsorption is commonly linked to a number of facets of human life, including the storage of clean gas oil, greenhouse gas management, and toxic gas monitoring. Hydrogen with methane, in particular, is important for the utilization of renewable energy; carbon dioxide is critical to lower down the effect of greenhouse; and hazardous gas management is critical for environmental conservation [28]. Porous polymers provide additional benefits when it comes to adsorption of gas:

- Constant porosity is present in the elastic building components.
- Porous polymers can be synthesized in a variety of ways, resulting in a wide range of structural configurations and pore sizes.
- Porous fibers having large surface areas may be made from porous polymer.
- The lightweight frameworks, which are usually made up of N, H, C, O along with B elements, provide an advantage of mass which facilitates the capacity of adsorption per gram to a higher extent.
- Because of the high stability, the allowance for several use cycles in case of covalent bonds.

The conventional gas separation techniques are (i) cryogenic distillation of air, (ii) condensation to remove condensable organic vapors from gas mixtures and (iii) amine absorption to remove acid gases such as carbon dioxide from natural gas. In all these processes a gas-to-liquid phase change is required. Gas separation through porous membranes is advantageous over those conventional gas separation techniques because it is economic as the energy cost is nil here as it does not involve any phase change. The simple and small structure of gas separation membrane units is another advantage of this process.

Gas separation through porous polymeric membranes is applicable in industry:

- In ammonia plants for hydrogen/nitrogen separation;
- In petrochemical industry for hydrogen/hydrocarbon separations;
- Nitrogen separation from air;
- Removal of CO_2 and water from natural gas;
- Removal of volatile organic components from air or nitrogen streams.

At the moment, the most widely used membrane materials for gas separation are polymers. Applications of different porous polymeric membranes to separate different gases in the petrochemical industry are presented in Table 8.2.

The chemical functionalities for gas adsorption may be planned using either a postsynthetic or a presynthetic alteration approach. Post-synthetic alteration of functional groups on the pore walls of porous polymers is rather simple to do synthetically, allowing for precise control of the chemical nature of the accessible surfaces [36]. Zhou and colleagues developed a class of polyamine-tethered porous polymer networks to improve CO_2 capture efficiency [37]. The decreasing order of CO_2 adsorption capacities of different attached polyamine groups is diethylenetriamine, triaminotriethylamine, triethylenetriamine and ethanediamine. This is very interesting as the surface areas of these compounds are completely opposite. Using a sulfonation approach, Zhang and colleagues discovered a new family of effective microporous polyimides on the basis of sulfonated binaphthalene dianhydride, which was used for the selective removal of hazardous organic vapors [38]. By steadily the sulfonation degree, the porous polymers transition from hydrophobic to hydrophilic, allowing for the regulation of nonpolar/polar organic vapor absorption. Guiver and colleagues proposed the synthesis of microporous polymers with the decoration of tetrazole, where an azide with an aromatic nitrile group was to the polymer by cycloaddition reaction. It gave them a very good uptake of CO_2 and high CO_2/N_2 mixed gas selectivity [39]. Patel and colleagues produced a new kind of nanoporous polymer containing high percentage of nitrogen [40]. The covalent organic polymers (COPs) containing the azo groups, obtained an upper threshold selectivity of ~290 for CO_2/N_2 at a temperature of 323 K, which corresponds to an industrially significant temperature above 40 °C. The inclusion of (NN) groups in azo-COPs is critical for rejecting N_2 gas and thereby increasing selectivity [41]. Lavigne and colleagues identified a variety of alkyl (methyl to propyl) indoctrinated microporous

TABLE 8.2
Different Kinds of Gas Separations Using Varieties of Porous Polymers

Sl. no	Gas Separation	Mixture of Gases	Name of the Polymer
1.	Hydrogen sulphide gas separation	CH_4/H_2S	Polyether-block-amide Cellulose acetate (UOP) Polyamide Cellulose triacetate (Cynara) Polyether-urethane-urea
2.	Hydrogen gas recovery	H_2/CO H_2/CH_4 H_2/N_2	Polysulfone (PRISM) Cellulose acetate (Separex) Polyetherimide Polyimide (UBE) Polyaramid (MEDAL)
3.	Carbon dioxide gas adsorption	CO_2/N_2 CO_2/CH_4	Polyimide (UBE) Cellulose acetate (UOP) Polysulfone Polyether-block-amide Polyether-urethane-urea polyamide
4.	Separation of air	O_2/N_2	Polysulfone (UOP) Cellulose acetate Poly(4-methyl-1-pentene) Polydimethylsiloxane (PRISM) (GENERON) Polyamide Polyimide Polyetherimide
5.	Recovery of VOC	VOC/N_2 VOC/air	Polyimide (Upjohn) Polydimethylsiloxane (GKSS)
6.	Removal of water	CH_4/H_2O	Cellulose acetate Polydimethylsiloxane (PRISM) Polyether-block-amide Polysulfone Polyimide
7.	LPG	$CH_4/(C_3–C_4)$	Polydimethylsiloxane Poly[1-(trimethylsilyl)-1-propyne]

polymers with varying pore diameters [42]. The added bulk would fill the pores as the alkyl substituents become longer, reducing the size and surface area. The findings show that changing the size of the alkyl groups in the pore interior induces the adsorption of hydrogen with the molar amount increment into the channel. Cooper and his collaborators have created a class of CMP networks [43]. Depending on the monomer composition, their BET surface areas can exceed 1000 m^2/g. For example, one of their obtained products has the shortest strut range, the largest micropore capacity, and the largest BET surface area, resulting in the maximum H_2 uptake (above 150 cm^3/g). Wu and colleagues were the first to make monodisperse porous polymer yolk-shell-shaped nanoparticles with recurring composites laminated by large-sized mesoporosity. They exhibit absorption of methanol vapor compared to standard hollow maladaptive microporous polymer nanoparticles because their periodic mesopores are significantly more sensitive to the adsorbate ions [44].

There is a wide range of porous polymers used for the gas separation process. Some of them are mentioned in Table 8.3 mentioning their adsorption properties.

TABLE 8.3
Surface Area of Different Porous Polymers to Adsorb Different Gases

Porous polymer	H2 (wt %)	Surface area (m^2/g)	Test conditions: Temperature (K)/ Pressure (bar)	CH_4 (wt.%)	Test conditions: Temperature (K)/ Pressure (bar)	CO_2 (wt.%)	Test conditions: Temperature (K)/ Presuure (bar)	References
ALP-1	2.2	1235	77/1.0	2.6	273/1.0	23.6	273/1.0	[45]
PAF-60	1.5	1094	77/1.0	1.9	273/1.0	21.9	273/1.0	[46]
CTF-TPC	1.8	1668	77/1.0	2.3	273/1.0	18.6	273/1.0	[47]
TNP4	1.8	1348	77/1.0	2.4	273/1.0	19.6	273/1.0	[48]
FCDTPA-K-700	2.61	2065	77/1.13	3.80	273/1.13	28.60	273/1.13	[49]
PPN-4	15.8	6461	77/ 90	38.9	295/55	212.1	295/50	[50]
Ni-Por-1	3.5	1711	77/40	12.3	273/1.08	12.8	273/35	[51]
PINK	2.48	2090	77/1.0	-	-	16.0	273/1.0	[52]
BILP-6-NH2	-	1185	-	2.8	273/1.0	24.5	273/1.0	[53]
CPOP-13	1.25	890	77/1.0	5.20	273/1.0	16.8	273/1.0	[54]
PCTF-1	1.86	2235	77/1.0	1.82	273/1.0	14.40	273/1.0	[55]
COP-1	0.98	827	77/1.13	2.9	298/18	19.4	298/18	[56]
BILP-2	1.3	708	77/1.0	1.4	273/1.0	14.9	273/1.0	[57]
SHCP-3b	2.48	3002	77/1.13	-	-	21.2	273/1.0	[58]

Some selective gas separations based on porous polymers can be summarized as follows:

- In dry air containing Ar, H_2, N_2 and O_2, PAF-1, PAF-3 and PAF-4 may detect greenhouse gases and isolate them. At atmospheric pressure, PAF-1 uptakes more than 37 times more CO_2 than N_2 and adsorbs almost 15 times more CH_4 than N_2, while PAF-3 has 85 times selectivity in the case of CO_2 vs. N_2. The heavy associations between CO_2 and the polymer's Si core may account for PAF-3's remarkable selectivity [59, 60].
- When using PAF-1, the selectivity of adsorption for CO_2/N_2 can be as high as over 205 at a CO_2 to N_2 ratio of 15:85. Also, at a ratio of 15:85 for CO_2 to CH_4 at a pressure between 0 and 4 bar, the CO_2/CH_4 adsorption selectivity has to be in the range of 8–10, which is ideal for waste gas separation. For the 20:80 CO_2:H_2 mixture, the measured CO_2 to H_2 selectivity of adsorption may be about 392 at a temperature of 273 K and pressure of 1 bar [61, 62].
- The thermal heats of adsorption decrease from PAF-301 to PAF-304 as the pore size increases in those PAFs, as does the selectivity. PAFs with narrower pores and higher thermal heats of adsorption, such as PAF-301, have a high CO_2 adsorption potential and specificity for CO_2/H2, CO_2/N_2, CO_2/CH_4 and CH_4/H_2 mixtures at low pressure. PAF-303 and PAF-304, on the other hand, have greater pores and lower isosteric heats of adsorption, resulting in elevated CO_2 adsorption at higher pressure [63, 64].
- The carbon scaffold of the network and the tetrahedral monomers can be added to PPN-6 to provide free and interconnected pores. When polar groups are incorporated into the PPN network, the electric fields generated on the network's surface interfere strongly with the quadrupole moment of CO_2, increasing the uptake potential. The frameworks' pore volume is essential for selective greenhouse gas absorption, as polymers with the least pore size have the best specificity. The pore volumes as determined for PPN-6-SO3H, PPN-6-SO3Li and PPN-6 are 0.58, 0.52 and 2.44 cm^3/g, respectively, with the lithiated moiety having the highest adsorption selectivity for CO_2 due to its limited pore volume [65, 66].
- Zhou et al. discovered porous polymer networks (PPNs), such as PPN-1, PPN-2 and PPN-3 that have similar affinity for CO_2 and CH_4. PPN-1 was found to be the best choice among them, owing to its high surface area density in terms of volume. Furthermore, the system has the greatest influence on the heat of adsorption between CO_2 and CH_4. They used a dual-site Langmuir-Freundlich model to match the basic isotherm models of CO_2 and CH_4 for the three PPNs and projected selectivity for CO_2 over CH_4 using the ideal adsorption solution theory. The result shows that CO_2 has a greater saturation capacity, allowing it to be more selective under higher pressures [67, 68].

8.4 WATER TREATMENT

Water pollution is becoming more widespread across the world as a result of increased industrialization and massive urbanization. Industrial and urban processes, e.g. trigger chemical micropollutants, poisonous heavy metal ions, synthetic dyes and other toxins in water supplies, both of which have negative impacts on wildlife and health impacts. Adsorption is more efficient at removing soluble and insoluble chemical, inorganic and biological toxins from polluted water, thanks to a porous adsorbent's ability to preferentially adsorb certain specific chemicals from polluted water. Because pollutants of a given size and shape have easy access to the openings of porous adsorbents, adsorption may readily remove them [69].

Waste treatment is based on the porous membrane separation process which is an advanced separation process. A membrane consists of a thin layer of semi-permeable material and separates influents due to driving force mainly pressure difference. With advancements in membrane production technology, membrane separation process costing is continuing to decline. The processes can be classified as microfiltration (MF), ultrafiltration (UF), nanofiltration (NF) and reverse osmosis (RO) depending on its pore size.

Membrane separation is an influent filtration by virtue of the pores size variation of the membrane [70]. Figure 8.1 presents different types of membrane separation processes based on the pore size ranges of membranes constituents found in water and the effective pore size of the membrane [71].

The characteristics of different filtration processes are given in Table 8.4 – including their pore size, operating condition, nature influents and consisting polymer.

By incorporating heavy mercury chelating thiol groups into PAF-1, Ma and colleagues created a mercury trap method [72]. The obtained material has a large surface area and is well-scattered across the single-walled pore surface, with heavily populated but readily beneficial Hg(II) binding sites. Mercury levels in polluted water drop from 10 parts per million to 0.4 parts per billion, well below the permissible limit set by drinking water regulations (2 ppb). Fang and colleagues published a 3D carboxyl-functionalized COF made by post-synthetic modification of hydroxyl groups [73]. Because of the heavy interaction between Nd and the carboxyl groups in Nd/3D-COOH-COF, the substance has a high metal loading potential as well as excellent selectivity of adsorption for

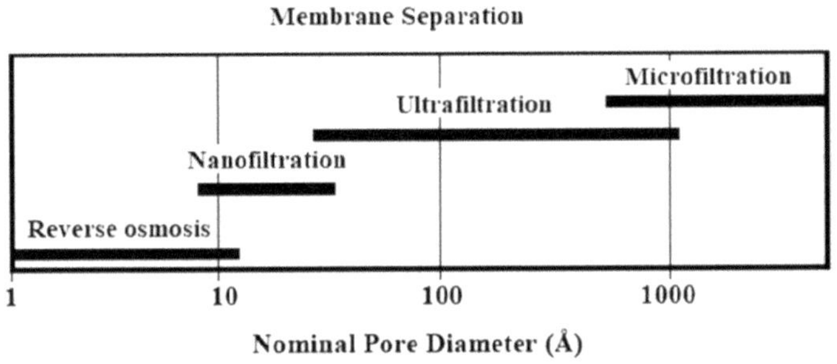

FIGURE 8.1 Range of pore diameters in different membrane separation processes.

TABLE 8.4
Characteristics of Different Filtration Processes [70]

Process	Pore Size	Molecular Cut-off (MWCO)	Operating Pressure (psi)	Influent	Polymeric Membrane Used
MF	~ 0.03–10 micron	>1000,000 Da	15–60 psi	Crypotosporidium Giardia lamblia, algae, cysts, clays, silt, sand and many bacterial species.	Polyvinylidene fluoride, cellulose acetate, polypropylene, polyacrylonitrile, polyethersulfone, polysulfone and others
UF	~ 0.002–0.1 microns	Approximately 10,000–100,000 Da	30–100 psi	UF successfully eliminates all the microbiological species which can be removed by MF also some viruses (but this is not a complete barrier to viruses) and humic materials.	Do
NF	~0.001 microns	1,000–100,000 Da	90–150 psi	Includes all bacteria, cysts, viruses, and humic materials	Polyamide-based materials or cellulose acetate
RO	~ 0.0001 micron	100–1000 Da	>250 psi	Effectively takes away radium, pesticides, natural organic materials, cysts, viruses and bacteria.	Do

neodymium (III) against iron (III) and strontium(II) ions [74]. Byun et al. developed a novel family of fluorinated polymers with limited microporosity for water handling [75]. Only methylene blue, whose hydrodynamic dimension goes beyond the COP-99's useable pore size, is size-selective over chargeable dye molecules including rhodamine B, brilliant blue G, and methylene blue, thereby eliminating these from water.

In another work, for the development of polyacrylonitrile, Dai and colleagues introduced an open structure containing chloride groups, which was then transformed into polyamidoxime for the extraction of uranium from seawater, using a dynamic micro-/meso-porous as the initiator for the transfer in radical polymerization [76]. Since most of the porous polymers have structures with cross-linking toward hydrophobicity and are unable to react to environmental triggers, their use in diverse aqueous media is severely restricted. By investigating this limitation, Wu and coworkers facilitated the simple and flexible emergence of a new type of water-dispersible, tactile stimulation fuzzy microporous polymeric nanospheres using a mixture of hyper-crosslinking chemistry dealt with hyper-crosslinking and radical polymerization with surface-initiated atom transfer [77]. To research the adsorption properties in an aqueous phase, they used water-soluble alizarin red as a model molecule. As compared to traditional microporous polymers, the MMP has a very high adsorption rate which boosts up a greater real adsorption potential per unit surface area as the water permeation allows the pores even more available to the adsorbates in the aqueous phase. These results can pave the way for the development and fabrication of advanced multifunctional porous polymers, pointing to a new avenue for the development of novel smart water purification adsorbents [78]. Dichtel and workmates recently developed a class of porous polymer networks based on cyclodextrin with the replacement of nucleophilic hydroxyl groups of β-CD by various electrophiles. The material, Bisphenol A, a part of plastics that disrupts human endocrine processes, is quickly removed using a tetrafluoroterephthalonitrile-crosslinked β-CD polymer by about 95% in just 10 min [79] which denotes a much higher adsorption rate compared to activated carbons against nonporous adsorbent materials like β-CD. The practical nanonetwork-structured polymer, whose network unit is a special hollow microporous nanosphere with an inbuilt poly(acrylic acid) lining, was stated by the Wu group [80]. The content displays the comparative accumulation of methyl violet along with malachite green which is appreciably greater than traditional polystyrene with nanostructured framework. It can be attributed to the influence of the interactions among simple dyes, PAA linings as well as unusual structures based on 3D nano-network.

The low-pressure membrane processes (MF, UF) may be associated with other conventional processes like the use of activated carbon, coagulation, etc., to remove particulate matter, protozoa, bacteria, etc., from the ground [81]. Xiangli et al. coagulation with UF to make highly turbid water of Taihu river potable. Another advanced separation process that includes porous membrane is membrane distillation which is an underdeveloped separation process [82]. Heat is used in this case as the driving force instead of pressure. Membranes specific for the MD process are yet not available commercially. The microporous hydrophobic polymeric membranes which are used for microfiltration purposes are generally applicable for this purpose. These membranes are characterized by their pore size and its' distribution, hydrophobicity and surface roughness. The common polymers used in this process are polytetrafluoroethylene, polypropylene, polyvinylidene fluoride and polydimethyl siloxane to a lesser extent [83]. There are three configurations of these polymeric membranes: flat, electrospun and hollow fiber membranes. Flat and hollow fiber. Polymer composites and nanocomposites are also used as membranes. In the case of nanocomposites, incorporation of inorganic nanoparticles like carbon nanoparticles, nanosilica or nanotitania increases hydrophobicity of the pristine polymer [84]. Different types of membrane distillation processes are there using porous membrane and the classification is presented in Figure 8.2.

This membrane technology is continuously growing and gaining attraction in water and wastewater treatment. There is enough scope for research in this field though continuous research is going on. The development of a new membrane will make the membrane distillation process more widely applicable.

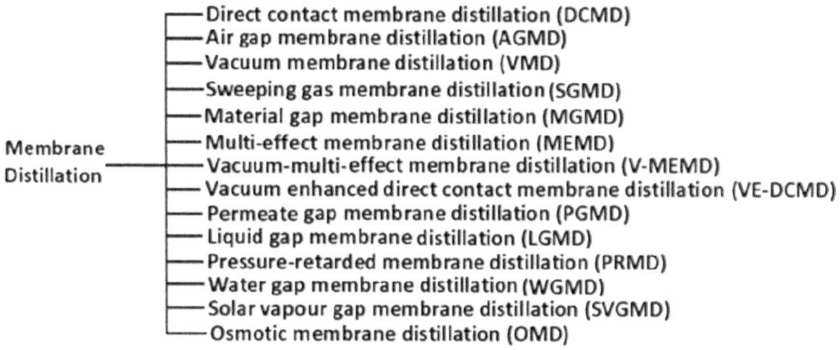

FIGURE 8.2 Various membrane distillation methods.

8.5 CONCLUSION

Porous polymers find numerous applications in separation. This chapter presented a summarized picture of that including the most important processes and the newly developed processes. It also highlighted their advantages and disadvantages. Probably, this chapter will be helpful in guiding further research into separation technology and the application of porous polymer.

ABBREVIATIONS

PIM	Polymer intrinsic microporosity
PVDF	Poly(vinylidene fluoride)
PAA-g-PVDF	Poly(acrylic acid) grafted PVDF
VOC	Volatile organic compounds
CMP	Conjugated microporous polymers
PAF	Porous aromatic frameworks
PPN	Porous polymer networks
MMP	Metal-induced ordered microporous polymers
ALP	Azo-linked polymersCTF-
TPC	Covalent triazine-based frameworks with triptycene
TNP	Triazole linked network polymers
PINK	Indole-based microporous organic polymersBILP-
6-NH2	Benzimidazole-linked polymers
CPOP	Carbazole-based porous organic polymers
COP	Covalent organic polymers
SHCP	Solvent knitting hyper-crosslinked microporous polymers
PCTF	Porous covalent triazine-based organic framework

REFERENCES

[1] Wu, D.; Xu, F.; Sun, B.; Fu, R.; He, H.; Matyjaszewski, K. Design and preparation of porous polymers. *Chem. Rev.*, 2012, 112, 3959–4015.
[2] Xu, L.-J.; Shi, X.-J.; Chai, M.-Y.; Ji, J.; Xu, Z.-K.; Wan, L.-S. Surface metallization of porous polymer materials for multifunctional applications. *Langmuir*, 2020, 36, 1454–1461.
[3] Sing, K.; Everett, D.; Haul, R.; Moscou, L.; Pierotti, R.; Rouquerol, J.; Siemieniewska, T. Reporting physisorption data for gas/solid systems with special reference to the determination of surface area and porosity. *Pure Appl. Chem.*, 1985, 57, 603–619.
[4] Fuoco, A.; Rizzuto, C.; Tocci, E.; Monteleone, M.; Esposito, E.; Budd, P.M.; Carta, M.; Comesaña-Gándara, B.; McKeown, N.B.; Jansen, J.C. The origin of size-selective gas transport through polymers of intrinsic microporosity. *J. Mater. Chem. A*, 2019, 7, 20121–20126.

[5] Qian, Q.; Asinger, P.H.; Jee, M.J.; Han, G.; Rodriguez, K.M.; Lin, S.; Benedetti, F.M.; Wu, A.X.; Chi, W.S.; Smith, Z.P. MOF-based membranes for gas separations. *Chem. Rev.*, 2020, 120, 8161–8266.

[6] Zou, L.; Sun, Y.; Che, S.; Yang, X.; Wang, X.; Bosch, M.; Wang, Q.; Li, H.; Smith, M.; Yuan, S.; Perry, Z.; Zhou, H.-C. Porous organic polymers for post-combustion carbon capture. *Adv. Mater.*, 2017, 29, 1700229.

[7] Lee, J.-S.M.; Cooper, A.I. Advances in conjugated microporous polymers. *Chem. Rev.*, 2020, 120, 2171–2214.

[8] Robeson, L.M.; The upper bound revisited. *J. Membr. Sci.*, 2008, 320, 390–400.

[9] Du, N.; Park, H.B.; Robertson, G.P.; Dal-Cin, M.M.; Visser, T.; Scoles, L.; Guiver, M.D. Polymer nanosieve membranes for CO_2 capture applications. *Nat. Mater.*, 2011, 10, 372–375.

[10] Che, H.; Huo, M.; Peng, L.; Fang, T.; Liu, N.; Feng, L.; Wei, Y.; Yuan, J. CO_2-responsive nanofibrous membranes with switchable oil/water wettability. *Angew. Chem. Int. Ed.*, 2015, 54, 8934–8938.

[11] Wang, Y.; Wang, J.; Ding, Y.; Zhou, S.; Liu, F. In situ generated micro-bubbles enhanced membrane anti-fouling for separation of oil-in-water emulsion. *J. Membr. Sci.*, 2021, 621, 119005.

[12] Carta, M.; Malpass-Evans, R.; Croad, M.; Rogan, Y.; Jansen, J.; Bernardo, P.; Bazzarelli, F.; McKeown, N. An efficient polymer molecular sieve for membrane gas separations. *Science*, 2013, 339, 303–307.

[13] Yi, S.; Ghanem, B.; Liu, Y.; Pinnau, I.; Koros, W.J. Ultraselective glassy polymer membranes with unprecedented performance for energy-efficient sour gas separation. *Sci. Adv.*, 2019, 5.

[14] Lv, Y.; Du, Y.; Qiu, W., Xu, Z.-K. Nanocomposite membranes via the codeposition of polydopamine/polyethylenimine with silica nanoparticles for enhanced mechanical strength and high water permeability. *ACS Appl. Mater. Interfaces*, 2017, 9, 2966–2972.

[15] Zhang, C.; Ou, Y.; Lei, W.X.; Wan, L.; Ji, J.; Xu, Z.K. $CuSO_4/H_2O_2$-induced rapid deposition of polydopamine coatings with high uniformity and enhanced stability. *Angew. Chem. Int. Ed.*, 2016, 55, 3054–3057.

[16] Zhou, H.; Tao, F.; Liu, Q.; Zong, C.; Yang, W.; Cao, X.; Jin, W.; Xu, N. Microporous polyamide membranes for molecular sieving of nitrogen from volatile organic compounds. *Angew. Chem. Int. Ed.*, 2017, 129, 5849–5853.

[17] Gao, X.; Xu, L.P.; Xue, Z.; Feng, L.; Peng, J.; Wen, Y.; Wang, S.; Zhang, X. Dual-scaled porous nitrocellulose membranes with underwater superoleophobicity for highly efficient oil/water separation. *Adv. Mater.*, 2014, 26, 1771–1775.

[18] Song, Q.; Cao, S.; Pritchard, R.H.; Ghalei, B.; Al-Muhtaseb, S.A.; Terentjev, E.M.; Cheetham, A.K.; Sivaniah, E. Controlled thermal oxidative crosslinking of polymers of intrinsic microporosity towards tunable molecular sieve membranes. *Nat. Commun.*, 2014, 5, 4813.

[19] Song, Q.; Cao, S.; Pritchard, R.H.; Ghalei, B.; Al-Muhtaseb, S.A.; Terentjev, E.M.; Cheetham, A.K.; Sivaniah, E. Controlled thermal oxidative crosslinking of polymers of intrinsic microporosity towards tunable molecular sieve membranes. *Nat. Commun.*, 2014, 4, 4813.

[20] Jimenez-Solomon, M.F.; Song, Q.; Jelfs, K.E.; Munoz-Ibanez, M.; Livingston, A.G. Polymer nanofilms with enhanced microporosity by interfacial polymerization. *Nat. Mater.*, 2016, 15, 760.

[21] Zhang, W.; Shi, Z.; Zhang, F.; Liu, X.; Jin, J.; Jiang, L. Superhydrophobic and superoleophilic PVDF membranes for effective separation of water-in-oil emulsions with high flux. *Adv. Mater.*, 2013, 25, 2071–2076.

[22] Che, H.; Huo, M.; Peng, L.; Fang, T.; Liu, N.; Feng, L.; Wei, Y.; Yuan, J. CO_2-responsive nanofibrous membranes with switchable oil/water wettability. *Angew. Chem. Int. Ed.*, 2015, 54, 8934–8938.

[23] Samitsu, S.; Zhang, R.; Peng, X.; Krishnan, M.R.; Fujii, Y.; Ichinose, I. Flash freezing route to mesoporous polymer nanofibre networks. *Nat. Commun.*, 2013, 4, 2653.

[24] Jiang, B.; Sun, Z.; Zhang, L.; Sun, Y.; Zhang, H.; Yang, H. Synthesis of a hypercrosslinked, ionic, mesoporous polymer monolith and its application in deep oxidative desulfurization. *J. Appl. Polym. Sci.*, 2018, 135, 46280.

[25] Zhang, W.; Zhu, Y.; Liu, X.; Wang, D.; Li, J.; Jiang, L.; Jin, J. Salt-induced fabrication of superhydrophilic and underwater superoleophobic PAA-g-PVDF membranes for effective separation of oil-in-water emulsions. *Angew. Chem. Int. Ed.*, 2014, 53, 856–860.

[26] Kota, A.K.; Kwon, G.; Choi, W.; Mabry, J.M.; Tuteja, A. Hygro-responsive membranes for effective oil–water separation. *Nat. Commun.*, 2012, 3, 1025.

[27] Ghanem, B.S.; Swaidan, R.; Litwiller, E.; Pinnau, I. Ultra-microporous triptycene-based polyimide membranes for high-performance gas separation. *Adv. Mater.*, 2014, 26, 3688–3692.

[28] Tsai, C.-Y.; Tai, H.-C.; Su, C.-A.; Chiang, L.-M.; Li, Y.-Y. Activated microporous carbon nanospheres for use in supercapacitors. *ACS Appl. Nano Mater.*, 2020, 3, 10380–10388.

[29] Bondar, V.I.; Freeman, B.D.; Pinnau, I. Gas transport properties of poly(ether-b-amide) segmented block copolymers. *J. Polym. Sci. Part B: Polym. Phys.*, 2000, 38, 2051–2062.
[30] Saeed, U.; Khan, A.L., Gilani, M.A.; Bilad, M.R.; Khan, A.U. Supported liquid membranes comprising of choline chloride based deep eutectic solvents for CO_2 capture: influence of organic acids as hydrogen bond donor. *J. Mol. Liq.*, 2021, 335, 116155.
[31] Khanipour, M.; Mirvakili, A.; Bakhtyari, A.; Farniaei, M.; Rahimpour, M.R. A membrane-assisted hydrogen and carbon oxides separation from flare gas and recovery to a commercial methanol reactor. *Int. J. Hydrog. Energy*, 2020, 45, 7386–7400.
[32] Alqaheem, Y.; Alomair, A.; Vinoba, M.; Pérez, A. Polymeric gas-separation membranes for petroleum refining. *Int. J. Polym. Sci.*, 2017, article ID 4250927.
[33] Shibasaki, Y.; Mori, T.; Fujimori, A.; Jikvi, M.; Sawada, H.; Oishi, Y. Poly(amide–ether) thermoplastic elastomers based on monodisperse aromatic amide hard segments as shape-memory and moisture-responsive materials. *Macromolecules*, 2018, 51, 9430–9441.
[34] U.S. Department of Energy, Membrane system for the recovery of volatile organic compounds from remediation off-gases. Innovative Technology Summary Report, U.S. Department of Energy, 2001.
[35] Raharjo, R.D.; Freeman, B.D.; Paul, D.R.; Sarti, G.C.; Sanders, E.S. Pure and mixed gas CH_4 and n-C_4H_{10} permeability and diffusivity in poly(dimethylsiloxane). *J. Membr. Sci.*, 2007, 306, 75–92.
[36] Wu, J.L.; Xu, F.; Li, S.M.; Ma, P.W.; Zhang, X.C.; Liu, Q.H.; Fu, R.W.; Wu, D.C. Porous polymers as multifunctional material platforms toward task-specific applications. *Adv. Mater.*, 2019, 31, 1802922.
[37] Lu, W.; Sculley, J.P.; Yuan, D.; Krishna, R.; Wei, Z.; Zhou, H.C. Polyamine-tethered porous polymer networks for carbon dioxide capture from flue gas. *Angew. Chem. Int. Ed.*, 2012, 51, 7480–7484.
[38] Yang, Y.; Zhang, Q.; Zhang, Z.; Zhang, S. Functional microporous polyimides based on sulfonated binaphthalene dianhydride for uptake and separation of carbon dioxide and vapors. *J. Mater. Chem. A*, 2013, 1, 10368–10374.
[39] Du, N.; Park, H.B.; Robertson, G.P.; Dal-Cin, M.M.; Visser, T.; Scoles, L.; Guiver, M.D. Polymer nanosieve membranes for CO_2-capture applications. *Nat. Mater.*, 2011, 10, 372–375.
[40] Patel, H.A.; Je, S.H.; Park, J.; Chen, D.P.; Jung, Y.; Yavuz, C.T.; Coskun, A. Unprecedented high-temperature CO_2 selectivity in N2-phobic nanoporous covalent organic polymers. *Nat. Commun.*, 2013, 4, 1357.
[41] Alam, A.; Mishra, S.; Hassan, A.; Bera, R.; Dutta, S.; Das Saha, K.; Das, N. Triptycene-based and Schiff-Base-linked porous networks: efficient gas uptake, high CO_2/N_2 selectivity, and excellent antiproliferative activity. *ACS Omega.*, 2020, 21, 4250–4260.
[42] Tilford, R.W.; Mugavero, S.J.; Pellechia, P.J.; Lavigne, J.J. Tailoring microporosity in covalent organic frameworks. *Adv. Mater.*, 2008, 20, 2741–2746.
[43] Jiang, J.X.; Su, F.; Trewin, A.; Wood, C.D.; Niu, H.; Jones, J.T.A.; Khimyak, Y.Z.; Cooper, A.I. Synthetic control of the pore dimension and surface area in conjugated microporous polymer and copolymer networks. *J. Am. Chem. Soc.*, 2008, 130, 7710–7720.
[44] Liang, Y.; Mai, W.; Huang, J.; Huang, Z.; Fu, R.; Zhang, M.; Wu, D.; Matyjaszewski, K. Novel hollow and yolk–shell structured periodic mesoporous polymer. *Chem. Commun.*, 2016, 52, 2489–2492.
[45] Taylor, D.; Dalgarno, S.J.; Xu, Z.; Vilela, F. Conjugated porous polymers: incredibly versatile materials with far-reaching applications. *Chem. Soc. Rev.*, 2020, 49, 3981–4042.
[46] Yuan, Y.; Cui, P.; Tian, Y.; Zou, X.; Zhou, Y.; Sun, F.; Zhu, G. Coupling fullerene into porous aromatic frameworks for gas selective sorption. *Chem. Sci.*, 2016, 7, 3751–3756.
[47] Dey, S.; Bhunia, A.; Esquivel, M.D.; Janiak, C. Covalent triazine-based frameworks (CTFs) from triptycene and fluorene motifs for CO_2 adsorption. *J. Mater. Chem. A*, 2016, 4, 6259–6263.
[48] Mondal, S.; Das, N. Triptycene based 1,2,3-triazole linked network polymers (TNPs): small gas storage and selective CO_2 capture. *J. Mater. Chem. A*, 2015, 3, 23577–23586.
[49] Tan, L.; Tan, B. Hypercrosslinked porous polymer materials: design, synthesis, and applications. *Chem. Soc. Rev.*, 2017, 46, 3322–3356.
[50] Zheng, J.; Wang, C.-G.; Zhou, H.; Ye, E.; Xu, J.; Li, Z.; Loh, X.J. Current research trends and perspectives on solid-state nanomaterials in hydrogen storage. *Research*, 2021, article ID 3750689.
[51] Wang, Z.; Yuan, S.; Mason, A.; Reprogle, B.; Liu, D.J.; Yu, L. Nanoporous porphyrin polymers for gas storage and separation. *Macromolecules*, 2012, 45, 7413–7419.
[52] Chang, G.; Shang, Z.; Yu, T.; Yang, L. Rational design of a novel indole-based microporous organic polymer: enhanced carbon dioxide uptake via local dipole–π interactions. *J. Mater. Chem. A*, 2016, 4, 2517–2523.

[53] Islamoglu, T.; Behera, S.; Kahveci, Z.; Tessema, T.D.; Jena, P.; El-Kaderi, H.M. Enhanced carbon dioxide capture from landfill gas using bifunctionalized benzimidazole-linked polymers. *ACS Appl. Mater. Interfaces*, 2016, 8, 14648–14655.
[54] Zhu, J.H.; Chen, Q.; Sui, Z.Y.; Pan, L.; Yu, J.; Han, B.H. Preparation and adsorption performance of cross-linked porous polycarbazoles. *J. Mater. Chem. A*, 2014, 2, 16181–16189.
[55] Wang, H.; Jiang, D.; Huang D. et al. Covalent triazine frameworks for carbon dioxide capture. *J. Mater. Chem. A*, 2019, 7, 22848–22870.
[56] Chen, S.; Lucier, B.E.G.; Luo, W.; Xie, X.; Feng, K.; Chan, H.; Terskikh, V.V.; Sun, X.; Sham, T.-K.; Workentin, M.S.; Huang, Y. Loading across the periodic table: introducing 14 different metal ions to enhance metal–organic framework performance. *ACS Appl. Mater. Interfaces*, 2018, 10, 30296–30305.
[57] Sadak, A.E.; Karakuş, E.; Chumakov, Y.M.; Dogan, N.A.; Yavuz, C.T. Triazatruxene-based ordered porous polymer: high capacity CO_2, CH_4, and H_2 capture, heterogeneous Suzuki–Miyaura catalytic coupling, and thermoelectric properties. *ACS Appl. Energy Mater.*, 2020, 3, 4983–4994.
[58] Wang, S.; Zhang, C.; Shu, Y.; Jiang, S.; Xia, Q.; Chen, L.; Jin, S.; Hussain, I.; Cooper, A.I.; Tan, B. Layered microporous polymers by solvent knitting method. *Sci. Adv.*, 2017, 3, e1602610.
[59] Ben, T.; Pei, C.; Zhang, D.; Xu, J.; Deng, F.; Jing, X.; Qiu, S. Gas storage in porous aromatic frameworks (PAFs). *Energy Environ. Sci.*, 2011, 4, 3991–3999.
[60] Xiang, Z.; Cao, D. Porous covalent–organic materials: synthesis, clean energy application and design. *J. Mater. Chem. A*, 2013, 1, 2961.
[61] Ben, T.; Li, Y.; Zhu, L.; Zhang, D.; Cao, D.; Xiang, Z.; Yao, X.; Qiu, S. Selective adsorption of carbon dioxide by carbonized porous aromatic framework (PAF). *Energy Environ. Sci.*, 2012, 5, 8370–8376.
[62] Chen, S.; Li, P.; Xu, S.; Pan, X.; Fu, Q.; Bao, X. Carbon doping of hexagonal boron nitride porous materials toward CO_2 capture. *J. Mater. Chem. A*, 2018, 6, 1832–1839.
[63] Yang, Z.; Peng, X.; Cao, D. Carbon dioxide capture by PAFs and an efficient strategy to fast screen porous materials for gas separation. *J. Phys. Chem. C*, 2013, 117, 8353–8364.
[64] Kupgan, G.; Abbott, L.J.; Hart, K.E.; Colina, C.M. Modeling amorphous microporous polymers for CO_2 capture and separations. *Chem. Rev.*, 2018, 118, 5488–5538.
[65] Tian, Y.; Zhu, G. Porous Aromatic Frameworks (PAFs). *Chem. Rev.*, 120, 8934–8986.
[66] Ben, T.; Qiu, S. Porous polymers: design, synthesis and applications. *RSC*, 2016, 256–257.
[67] Xiang, Z.; Zhou, X.; Zhou, C.; Zhong, S.; He, X.; Qin, C.; Cao, D. Covalent-organic polymers for carbon dioxide capture. *J. Mater. Chem.*, 2012, 22, 22663–22669.
[68] Reddy, M.S.B.; Ponnamma, D.; Sadasivuni, K.K.; Kuman, B.; Abdullah, A.M. Carbon dioxide adsorption based on porous materials. *RSC Adv.*, 2021, 11, 12658–12681.
[69] Mohan, D.; Sarswat, A.; Ok, Y.S.; Pittman, Jr. C.U. Organic and inorganic contaminants removal from water with biochar, a renewable, low cost and sustainable adsorbent – A critical review. *Bioresour. Technol.*, 2014, 160, 191–202.
[70] Ezugbe, E.O.; Rathilal, S. Membrane technologies in wastewater treatment: a review. *Membrane*, 2020, 10, 89.
[71] Singh, R.; Hankins, N. *Emerging membrane technology for sustainable water treatment*, Elsevier, Amsterdam, The Netherlands, 2016.
[72] Li, B.; Zhang, Y.; Ma, D.; Shi, Z.; Ma, S.; Mercury nano-trap for effective and efficient removal of mercury(II) from aqueous solution. *Nat. Commun.*, 2014, 5, 5537.
[73] Lu, Q.; Ma, Y.; Li, H.; Guan, X.; Yusran, Y.; Xue, M.; Fang, Q.; Yan, Y.; Qiu, S.; Valtchev, V.; Postsynthetic functionalization of three-dimensional covalent organic frameworks for selective extraction of lanthanide ions. *Angew. Chem. Int. Ed.*, 2018, 57, 6042–6048.
[74] Ruoyang, L.; Ke, T.T.; Yifan, G.; Yongzhi, C.; Zhuoer, L.; Shuailei, X.; Ting, H.; Zhen, L.; Hao, Y.; Donglin, J. Covalent organic frameworks: an ideal platform for designing ordered materials and advanced applications. *Chem. Soc. Rev.*, 2021, 50, 120–242.
[75] Byun, J.; Patel, H.A.; Thirion, D.; Yavuz, C.T. Charge-specific size-dependent separation of water-soluble organic molecules by fluorinated nanoporous networks. *Nat. Commun.*, 2016, 7, 13377.
[76] Yue, Y.; Mayes, R.T.; Kim, J.; Fulvio, P.F.; Sun, X.G.; Tsouris, C.; Chen, J.; Brown, S.; Dai, S. Seawater uranium sorbents: preparation from a mesoporous copolymer initiator by atom-transfer radical polymerization. *Angew. Chem. Int. Ed.*, 2013, 52, 13458–13462.
[77] Mai, W.; Sun, B.; Chen, L.; Xu, F.; Liu, H.; Liang, Y.; Fu, R.; Wu, D.; Matyjaszewski, K. Water-dispersible, responsive, and carbonizable hairy microporous polymeric nanospheres. *J. Am. Chem. Soc.*, 2015, 137, 13256–13259.

[78] Zhang, W.; Zhao, Q.; Yuan, J. Porous polyelectrolytes: the interplay of charge and pores for new functionalities. *Angew. Chem. Int. Ed. Engl.*, 2018, 57, 6754–6773.
[79] Xiao, L.; Ling, Y.; Alsbaiee, A.; Li, C.; Helbling, D.E.; Dichtel, W.R. β-cyclodextrin polymer network sequesters perfluorooctanoic acid at environmentally relevant concentrations. *J. Am. Chem. Soc.*, 2017, 139, 7689–7692.
[80] Mai, W.; Zuo, Y.; Li, C.; Wu, J.; Leng, K.; Zhang, X.; Liu, R.; Fu, R.; Wu, D. Functional nanonetwork-structured polymers with inbuilt poly(acrylic acid) linings for enhanced adsorption. *Polym. Chem.*, 2017, 8, 4771–4775.
[81] Stoquart, C.; Servais, P.; Bérubé, P.R.; Barbeau, B. Hybrid membrane processes using activated carbon treatment for drinking water: a review. *J. Membr. Sci.*, 2012, 411–412, 1–12.
[82] Xiangli, Q.; Zhenjia, Z.; Nongcun, W.; Wee, V.; Low, M.; Loh, C.S.; Teck Hing, N. Coagulation pretreatment for a large-scale ultrafiltration process treating water from the Taihu River. *Desalination*, 2008, 230, 305–313.
[83] Mohammaad, M.A.; Shirazi, D.B.; Ali, K.; Meisam, T. Characterization of polymeric membranes for membrane distillation using atomic force microscopy. *Desalin. Water Treat.*, 2013, 51, 6003–6008.
[84] Ravi, J.; Othman, M.H.D.; Matsuura, T.; Bilad, M.R.; El-Badawy, T.H.; Aziz, F.; Ismail, A.F.; Rahman, M.A.; Jaafar, J. Polymeric membranes for desalination using membrane distillation: a review. *Desalination*, 2020, 490, 114530.

9 Porous Polymers for Membrane Applications

Aneela Sabir, Humaira Idrees, Muhammad Shafiq, and Rafi Ullah Khan
Institute of Polymer& Textile Engineering, University of the Punjab, Lahore, Pakistan

CONTENTS

9.1 INTRODUCTION

The materials technology has always been an essential field from ancient times and helpful in the great progress of civilization. The development of nanomaterials introduces huge applications in different fields of life such as paint industries, environmental field, corrosion-resistant field, as coatings, information technology field, etc., because of its structural and thermal properties, ease of transportation, energy storage, sequestration of industrial waste, etc., [1–6]. The novel properties of nanomaterials can be obtained by mixing them with other materials [7–9]. The main matrix of porous material is often called the skeleton of material, mostly gas or aqueous solution that fills the pore spaces. Mostly the surrounding material is in the form of solid to show the best interaction with inducing materials [1].

Over time, porous polymers attained remarkable attention because of their huge applications in the field of biomedicine, catalytic field, energy storage, separation, etc. [10–14]. Porous polymers

DOI: 10.1201/9781003169604-9

have exclusive properties like greater surface area, greater porosity, excellent stability, freely functionalizable scaffolds [10, 15, 16]. Porosity can be helpful to recognize the nature of synthesized novel structures. Some unique examples of such structures in nature are honeycomb structure, the structure of alveoli inside lungs, bamboo sticks, etc. Researchers gain an idea from nature to synthesize materials that mimic naturally existing porous things; their size varies from macro to nano [17]. Porous polymers can be easily synthesized having great advantages and applications in different fields of life.

The porous polymers include polymers with intrinsic microporosity (PIMs), covalent organic frameworks (COFs), polyacetylenes, covalent triazines framework, hyper-crosslinked polymers, porous aromatic framework, conjugated microporous polymers (CMPs), etc. These polymers have crosslinked structure and show better properties as compared to the linear polymers. The solubility of porous polymers depends upon the type of organic solvent. PIMs are easily soluble while others have insoluble nature. Hyper-crosslinked polymers are prepared by different chemistry with different properties, though they have a lack of structural diversity but are more accessible and cheaper. PIMs are amorphous in nature with rigid packing abilities [10, 18]. Some of the porous polymers can be made easily soluble by using solvent-based processes without damaging the porosity and this property is nearly impossible to develop in other porous polymers such as zeolites, carbons and porous silica.

Availability of numerous ways of synthesis and preparation of various porous polymers help in the incorporation of many chemical modifications into the porous skeleton structure [19–24]. The material can be modified to prepare stimuli-responsive properties responsible for reversibly altering the pore structure [25–28], switching the closing and opening of porous state when exposed to environment stimulus [29–32]. Such exceptional properties are usually not present in other porous materials. Furthermore, because of its organic nature, the polymeric network is comprised of elements that show importance in various applications [17].

Porous polymers can be used in separation processes and also in gas storage processes. They are also used in controlled drugs release, catalysts, used as scaffolds, in packaging, photonic band gap materials, membrane filtration process, used in lithography, used in chromatography, used as energy storage material in electrode, used as structure template for structure duplication, as coating materials, etc. [33–41]. So, these wide range of applications give a researcher a gateway to develop novel and reliable porous polymeric materials with great modification and also with modified and economical features. Porous polymers with a single pore are usually called hollow polymers such as in the form of tubes and spheres [42–50].

Moreover, the use of porous polymers started after the First World War. According to IUPAC, microporous polymers are defined as porous polymers with a pore size of 2 nm in diameter, mesoporous polymers have a pore size of 2–50 nm and macroporous polymers have a pore size of 50 nm [17, 51, 52]. Polymers with pore size in micrometer are also called microporous polymers [53–55]. Dingaci et al. (2012) described in the review that there are many structural properties of porous polymers such as surface modification, the geometry of pores, pore size, polymer network such as functionality, topography and composition of polymeric material. Figure 9.1 shows the pore size, geometry and framework.

Geometry of porous polymers are usually tubular, spherical of in form of network that are mostly in form of order and in irregular format. Porous polymers are present in various shapes such as fibers, particles, wires, foams, membranes, etc. Due to important properties of porous polymers such as size, shape, density, huge surface to volume ratio which allows a greater number of functional groups to incorporate into the pores of the material surface. Porous materials are usually alumina, mesoporous silica and inorganic polymers which show an extensive level of applications.

Polymeric membranes are the main study field with experiments. Many processes are present for membrane synthesis as mentioned in Figure 9.2. Top down and bottom up synthesis method [56]. Membranes show substantial performance in various fields of life; in order to get a membrane with successful properties, it will merge with the polymeric material. The use of a polymeric membrane

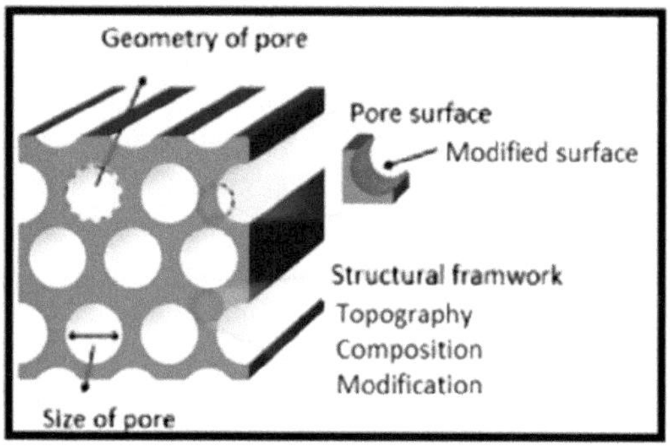

FIGURE 9.1 Shows the pore size, geometry and framework of materials.

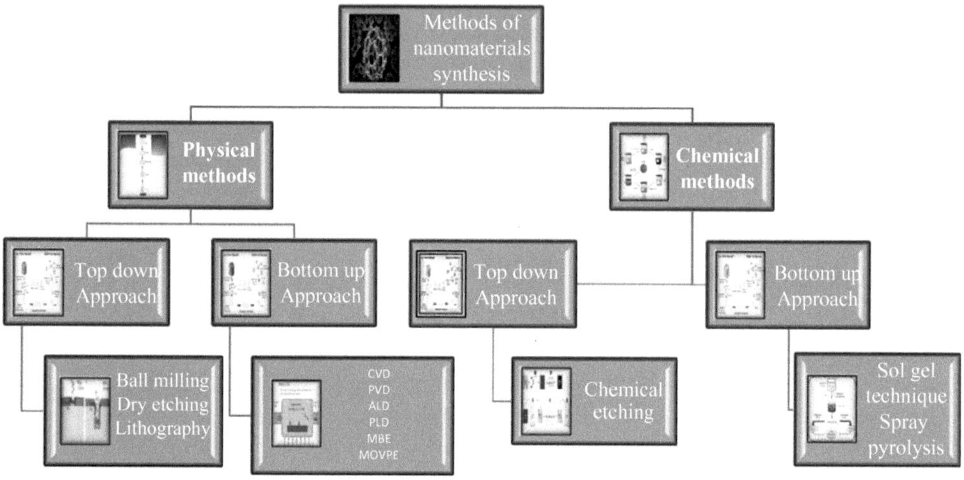

FIGURE 9.2 Shows the different methods of nanomaterials synthesis.

is economical due to its cost, simplicity of synthesis and mechanical reliability [57, 58]. Membrane technologies are applicable in various fields of life such as pharmaceutical, biotechnology, drug release, packaging and food industries, in renewable energy fields as conducting membranes, used in membrane redox flow battery, hemodialysis, as scaffolds, used in the artificial photosynthetic system for water filtration and media filtration [58]. The present chapter describes routes and schemes for porous membrane synthesis and their application in various fields of life.

9.2 INTRODUCTION TO SYNTHESIS OF POROUS POLYMERIC PARTICLES

Form last many years' researchers used physical principle plan their reactors as compared to chemical methods. The important factors which are considered for the synthesis of polymeric particles are pressure, temperature, viscosity and fluid dynamics; stirring and liquid immiscibility are also points of interest for researchers for particle synthesis. As we know that oil and water are immiscible for their solubility stabilizer are added in it, emulsifier, surfactant, surface activated substance are the terminologies used. The addition of emulsifiers reduces the surface tension and produces an interaction between oil and water molecules. Milk is a known example of the emulsion process. In the early 19's scientists realized that emulsion is the process to produce polymer particles [59]. In the emulsion process, water is the continuous phase and oil droplets are the discontinuous phase which are also hydrophobic in nature. After free radical polymerization they are converted into particles by the addition of emulsifiers. Stabilizers are usually soap, detergents, gum, starch or gelatin used. Inverse heterogeneous polymerization is a type of polymerization in which water droplets are the discrete phase in the secondary (oil) phase. The formation of spherical shape of polymers can be formed due to interfacial tension in two phases, that's why in most cases the shape of polymeric particles is spherical. The spherical shape shows less surface tension and surface-to-volume ratio. Techniques usually used for the synthesis of polymeric particles are microemulsion, suspension, precipitation, microfluidics polymerization and membrane emulsification.. Due to some limitations of the emulsion process, microemulsion and miniemulsion are not used for the synthesis of porous polymeric particles [60].

Soluble porous polymers such as PIMs) are used in membrane formation for the separation process. Membranes synthesized with PIM-1 and PIM-7 show a superb gas flux rate with outstanding discrimination, exceeding the permeability–selectivity trade-off growing in polymer membranes [61]. Intrinsic microporous polymers have less intermolecular cohesive interaction with a large amount of free volume due to which they show the best separation efficiencies [62]. Development of permeability and selectivity modification of physical durability of PIM membranes are promising including many steps such as copolymerization, post-modification [45, 63], etc. Although intrinsic microporous membranes have the best mechanical strength and permeability, they do not show targeted selectivity. Introducing fillers into the porous polymeric membranes induces unique selectivity properties. example of polyacetylenes are intrinsic microporous polymers are soluble and film forming abilities [10].

9.3 PREPARATION OF POROUS POLYMERIC MEMBRANE

Different techniques are used for the synthesis of membranes and films from soluble COFs and CMPs. Different synthesis methods are also introduced to prepare the insoluble porous polymeric membranes. Techniques used for the synthesis of the membrane are: exfoliation, solvothermal, interfacial polymerization, layer-by-layer assembly, these methods successfully synthesized the membranes [64–68]. Composite membranes have multiple layers of porous polymers with extraordinary properties with novel interlayer properties [71–73]. Solvothermal process is used for the synthesis of composite membrane that shows great chemical bonding in between them [69].

9.4 MORPHOLOGY OF MEMBRANE AND ITS PARAMETERS

Generally, the parameters significant for membranes are tortuosity, gravimetric porosity, pore size distribution, molecular weight cutoff and thickness, surface roughness. Tortuosity is the reciprocal of the average ratio between the connecting two points (membrane thickness) and the real path length inside the porous medium (on average). Tortuosity provides some data on the change between the real internal pore shape and an ideal structure [70]. The measurement of the total quantity of void in a membrane is the gravimetric porosity and it will show the ratio between density of the neat matrix and membrane density. Size of pore characterizes the dimensions of the pores, which are networks of adjustable cross-section [71]. If some irregularity is present in pores, then approximations is made to account an average size of pore. Mostly, equal spherical shape is predictable, but many illustrations were suggested such as non-spherical particles, cage-like/mesoporous structure, etc. Some statistical analyses such as Monte Carlo integration and non-linear optimization are applied to find the pore size distribution of pores for irregular pores [72].

The molecular weight cutoff (MWCO) is the lowest molecular weight solute for which 90% of the solute is retained by the membrane [73]. Roughness of the Surface is counted by the non-conformities in the direction of the normal vector of an actual surface from its perfect geometric flat shape and dimensions. The thickness shows the distance between both surfaces of a membrane. On the basis of application and characteristics membranes are divided into four types such

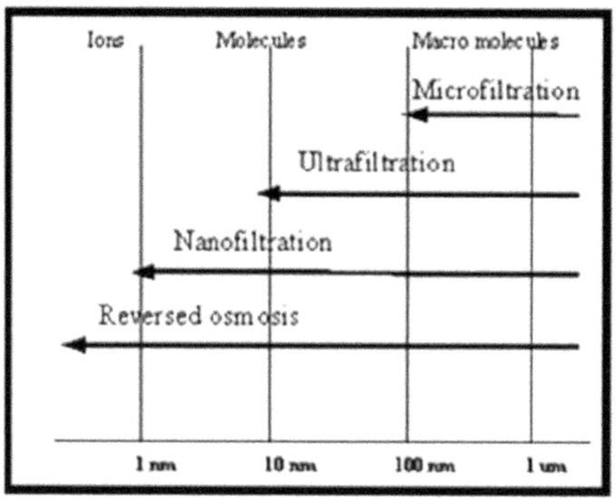

FIGURE 9.3 Show types of membrane with pore size.

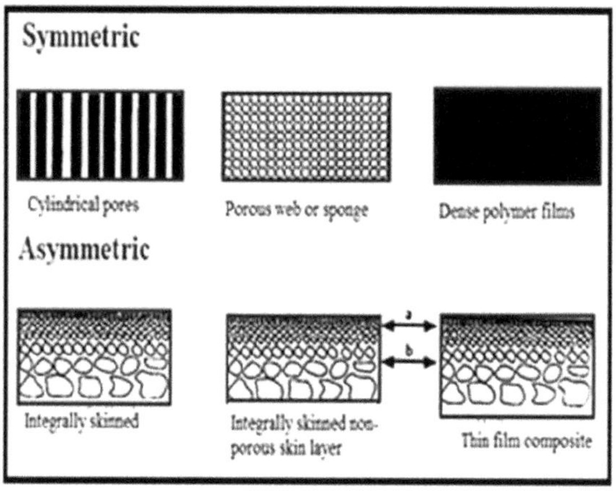

FIGURE 9.4 Show the types asymmetric and symmetric membranes.

as microfiltration, ultrafiltration, nanofiltartion and reverse osmosis. Figure 9.3 represents the membrane type with pore size [74].

On the basis of the structure, membranes are divided into two categories: asymmetric and symmetric membranes. In asymmetric membranes, pore size is not uniform and having a dense surface layer functional porous layer present on the top and the remaining are below and in symmetric membranes pore size is uniformly distributed. Figure 9.4 shows the types of asymmetric and symmetric membranes

9.5 EMERGING APPLICATIONS OF POROUS POLYMER MEMBRANES

These are used in various fields of life, for example, organic solvent nanofiltartion, desalination, separation of gases, removal of dyes, etc. [75–78].

COFs, CMPs and PIMs are used in electronic and photonic applications. Khayum et al. [79] reported that COFs and PIMs are laid toward the application of electronic wearables. Due to extensive mechanical application, a large contact area and more flexible covalent organic nanosheets are applicable for biomedical applications [80]. Drug delivery is the extensive application of these polymers used as a therapeutic for cancer cell treatment [81]. Modification of COF with photosynthetic procedure allowed the improvement of COF for cancer cell therapy. Zhnag et al. reported that the use of CONs shows a promising application in sensing DNA sequence, as bioinformatics and genomic analysis [84]. Ionic CONs (iCONs) may find sufficient opportunity in biomedicine in consideration of the ionic nano–bio interfaces, used as protein transport and fluorescent detection in DNA [85, 86]. Mitra et al. reported the anti-microbial property of iCONs against *Escherichia coli* bacteria and *Staphylococcus aureus*. The biomedical applications of CON are also reported such as cellular imaging, wound dressing and artificial enzymes [67].

9.6 POLYSULFONE AND POLYVINYLIDENE FLUORIDE USED AS POROUS POLYMERS FOR MEMBRANE APPLICATION

9.6.1 Polysulfone Membranes

Polysulfone (PS) is a significant thermoplastic beneficial for the synthesis of polymer membranes. This is related to the structure of its repeating unit $[OC_6H_4OC_6H_4SO_2C_6H_4]n$ providing outstanding properties such as chemical steadiness, pH sensitivity, mechanical stability, thermal stability, chemical inertness over the whole pH range, great process ability factor and process ability. PS membranes have great application in membrane oxygenators, membrane desalination, wastewater treatment, pervaporation, gas separation, support for gas separation and as separators in lithium-ion battery. The methods used for the synthesis of PS membranes are electrospinning and phase inversion method [58].

9.6.2 Polyvinylidene Fluoride Membranes

Polyvinylidene fluoride (PVDF) is semi-crystalline polymer used for the synthesis of membranes; this polymer has CF_2–CH_2 repeating units. Based on its excellent mechanical, chemical, thermal and hydrophobicity stability, PVDF membranes are preferred more as compared to other polymeric membranes [58, 82]. Properties of PVDF are given in Table 9.1.

9.7 USE OF POROUS POLYMERIC MEMBRANES FOR SENSING APPLICATION

Brijesh Prasad et al. have reported that in the last few decades, very little work has been done in the field of polymer sensors. Different techniques are used to make the material piezo-resistive and few of the researches done on analyzing and morphological changes occur in the material. Brijesh and his co-workers synthesized the porous structure by using PVDF and making its composite as PVDF/CNF/IL as a sensing application in both wet and dry conditions. Soft polymer sensors are valued for properties of sensors, electronic devices, human health and structural health uses. PVDF has been mostly used by many scientists for the improvement of sensing applications. In the case of sensing behavior, surface morphology is important because the surface acts as receptors. Stress and strain are important parameters for sensing applications as a smaller change in the shape of polymer can affect the performance of sensors. They introduced the PVDF/CNF/IL model sensing membranes with improved properties such as increased formation of pores and improved stress-strain behavior. With increase in pores formation, the concentration of stress also increases [83].

TABLE 9.1
Shows the Properties of PVDF Membranes

Properties	Range
Melting point °C	140–170
Crystallinity (%)	35–75
Glass transition temperature °C	–41–38

9.8 USE OF POROUS POLYMERIC ELECTROLYTIC MEMBRANES APPLICATION

Porous polymers are also used for the formation of photoelectrochemical (PEC) cell applications.

The conventional PEC cells depend upon planar photoelectrodes reinforced on glass substrates and liquid electrolytes. Very few studies have been reported which are scalable as an alternate of PEC. The key element use for the synthesis of successful PEC having porous polymeric electrolytic membranes (PEMs) and photoelectrodes. G. Zafeiropoulos et al. proposed the novel cells based on titania photoelectrodes and proton and hydroxide ion-conducting membranes. Two families of photoelectrodes were synthesized on Ti porous substrate; one is the TiO_2 nanotubes developed by anodization and subsequent oxygen annealing, and the second layers of TiO_2 developed under oxygen annealing. Primary screening of the photoanodes for water splitting and (poly)alcohol photo-oxidation took place in conventional PEC cells. The annealing temperature affects the efficiency of photoanode; a monotonic increase in the performance of water photo-oxidation increases the annealing temperature. Moreover, it was established that the anatase phase is mainly active for the (poly) alcohol electro-oxidation. The most favorable and novel photoanodes of PEMs PEC (photoelectrochemical cell) were developed for water splitting during gas-phase process. During performance testing, it was investigated that (PEM-PEC) membranes are promising one as compared to the conventional PECs in both alkaline and acidic media [84].

9.9 USE OF POROUS POLYMERIC MEMBRANE FOR NUMERICAL MODELING AND OPTIMIZATION

Mashallah Rezakazemi et al. synthesized the porous membranes for ammonia removal from water. The novel model they synthesized consists of adaptive neuro-fuzzy interface system (ANFIS) along with hollow-fiber membranes to regulate the ammonia present in water. They proposed the method for ammonia striping by using the Taguchi method. The primary properties which are important for ammonia separation are initial concentration of ammonia, ammonia solution pH, presence of excess ions in ammonia solution and velocity of the solution act as input properties. Mass transfer coefficient is considered as output property. The prime thing of this method is the pH which plays an important role in the separation of ammonia from water. The ANFIS model is used specially for porous membranes [85].

9.10 USE OF POROUS POLYMERS FOR BIOMEDICAL APPLICATION

Porous polymers show extensive potential applications such as energy storage, catalytic reactions, wastewater treatment, etc. They also show great application in the biomedical field, for biomedical application biocompatibility and biodegradation, prevention of biofouling and many biomedical responses are important. Surface modification of the porous polymer is important to increase the functional properties of the membrane as a biomedical application. Surface modification of porous polymers means developing of covalent bond on the surface to get tremendous results as drug delivery [86].

9.11 USE OF POROUS POLYMERIC MEMBRANE IN TISSUE ENGINEERING

The use of porous polymeric membranes as scaffolds for tissue regeneration is in latest consideration among the scientific community. For this application of scaffolds, biocompatibility, space preservation ability and medical manageability have important values. Normally, natural or bio-polymers such as chitosan, collagen, are used for the biomedical application membranes, and synthetic polymers such as Polylactic acid (PLA), Polycaprolactone (PCL) and prolonged polytetrafluoroethylene (e-PTFE) are used. Both synthetic and natural porous polymeric membranes have great applications in tissue engineering as they show better biocompatibility and biodegradation properties. Collagen without showing foreign-body reaction shows good osteoblastic adhesion, good tissue integration and speedy vascularization. Collagen is also stimulated by inducing fibroblast and hemostasis processes as compared to polystyrene and polyethylene which are synthetic in nature. Despite having many applications collagen has some drawbacks due to less mechanical strength and greater biodegradation; improvement of mechanical strength is carried out by inducing reinforcement material in the surface of collagen. Crosslinkers such as glutaraldehyde, oligomeric proanthocyanidins and genipin are used to strength. By introducing a crosslinker, the strength of collagen increases from 0% to 20%. Synthetic polymers such as PEG, PCL, PLA and poly (glycolic acid) (PGA) due to their inflammatory behavior show limited properties in tissue engineering applications. PLA is a food and drug administration approved polymer used in tissue engineering because of its better mechanical and biocompatible properties. To overcome the hydrophilicity and degradation behavior of PLA, it is combined with polylactic acid/polyglycolic acid (PLGA).

Mixed matrix membrane of collagen and PLGA shows great importance in wound dressing, periodontal applications. Composite porous membranes of PLA and PLGA have some drawbacks such as they are inflammable and stiff. To reduce the stiffness of PLA and PLGA membrane, some softening agents such as N-methyl-2-pyrrolidone (NMP) and hydroxyapatite, are added. PCL is also another polymer which is approved by FDA for the tissue engineering application, PCL, PLA, PLGA porous polymeric membrane due not produce any acidic species during the degradation process for at least 2–3 years. Moreover, their degradation kinetics can be modified by adjusting the PEG composition [86].

9.12 USE OF POROUS POLYMERIC MEMBRANE IN WASTEWATER TREATMENT

Polymeric membranes are mostly applicable for wastewater treatment. However, the membranes used for pressure-driven membrane processes face unwanted procedure of biofouling or fouling, due to the productivity drop and rise in treatment costs. Forward osmosis membranes facing less fouling behavior, less permeation ability have restricted possibility in wastewater treatment. Many nanomaterials have been introduced which show anti-biofouling/antifouling, permeability. Amanda Grylewiz et al. presents a review which contains literature on polymer mixed-matrix membranes having modified halloysite nanotubes (HNTs). The effect of HNTs on the performance or properties of membranes of polymers such as PVDF, polysulfone, polystyrene and polyether sulfone are widely discussed. Moreover, the key issues related to research on the mixed-matrix membranes modified with HNTs were suggested [87].

9.13 USE OF POROUS POLYMERIC MEMBRANE FOR DYE REJECTION APPLICATION

Xiaocan Liu et al. proposed the hybrid porous polymeric membrane which is helpful in organic dye rejection and it's also economical. The hybrid polymers are amphiphilic such as sulfonated polyarylene ether nitrile by consuming polyethylene glycol (PEG) and titanium dioxide (TiO_2) used as pore-forming agents and photocatalysts. Moreover, in this study, PEG-TiO_2 showed the key role in surface wettability and morphology and their absorption capacity toward methyl blue dye increases. Also, the concentration of PEG-TiO_2 increases attained hybrid porous membrane showed primarily improved and reduced equilibrium porosity and water content, which show the deviation of adsorption capabilities near MB (Methyl blue). On the contrary, the increased loading content of PEG-TiO_2 renders the obtained hybrid porous membrane with greater pore size and more hydrophilic morphology, which leads toward 99.9% MB absorption by hybrid porous membrane [88].

9.14 POROUS POLYMERIC MEMBRANE ANTIFOULING APPLICATION

Fouling is the main difficulty in wastewater treatment membranes. Therefore, modification of the surface with a zwitterion is useful for antifouling. Yuhei Oshiba et al. proposed a method in which they modified the surface of the polymer by grafting by controlling the length and density of the polymer to show better fouling properties. They used the atomic transfer radical polymerization. Moreover, the polymer density was measured during immobilization reaction by changing the solvent type; high-density surface modification was attained using poor solvents, whereas, low-density surface functionalization was recognized using better solvents. This method, poly (polyMPC) (2-methacryloyloxyethyl phosphoryl choline) was effectively functionalized with measured densities on porous membrane surfaces. The dynamic and static antifouling properties of the polyMPC-functionalized membranes were estimated by consuming the bovine serum albumin solution, and the functionalized membrane having higher–molecular-weight polyMPC with a greater density was more operative for antifouling. This modification of polymers introduces the membrane with greater antifouling properties [89].

9.15 POROUS POLYMERIC MEMBRANE USED FOR FUEL CELL APPLICATION

Carbonaceous materials and porous metals and have greater application in the polymer electrolyte membrane fuel cells (PEMFCs) as the porous flow fields/flow suppliers because of required properties such as lightweight, high porosity, greater electrical conductivity and controllable structure. In comparison with the conventional beamed flow distributor, the porous flow distributor is authenticated to allocate reactant gases more evenly all over the whole electrode area and distribute superior electrochemical significance of hydrogen PEMFCs. Also, the result of the critical morphological properties of porous flow distributors and the operational conditions on the fuel cell performance in case of water management, power density, gas distribution and pressure drop, etc., are investigated [90].

9.16 CONCLUSION

In this chapter, the basic description about the porous polymers and some methods of synthesis and application of porous polymeric membranes are mentioned. Developing advanced porous polymeric membranes is of great importance in the field of science and technology. Here, the different applications of porous polymeric membranes are well-known due to their pore structure, adjustable organic and chemical components in sensing, separation, fuel cell and biomedical application. In the last few years, we have seen marvelous research innovations in these fields, exceptional pore structures and adaptable skeletons of porous polymers.

REFERENCES

1. Sadasivuni, K.K., et al., A review on porous polymer composite materials for multifunctional electronic applications. *Polymer-Plastics Technology and Materials*, 2019. **58**(12): pp. 1253–1294.
2. Ashok, C.H.R., V. K. Rao, and C. H. S. Chakra, CuO/TiO_2 metal oxide nanocomposite synthesis via room temperature ionic liquid. *Journal of Nanomaterials & Molecular Nanotechnology*, 2016. **5**: pp. 1–4.
3. Ranjith Reddy, P., et al., Structural and morphological studies of ZnO-CeO_2 nanocomposites. *Materials Today: Proceedings*, 2016. **3**(10, Part B): pp. 4146–4150.
4. Jana, A., et al., Tunable charge transport through n-ZnO nanorods on Au coated macroporous p-Si. *Journal of Materials Chemistry C*, 2014. **2**(45): pp. 9613–9619.
5. Bykkam, S., et al., Few layered graphene sheet decorated by ZnO nanoparticles for anti-bacterial application. *Superlattices and Microstructures*, 2015. **83**: pp. 776–784.

6. Rajendar, V.D., T. Dayakar, B. Satish, K. Subramanyam, and Y. Prashanthi, Synthesis and characterization of CuInS2 nanoparticles as potential candidates for photocatalyst and photovoltaic materials. *Chalcogenide Letters*, 2016. **13**(10): pp. 467–475.
7. Sadasivuni, K.K., et al., Effect of molecular interactions on the performance of poly(isobutylene-co-isoprene)/graphene and clay nanocomposites. *Colloid and Polymer Science*, 2013. **291**(7): pp. 1729–1740.
8. Ponnamma, D., et al., Carbon nanotube based elastomer composites – An approach towards multifunctional materials. *Journal of Materials Chemistry C*, 2014. **2**(40): pp. 8446–8485.
9. Ossipov, D.A., Nanostructured hyaluronic acid-based materials for active delivery to cancer. *Expert Opinion on Drug Delivery*, 2017. **7**(6):681–703.
10. Das, S., T. Ben, and S. Qiu, Shaping of porous polymers. *Polymer*, 2020. **207**: p. 122928.
11. Xu, D., J. Guo, and F. Yan, Porous ionic polymers: Design, synthesis, and applications. *Progress in Polymer Science*, 2018. **79**: pp. 121–143.
12. Dawson, R., A.I. Cooper, and D.J. Adams, Nanoporous organic polymer networks. *Progress in Polymer Science*, 2012. **37**(4): pp. 530–563.
13. Das, S., et al., Porous organic materials: Strategic design and structure–function correlation. *Chemical Reviews*, 2017. **117**(3): pp. 1515–1563.
14. Tsyurupa, M.P., et al., Sorption of organic compounds from aqueous media by hypercrosslinked polystyrene sorbents 'Styrosorbrs. *Reactive Polymers*, 1995. **25**(1): pp. 69–78.
15. Masuda, T., et al., Poly[1-(trimethylsilyl)-1-propyne]: A new high polymer synthesized with transition-metal catalysts and characterized by extremely high gas permeability. *Journal of the American Chemical Society*, 1983. **105**(25): pp. 7473–7474.
16. McKeown, N.B., et al., Porphyrin-based nanoporous network polymers. *Chemical Communications*, 2002(23): pp. 2782–2783.
17. Wu, D., et al., Design and preparation of porous polymers. *Chemical Reviews*, 2012. **112**(7): pp. 3959–4015.
18. Guo, H., et al., Eyes of covalent organic frameworks: Cooperation between analytical chemistry and COFs. *Reviews in Analytical Chemistry*, 2019. **38**(1). doi:10.1515/revac-2017-0023
19. Ben, T., et al., Targeted synthesis of a porous aromatic framework with high stability and exceptionally high surface area. *Angewandte Chemie International Edition*, 2009. **48**(50): pp. 9457–60.
20. El-Kaderi, H.M., et al., Designed synthesis of 3D covalent organic frameworks. *Science*, 2007. **316**(5822): pp. 268–72.
21. Beiler, B., et al., Poly(2-hydroxyethyl acrylate-co-ethyleneglycol dimethacrylate) monoliths synthesized by radiation polymerization in a mold. *Polymer*, 2007. **48**: pp. 3033–3040.
22. Rose, M., et al., A new route to porous monolithic organic frameworks via cyclotrimerization. *Journal of Materials Chemistry*, 2011. **21**(3): pp. 711–716.
23. Du, N., et al., Polymers of intrinsic microporosity containing trifluoromethyl and phenylsulfone groups as materials for membrane gas separation. *Macromolecules*, 2008. **41**(24): pp. 9656–9662.
24. Hu, X., et al., Imprinted photonic polymers for chiral recognition. *Angewandte Chemie International Edition*, 2006. **45**(48): pp. 8145–8148.
25. Rzayev, J. and M.A. Hillmyer, Nanoporous polystyrene containing hydrophilic pores from an ABC triblock copolymer precursor. *Macromolecules*, 2005. **38**(1): pp. 3–5.
26. Li, G., et al., pH-responsive hollow polymeric microspheres and concentric hollow silica microspheres from silica–polymer core–shell microspheres. *Langmuir*, 2008. **24**(16): pp. 9050–9055.
27. Zhang, Y., et al., A novel route to thermosensitive polymeric core–shell aggregates and hollow spheres in aqueous media. *Advanced Functional Materials*, 2005. **15**(4): pp. 695–699.
28. Schacher, F., M. Ulbricht, and A.H.E. Müller, Self-supporting, double stimuli-responsive porous membranes from polystyrene-block-poly(N,N-dimethylaminoethyl methacrylate) diblock copolymers. *Advanced Functional Materials*, 2009. **19**(7): pp. 1040–1045.
29. Du, J. and S.P. Armes, pH-responsive vesicles based on a hydrolytically self-cross-linkable copolymer. *Journal of the American Chemical Society*, 2005. **127**(37): pp. 12800–12801.
30. Chen, Z., et al., Responsive micellar films of amphiphilic block copolymer micelles: Control on micelle opening and closing. *Langmuir*, 2010. **26**(11): pp. 8869–8874.
31. Hyuk Im, S., U. Jeong, and Y. Xia, Polymer hollow particles with controllable holes in their surfaces. *Nature Materials*, 2005. **4**(9): pp. 671–675.

32. Li, B., et al., Synthesis of uniform microporous polymer nanoparticles and their applications for hydrogen storage. *Journal of Materials Chemistry*, 2010. **20**(35): pp. 7444–7450.
33. Han, S.S., et al., Covalent organic frameworks as exceptional hydrogen storage materials. *Journal of the American Chemical Society*, 2008. **130**(35): pp. 11580–11581.
34. Furukawa, H. and O.M. Yaghi, Storage of hydrogen, methane, and carbon dioxide in highly porous covalent organic frameworks for clean energy applications. *Journal of the American Chemical Society*, 2009. **131**(25): pp. 8875–8883.
35. Doonan, C.J., et al., Exceptional ammonia uptake by a covalent organic framework. *Nature Chemistry*, 2010. **2**(3): pp. 235–238.
36. Ghanem, B.S., et al., A triptycene-based polymer of intrinsic microposity that displays enhanced surface area and hydrogen adsorption. *Chemical Communications*, 2007(1): pp. 67–69.
37. Yuan, S., et al., Microporous polyphenylenes with tunable pore size for hydrogen storage. *Chemical Communications*, 2010. **46**(25): pp. 4547–4549.
38. Yuan, S., et al., Nanoporous polymers containing stereocontorted cores for hydrogen storage. *Macromolecules*, 2009. **42**(5): pp. 1554–1559.
39. Lee, J.-Y., et al., Hydrogen adsorption in microporous hypercrosslinked polymers. *Chemical Communications*, 2006(25): pp. 2670–2672.
40. Germain, J., et al., High Surface Area Nanoporous Polymers for Reversible Hydrogen Storage. *Chemistry of Materials*, 2006. **18**(18): pp. 4430–4435.
41. Budd, P.M., et al., The potential of organic polymer-based hydrogen storage materials. *Physical Chemistry Chemical Physics*, 2007. **9**(15): pp. 1802–1808.
42. Makhseed, S. and J. Samuel, Hydrogen adsorption in microporous organic framework polymer. *Chemical Communications*, 2008(36): pp. 4342–4344.
43. Li, B., et al., A new strategy to microporous polymers: Knitting rigid aromatic building blocks by external cross-linker. *Macromolecules*, 2011. **44**.
44. Shi, Z., Y. Zhou, and D. Yan, Preparation of robust poly(ε-caprolactone) hollow spheres with controlled biodegradability. *Macromolecular Rapid Communications*, 2006. **27**(15): pp. 1265–1270.
45. Li, G., et al., Monodisperse temperature-responsive hollow polymer microspheres: Synthesis, characterization and biological application. *Polymer*, 2008. **49**(16): pp. 3436–3443.
46. Yang, X., et al., Synthesis of pH-sensitive hollow polymer microspheres and their application as drug carriers. *Polymer*, 2009. **50**(15): pp. 3556–3563.
47. Abidian, M.R., D.H. Kim, and D.C. Martin, Conducting-polymer nanotubes for controlled drug release. *Advanced Materials*, 2006. **18**(4): pp. 405–409.
48. Duan, H., et al., pH-responsive capsules derived from nanocrystal templating. *Langmuir*, 2005. **21**(24): pp. 11495–11499.
49. Zhang, Y., et al., Synthesis and catalytic applications of mesoporous polymer colloids in olefin hydrosilylation. *Advanced Synthesis & Catalysis*, 2008. **350**(5): pp. 662–666.
50. Chan-Thaw, C.E., et al., Covalent triazine framework as catalytic support for liquid phase reaction. *Nano Letters*, 2010. **10**(2): pp. 537–41.
51. Ishizu, K., Y. Tokuno, and M. Makino, Microporous films from diblock copolymer micelles based on solvent-induced mechanism by temperature control of casting solution. *Macromolecules*, 2007. **40**(3): pp. 763–765.
52. Joo, W., M.S. Park, and J.K. Kim, Block copolymer film with sponge-like nanoporous strucutre for antireflection coating. *Langmuir*, 2006. **22**(19): pp. 7960–7963.
53. Ishizu, K., N. Kobayakawa, and S. Uchida, Ordered microporous surface films formed by core-shell-type nanospheres. *Macromolecular Rapid Communications*, 2006. **27**(12): pp. 961–965.
54. Choi, K.Y., et al., Self-assembled hyaluronic acid nanoparticles for active tumor targeting. *Biomaterials*, 2010. **31**(1): pp. 106–114.
55. Matyjaszewski, K., Architecturally complex polymers with controlled heterogeneity. *Science*, 2011. **333**(6046): p. 1104.
56. Remanan, S., et al., Recent advances in preparation of porous polymeric membranes by unique techniques and mitigation of fouling through surface modification. *ChemistrySelect*, 2018. **3**(2): pp. 609–633.
57. Zhang, J., et al., Porous structure design of polymeric membranes for gas separation. *Small Methods*, 2017. **1**(5): p. 1600051.

58. Tan, X. and D. Rodrigue, A review on porous polymeric membrane preparation. Part I: Production techniques with polysulfone and poly (Vinylidene Fluoride). *Polymers*, 2019. **11**(7).
59. Antonietti, M. and K. Tauer, 90 years of polymer latexes and heterophase polymerization: More vital than ever. *Macromolecular Chemistry and Physics*, 2003. **204**(2): pp. 207–219.
60. Gokmen, M.T. and F.E. Du Prez, Porous polymer particles—A comprehensive guide to synthesis, characterization, functionalization and applications. *Progress in Polymer Science*, 2012. **37**(3): pp. 365–405.
61. Li, L., et al., Synthesis and characterization of suspension polymerized styrene-divinylbenzene porous microsphere using as slow-release-active Carrier1 1Supported by natural science foundation of guangdong province (No.05006556). *Chinese Journal of Chemical Engineering*, 2006. **14**(4): pp. 471–477.
62. Li, W.-H. and H.D.H. Stöver, Porous monodisperse poly(divinylbenzene) microspheres by precipitation polymerization. *Journal of Polymer Science Part A: Polymer Chemistry*, 1998. **36**(10): pp. 1543–1551.
63. Saraçoğlu, B., et al., Synthesis of monodisperse glycerol dimethacrylate-based microgel particles by precipitation polymerization. *Industrial & Engineering Chemistry Research*, 2009. **48**(10): pp. 4844–4851.
64. Fan, H., et al., High-flux membranes based on the covalent organic framework COF-LZU1 for selective dye separation by nanofiltration. *Angewandte Chemie International Edition*, 2018. **57**(15): pp. 4083–4087.
65. Colson, J.W., et al., Oriented 2D covalent organic framework thin films on single-layer graphene. *Science*, 2011. **332**(6026): pp. 228–31.
66. Berlanga, I., et al., Delamination of layered covalent organic frameworks. *Small*, 2011. **7**(9): pp. 1207–11.
67. Mitra, S., et al., Self-exfoliated guanidinium-based ionic covalent organic nanosheets (iCONs). *Journal of the American Chemical Society*, 2016. **138**(8): pp. 2823–8.
68. Wang, S., et al., Exfoliation of covalent organic frameworks into few-layer redox-active nanosheets as cathode materials for lithium-ion batteries. *Journal of the American Chemical Society*, 2017. **139**(12): pp. 4258–4261.
69. Fan, H., et al., Covalent organic framework-covalent organic framework bilayer membranes for highly selective gas separation. *Journal of the American Chemical Society*, 2018. **140**(32): pp. 10094–10098.
70. Nagai, K., et al., Poly[1-(trimethylsilyl)-1-propyne] and related polymers: Synthesis, properties and functions. *Progress in Polymer Science*, 2001. **26**(5): pp. 721–798.
71. Guan, X., et al., Design and applications of three dimensional covalent organic frameworks. *Chemical Society Reviews*, 2020. **49**(5): pp. 1357–1384.
72. Krishnaraj, C., et al., Covalent triazine frameworks – a sustainable perspective. *Green Chemistry*, 2020. **22**(4): pp. 1038–1071.
73. Xu, Y., et al., Conjugated microporous polymers: Design, synthesis and application. *Chemical Society Reviews*, 2013. **42**(20): pp. 8012–8031.
74. Tian, Y. and G. Zhu, Porous aromatic frameworks (PAFs). *Chemical Reviews*, 2020. **120**(16): pp. 8934–8986.
75. Shinde, D.B., et al., Crystalline 2D covalent organic framework membranes for high-flux organic solvent nanofiltration. *Journal of the American Chemical Society*, 2018. **140**(43): pp. 14342–14349.
76. Hao, Q., et al., Confined synthesis of two-dimensional covalent organic framework thin films within superspreading water layer. *Journal of the American Chemical Society*, 2018. **140**(38): pp. 12152–12158.
77. Sasmal, H.S., et al., Superprotonic conductivity in flexible porous covalent organic framework membranes. *Angewandte Chemie International Edition*, 2018. **57**(34): pp. 10894–10898.
78. Das, S. and T. Ben, A [COF-300]-[UiO-66] composite membrane with remarkably high permeability and H_2/CO_2 separation selectivity. *Dalton Transactions*, 2018. **47**(21): pp. 7206–7212.
79. Khayum, M.A., et al., Convergent covalent organic framework thin sheets as flexible supercapacitor electrodes. *ACS Applied Materials & Interfaces*, 2018. **10**(33): pp. 28139–28146.
80. Takeoka, S., et al., Development of biodegradable nanosheets as nanoadhesive plaster. *Pure and Applied Chemistry*, 2008. **80**(11): pp. 2259–2271.
81. Mitra, S., et al., Targeted drug delivery in covalent organic nanosheets (CONs) via sequential postsynthetic modification. *Journal of the American Chemical Society*, 2017. **139**(12): pp. 4513–4520.
82. Ma, J., et al., Preparation and characterization of PVDF-PFSA flat sheet ultrafiltration membranes. *Frontiers of Chemical Science and Engineering*, 2012. **6**(3): pp. 301–310.
83. Prasad, B., et al., Analysis of porous polymer membranes using FEM software for sensing applications. *Materials Today: Proceedings*, 2020. **28**: pp. 2133–2136.

84. Zafeiropoulos, G., et al., Porous titania photoelectrodes built on a Ti-web of microfibers for polymeric electrolyte membrane photoelectrochemical (PEM-PEC) cell applications. *Solar Energy Materials and Solar Cells*, 2018. **180**: pp. 184–195.
85. Rezakazemi, M., et al., Numerical modeling and optimization of wastewater treatment using porous polymeric membranes. *Polymer Engineering & Science*, 2013. **53**(6): pp. 1272–1278.
86. Shiohara, A., B. Prieto-Simon, and N.H. Voelcker, Porous polymeric membranes: Fabrication techniques and biomedical applications. *Journal of Materials Chemistry B*, 2021. **9**(9): pp. 2129–2154.
87. Grylewicz, A. and S. Mozia, Polymeric mixed-matrix membranes modified with halloysite nanotubes for water and wastewater treatment: A review. *Separation and Purification Technology*, 2021. **256**: p. 117827.
88. Liu, X., et al., Design of polymer composite-based porous membrane for in-situ photocatalytic degradation of adsorbed organic dyes. *Journal of Physics and Chemistry of Solids*, 2021. **154**: p. 110094.
89. Oshiba, Y., Y. Harada, and T. Yamaguchi, Precise surface modification of porous membranes with well-defined zwitterionic polymer for antifouling applications. *Journal of Membrane Science*, 2021. **619**: p. 118772.
90. Zhang, Y., Y. Tao, and J. Shao, Application of porous materials for the flow field in polymer electrolyte membrane fuel cells. *Journal of Power Sources*, 2021. **492**: p. 229664.

10 Porous Polymers in Solar Cells

Shah Imtiaz, Masrat Bashir, and Mohd. Imran Ahamed
Aligarh Muslim University, Aligarh, India

Syqa Banoo
Mangalayatan University, Beswan, India

Nimra Shakeel and Naushad Anwar
Aligarh Muslim University, Aligarh, India

CONTENTS

10.1 INTRODUCTION

The energy demand is increasing day by day with the increasing population and growing economy. To counter these energy demands, a large amount of fossil fuels are consumed leading to the destruction of resources and environmental pollution. These fossils that we look for as energy sources are limited and may replenish after some time. Therefore, there is a need for an energy source that

DOI: 10.1201/9781003169604-10

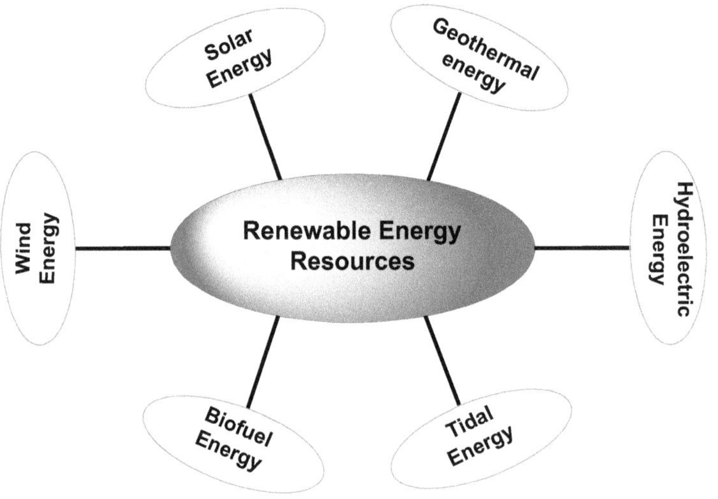

FIGURE 10.1 Different sources of renewable energy resources.

must be environment friendly as well as renewable. There are large numbers of renewable resources of energy that exist in nature some of which are windmills, hydroelectricity projects, ocean currents, solar energy, etc. [1]. Figure 10.1 describes different paths of renewable energy sources. These renewable resources may have some regional restrictions, large initial investments except utilization of solar energy that did not have any regional restriction, and are environment friendly [2]. Solar energy can be utilized by many approaches, of which solar cell is one of the best and most effective approaches. The solar cell commonly known as the photovoltaic cell is a photoelectric device that converts light energy into electrical energy via physical and chemical phenomenon. These solar cells are often the building blocks of solar panels. A single junction of silicon solar cell produces an open-circuit voltage of about 0.5–0.6 volts [3]. The solar cell is basically a p–n junction diode; however, its construction has a minor difference from a normal diode. The n-region in a solar cell is thin as compared to the p-region and is heavily doped, as a result when a light photon is incident on the n-surface it easily penetrates the surface [4]. The penetration also depends on energy and absorption coefficient. The p-region is comparatively thick and lightly doped because of which most of the depletion layer lies in it. Photon penetrates into the junction through the n-surface generating an electron-hole pair in the depletion region. Because of built-in potential and electric field, hole and electrons move to the p- and n-junctions respectively. When an external circuit is connected to this diode, electrons show movement in it from the n-junction to the p-junction externally, where they recombine with holes, neutralizing the circuit. The carriers (electrons and holes) are extracted using metal electrodes on both sides of the cell, i.e. front electric contact and back electric contact. Front electric contact is used as a finger electrode of small size, because of the need for a large surface to allow more penetration of light [5]. Figure 10.2 shows the basic structure and functioning of a solar cell.

In the present era, research related to solar cells mainly focuses on:

10.1.1 Si-Based Solar Cells

The majority of the solar cells available in the markets in the current era are made up of some type of Si material. The variation in silicon used in solar cells is due to its impurity, which depends on the alignment of silicon modules. The power conversion efficiency of Si-based solar cells directly depends on the purity of silicon in solar cells. More is the purity of silicon in solar cells more is the power conversion efficiency. Generally, two types of Si-based solar cells are mostly available in the market, i.e. monocrystalline and polycrystalline solar cells [6].

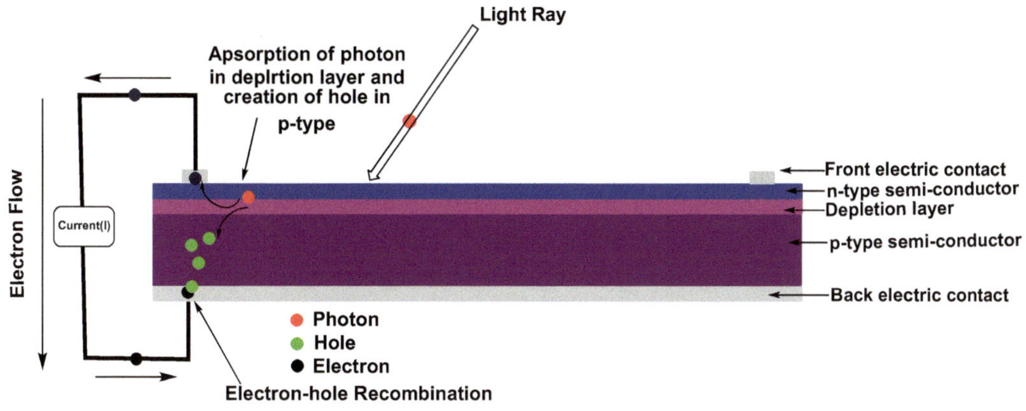

FIGURE 10.2 Basic structure and functioning of solar cell.

(https://www.electrical4u.com/solar-cell/).

Monocrystalline Si solar cells are easily recognized by colouring, their high purity makes them unique. They also show good power conversion efficiency of about 20% as compared to polycrystalline and thin-film solar cells [7]. They are long-lasting as well as more expensive because of the big loss of cylindrical crystals of silicon during designing. However, polycrystalline silicon cell does not require any designing, it is just melted and poured into a square mould to get square-shaped polycrystalline silicon [8]. These are less expensive as compared to monocrystalline silicon because of no loss during designing, but the system operates at about 13%–14% efficiency due to lack of purity. In summary, the laboratory efficiency of these cells is more than 25%, but due to their sophisticated manufacturing and high cost, their development becomes limited. High energy is required for their development.

10.1.2 Thin-film Solar Cells

These solar cells as present in name consists of one layer of semiconducting material over the other to form the films. The layer may consist of silicon as one of the semiconducting materials. They typically show variations in efficiencies between 7% and 13% and few research processes have shown that their efficiencies may climb up to a maximum of 16% [9]. The major draw for these film cells is their low cost, flexible nature, and easy mass production. Another advantage of these film technologies is that they show less negative effects occurring as a result of high heating and shadowing. In amorphous silicon solar cells, the method of stacking has increased their efficiencies to 8%. A few examples of thin-film solar cells are gallium arsenide (GaAs), copper indium gallium selenide (CuInGaSe), cadmium telluride (CdTe), etc. [10]. CdTe solar cells show an efficiency range of 9%–11% and CuInGaSe has shown a maximum efficiency of 10%–12%. One of the major drawbacks of these types of solar cells is that they require a lot of space for film technology [11]. In summary, this type of solar cell showed high efficiency as well as good stability compared to silicon-based ones. They are not considered good because of the toxicity of cadmium (Cd) and the low abundance of gallium (Ga) and indium (In).

10.1.3 Organic Solar Cells

These cells are also known as emerging solar cells and their few examples are organic photovoltaics, dye-sensitized solar cell (DSSC), perovskite solar cell, etc. Organic solar cells are easy to carry because of their lightweight and their production cost is low. However, because of a few problems their large-scale applications are limited. Some problems limit their large-scale application.

In summary, these cells need to be improved for efficiency, spectrum response, and stability [12]. For example, DSSCs show easy leakage of electrolyte and solvent evaporation, decreasing their stability. A little shift towards solid perovskite solar cells from DSSCs has attracted the attention of today's scientists as soon as they were reported [13]. The solid perovskite solar cells also work similar to DSSCs, but overcome a few shortcomings. Besides overcoming these shortcomings, the devices still show less stability resulting in limited applications. It is because the performance of the device depends on each component for their respective task. Therefore, investigating each component separately, their optimization can be of great significance. To improve these components, scientists are working on perovskite solar cells, bifacial solar cells, intermediate band, liquid inks, organic/polymer solar cells, etc. [14]. Various research areas of solar cells currently in progress have been seen in Figure 10.3.

The polymers have shown a lot of uses in each component of a solar cell because of their versatile chemical and physical properties (Figure 10.4). The 3D network of these polymers makes them usable in fabricating mesoporous materials or can be found their use as solid electrolytes [15]. The polymers showing electrolytic properties may be thermoplastic, thermosetting, or composite polymers. Because of their high catalytic activity, they also found their use as counter electrodes, e.g. polypyrrole, polyaniline, poly(3,4-ethylenedioxythiophene), etc. The polymers can regulate the morphology of perovskite because of the presence of diverse functional groups. As a result of their high carrier mobilities, they also act as hole and electron transfer materials in solar cells [16]. Triarylamine- and poly-3-hexylthiophene-based polymers and composites show hole carrier properties. The presence of different functional groups in polymers improves the performance of the device by adjusting the work function of metal electrodes and their use in interface layers. Polymers have found their use in the photoactive layer or buffer layer on organic photovoltaics due to variable

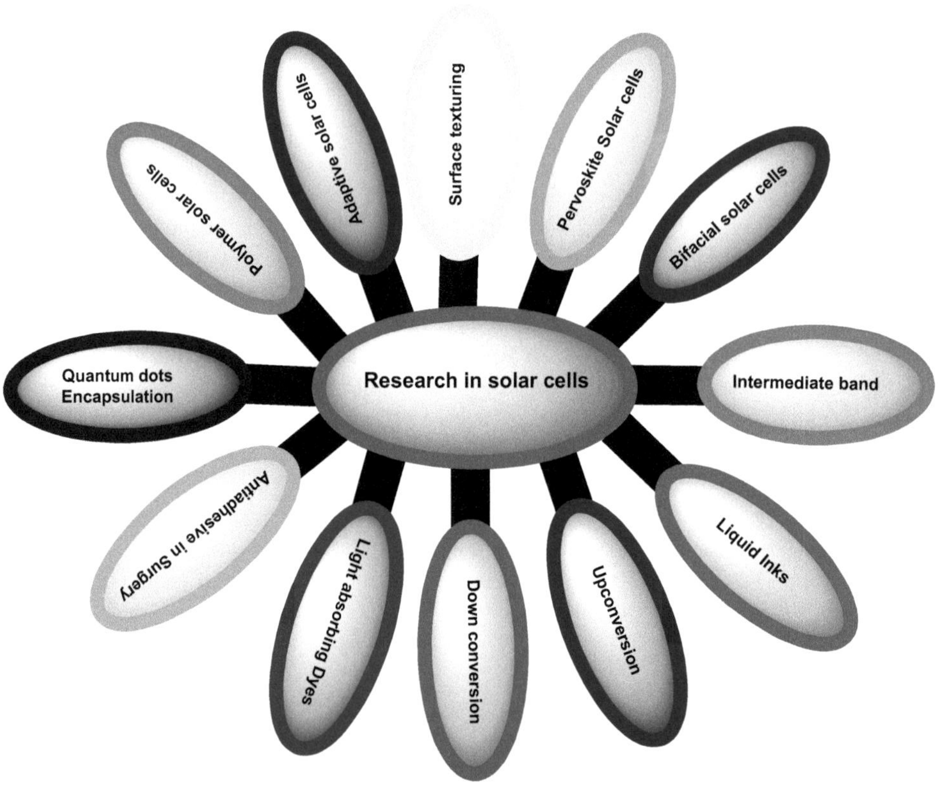

FIGURE 10.3 Different research areas of solar cells currently in progress.

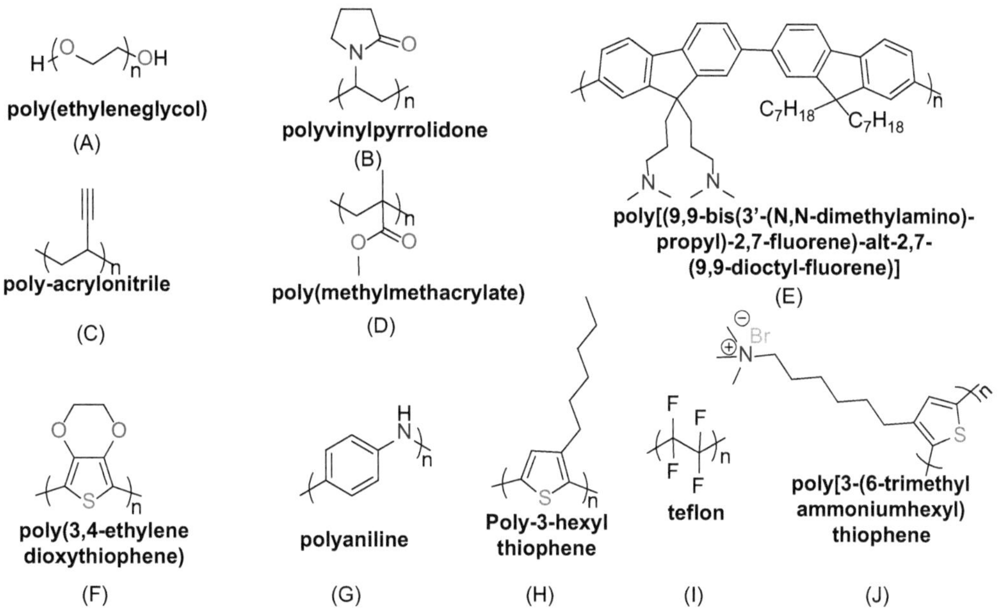

FIGURE 10.4 Few polymers having applications in solar cells; (a)–(e): Morphology regulation in perovskite cells. (f): Interlayer formation, Hole transport layer, counter electrode, Binary organic photovoltaics and buffer layers. (g): counter electrode and hole transport layer. (h): Hole transport layers, Interlayers and in Binary organic photovoltaics. (i)–(j): Interlayer formation.

optical adsorption behaviour and electron mobility. The modification of functional groups can provide us with diversified polymer structures which can be further used to obtain components of solar cells with good performance. The conductive polymers, i.e. polymers with good conductivity have also their use in many fields of solar cells. The diverse use of polymers in different solar cells is shown in the flowchart given in Figure 10.5 [17].

In addition to naturally occurring inorganic porous materials such as silicates and zeolites, research in the field of porous polymers has revealed that they show a lot of applications in solar cells, electrochemical capacitors, electrochromic devices, light-emitting diodes, etc. [18]. These potential applications of porous polymeric substances have drawn the significant attention of scientists. These porous polymeric materials include covalent organic frameworks (COFs), porous aromatic frameworks (PAFs), conjugated porous polymers (CPPs), microporous intrinsic polymers (MIPs), and hyper-crosslinked polymers (HIPs) [19, 20]. Each class of these polymers shows unique aspects in different fields, that vary depending on the method of synthesis, origin of porosity, degree of conjugation, etc., which impact the final properties and application of materials. For example, CPPs have high surface area, permanent porosity, and extended conjugation, as a result, these show various applications in separation and storage of gases, photoelectronic and energy storage, chemical sensors, photocatalysis, etc. [21]. In this chapter, we have discussed the applications of porous polymers in solar cells.

The porous polymers have found their use directly in different components of the solar cell as well as in improving efficiencies of its components. The increase in porosity decreases the crystallinity of components and increases the amorphous nature, which leads to better photovoltaic performance of solar cells, e.g. increased surface durability, modified anti-reflectiveness, increased hydrophobicity, increased light absorption, film formation in perovskite solar cells, etc. [22]. In one of the studies, a metal-organic framework (MOF) (Zn-CBOB) was found to be hydrophobic, which was combined with a hole transport material, Spiro-OMeTAD to make it porous. This porous material obtained improved the conductivity of Spiro-OMeTAD and hydrophobic Zn-CBOB, and was

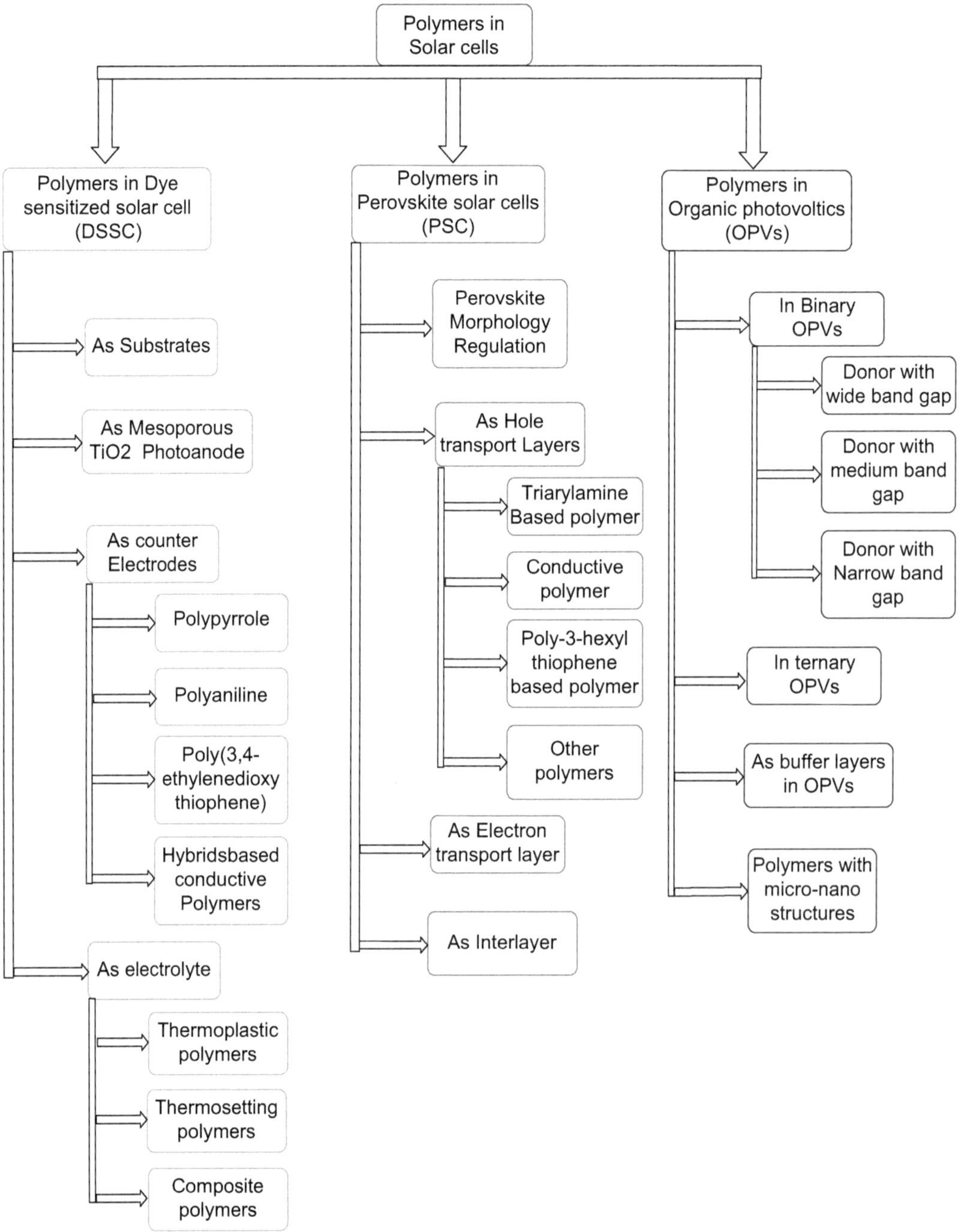

FIGURE 10.5 Flowchart showing uses of polymers in different components of solar cells.

found to prevent the perovskite layer, and hence increasing the stability of perovskite solar cells [23]. Moreover, the porous membrane structures have also shown the enhanced generation and storage of electricity in addition to energy conversion.

One of the great challenging tasks is to use concentrated solar energy in industrial processes, as a source of high-temperature heat. For this challenge to be successful, there is a need to design a solar receiver with great absorption frequency and less radiation loss, that can be achieved by porous materials. The porous materials when exposed to incident light enable penetration of light followed by absorption inside the volume of the pore, showing more absorption as compared to the exposed surface [24]. The heat is then efficiently transferred to air molecules present in the porous matrix to achieve the maximum required temperature and hence, increasing the efficiency of the solar cell. The porous structure of this surface material also reduces the re-radiation loss of heat back to the environment. Saleem et al. added lignocellulose to ZnO/TiO_2, as a result of this, the porosity of the material was increased which in turn, increased the surface area of the material, higher dye loading, and light-harvesting with a decrease in the backward scattering of electrons [25].

Several MOFs are reported recently which have found their use in many optoelectronic devices because of their unique properties that result in enhanced performance and stability in solar cells. A lot of evidence have shown them as promising materials in the electron transporting layers of perovskite solar cells which use flexible substrates because of their easy processing even at low temperature. Pavithra et al. in 2015, revealed that gel polymer electrolytes synthesized by incorporating acetamide in poly(ethylene oxide) with LiI/I_2 results in increased efficiency (η = 9.01%) as compared to the original gel polymer electrode without acetamide (5.7%) under 85 mW cm^{-2} solar irradiation. Further, it was seen that adding porosity led to an increase in open-circuit voltage that may be due to the decrease in electron-hole recombination and increase in the lifetime of electrons due to incorporation of acetamide on the poly(ethylene oxide). Similarly, Song et al. have compared porous plasticized poly(vinylidene fluoride-co-hexafluoropropylene) (PVDF/HFP)-based electrolytes, with conventional polypropylene separators, by different methods. The study found that electrolyte retention characteristics and ionic conductivity of these porous electrolyte membranes are much better as compared to conventional polypropylene separators. Moreover, PVDF/HFP-based electrolyte doesn't show wetting problems, whereas polypropylene separators show poor wetting and therefore showdecrease in their conductivity [26, 27]. In this chapter, we have briefly discussed the evolution and applications of various types of porous polymers which can be used in solar cell applications.

10.2 POROUS POLYMERS IN DSSCS

10.2.1 Porous Polymers in Electrodes

The porous photoelectrodes of ZnO/TiO_2 were prepared by adding lignocellulose on fluorine-doped tin oxide (FTO) substrate for DSSCs. The addition of lignocellulose to ZnO/TiO_2 results in an increase in porosity and surface area for loading more dye and increased light-harvesting and decreased backward scattering of electrons. These modifications increased the power conversion efficiency of ZnO (38%) and TiO_2 (104%) as compared to their pure form in DSSC. In summary, the addition of lignocellulose developed nanocomposites of highly porous metal oxide with enhanced porosity and thus enhancing the efficiency of DSSCs and may solve the electrolyte leakage issue [28]. A two-dimensional MOF (Cu-BHT) is a solution-processed transparent electrode having a conductivity of more than 2500 Sc $m-^{1}$ and a high value of transmittance (82%). Then three classic photovoltaic cells (quantum dots solar, organic solar cells, and perovskite solar cells) were then fabricated with this Cu-BHT electrode. The performances of all three devices were found to be improved as compared to the corresponding ITO electrode [29].

10.2.2 Porous Polymer as a Counter Electrode

A combination of MOF-derived selenide ((rGO/ZIF)@Se) with a mixture of polystyrene sulfonate-doped poly (3,4-ethylenedioxythiophene) is used in counter electrode for DSSC. The (rGO/ZIF)@ Se has a porous nanowires-like structure, encapsulated into a carbon matrix improves the counter electrode by inserting the porosity into it. The electrochemical studies for the counter electrodes based on different weight percentages showed that 3%wt (rGO/ZIF)@Se/PEDOT:PSS composite counter electrode has higher power conversion efficiency (8.60%) compared to the platinum counter electrode (7.14%) for DSSC. This was found to be an excellent composite to replace the platinum electrode in DSSCs [30].

In 2021, Li et al. synthesized a new type of counter electrode, bismuth molybdate/carbon nanofibres (Bi_2MoO_6/CNFs) composite utilizing hydrothermal techniques via electrospun and used it in DSSCs on Ti mesh substrate, showing strong stability and sustainability. The electrochemical studies for this electrode revealed that the DSSCs assembled with Bi_2MoO_6/CNFs Ti mesh counter electrode have good power conversion efficiency (9.02%) as compared to that of Pt counter electrode (7.47%) and pure carbon nanofibres (7.48%). The results also reveal that Bi_2MoO_6/CNFs Ti mesh counter electrodes can be seen as promising low-cost-effective and replaceable counter electrodes for DSSCs [31].

Zhou et al. synthesized Co/$Co_1P_1N_3$ (MOF) which was doped as a counter electrode in DSSCs, with power conversion efficiency (8.51%), as compared to Co/Co_1N_4 (6.62%) counter electrode and commercial Pt (7.88%) counter electrode [32]. Wang and co-workers developed nanotubes (CoNi alloy@carbon) which were further used to develop embedded carbon nanocages (CoNi@CNTs-C). These porous nanocages have found their use as the counter electrode of DSSCs. The power conversion efficiency of the CoNi@CNTs-C-200 counter electrode was found to be good with long-term electrochemical stability as compared to conventional Pt counter electrodes [33]. Another group of researchers in 2018 have developed stable zinc-metalloporphyrin thin films (ZnTCPP-C) that have found their use as counter electrodes with good performance in DSSCs. The power conversion efficiency was found to increase by 5.63% as compared to platinum electrodes [34].

10.2.3 Porous Polymers in TiO_2 Photoanode

The power conversion efficiency of titanium dioxide (TiO_2) photoanode in DSSCs was found to be enhanced by a MOF of zinc, i.e. Zinc(II)-benzene-1,3,5-tricarboxylate (Zn-BTC). The optical properties were enhanced due to a decrease in band gap values after combining TiO_2 powder and MOF. A comparison of the values of bandgap showed that TiO_2-Zn-BTC MOF (2%) has a lower bandgap of 2.9 eV, TiO_2-Zn-BTC MOF (4%) and TiO_2-Zn-BTC MOF (6%) had the same value of bandgap (3.1 eV). Therefore, a reduction in electron-hole pair recombination can be observed in TiO_2-Zn-BTC MOF (2%), which results in increased photoactivity. Analysis of I–V characteristics for TiO_2-Zn-BTC (2%) further confirmed that it can be used as an excellent electron transporter, hence enhances the photo efficiency of TiO_2 for their better use as photoanodes in DSSCs [35].

In one of the studies, Ramesh et al. have studied porous aerogel photoanode via modelling and simulation. These studies have shown that solid-state DSSCs with Titania aerogel photoanodes exhibited enhanced power conversion efficiency, which may be attributed to large short-circuit current density. This high short-circuit current density is achieved because of more dye absorption as a result of which electron-hole recombination decreases. A comparison of experimental and theoretical results showed that they are in close agreement with each other [36]. Lee et al. were the first to perform a study where a thin layer fabrication of Cu-based MOFs, i.e. copper(II) benzene-1,3,5-tricarboxylate was done using the layer-by-layer technique. This layer found its application in TiO_2-based solar cells as a light-absorbing layer [37]. More applications of porous polymers in TiO_2 photoanode can be found in detail in the work of Dou et al. [38], and Wang [39].

10.2.4 Porous Polymers in Electrolyte

Poly-3-(9H-carbazol-9-yl)propyl methacrylate (pCMA), a gel electrolyte is used in the fabrication of DSSC. The photovoltaic and electrochemical studies of pCMA, a gel electrolyte, showed that it is an efficient electrolyte and provides an open-circuit voltage (545 mV), current density (10 mA) for a cell having 0.25 cm^2 area. This besides showing a good efficiency (%) also performs like a good host polymer for redox couple in electrolyte [40].

Another polymer gel electrolyte with 4-tertbutylpyridine as additive and LiI/I_2 redox couple, which is based on polymethyl methacrylate (PMMA)/polyaniline nanotube (PANI-NT) is a good performing and efficient electrode. The electrochemical impedance spectroscopy showed that combining PANI-NT with PMMA-based gel electrolyte results in an increased conductivity of the electrolyte. The different studies for a few weight percentages were performed providing an optimized system with open-circuit voltage (0.75 V), short-circuit current density (9.80 mA cm^{-2}), and the photochemical conversion efficiency (5.11%) [41]. Vignesh et al. developed electrospun polyacrylonitrile (PAN)/cobalt sulphide (CoS) membranes by encapsulating CoS nanoparticles in PAN nanofibres. The prepared membranes (esCPMs) were immersed into an ionic electrolyte (liquid) (0.5 M LiI + 0.05 M I_2 + 0.5 M 4-tert-butylpyridine + 0.5 M 1-butyl-3-methylimidazolium iodide) in acetonitrile to obtain the activated nanocomposite membrane electrolytes (esCPMEs). The Tafel polarization and electrochemical impedance studies showed that the obtained esCPME have enhanced electrocatalytic activity as well as ionic conductivity. The 2 wt% CoS encapsulated esCPME was found to provide optimized photoconversion efficiency (7.41%) [42].

In 2013, Dong et al. have synthesized a block copolymer from poly(oxyethylene)-segmented diamine and 4,4′-oxydiphthalic anhydride by the process of oligomerization, which was further cured to obtain amide-imide cross-linked gels. An electrolyte was absorbed into a gel-like copolymer having 3D interconnected nanochannels. This elastomeric copolymer on its use as an electrolyte (PGE) in DSSCs showed extremely high photovoltaic performance. This polymer gel electrolyte that contains about 76.8 wt% liquid electrolyte, showed a power conversion efficiency (9.48%), short-circuit current density (19.50 mA cm^{-2}), open-circuit voltage (0.76 V), and fill factor (0.64). The studies reveal that the outstanding performance of this cell, as compared to normal (8.84%) is because of suppression of the back-electron transfer through polymer gel electrode [43].

One more study has reported quasi-solid-state DSSCs fabricated with the mixed-plasticizer, i.e. a modified polymer electrolyte. This is obtained by adding the original polymer electrolyte to a composite of poly(ethylene oxide)/poly(vinylidene fluoride-hexafluoropropylene) (PEO/P(VDF-HFP)) and KI/I_2 with methyl acrylate and ethylene glycol as plasticizers. The olefinic bonds in methyl acrylate and –OH groups in ethylene glycol improve the ionic conductivity of polymer electrolytes. The modification leads to the better performance of photovoltaics, e.g. power conversion efficiency (6.79%) and short current density (15.23 mA·cm^{-2}) [44]. A heterocyclic conjugated complex of cobalt complex with iodide ligand on an enriched graphene oxide sheet of nitrogen functional group was found to be an efficient charge transferable nanofiller and enhanced the overall performance of quasi-solid-state DSSC. The ionic liquid electrolyte was found to consist of an electrochemical window of more than 86% width and a charge transfer resistance lowered by 64% as compared to bare ionic liquid electrolyte. The cell using these as electrolyte showed a power conversion efficiency (7.29%) which is an increase of 70.37% compared to the cells utilizing bare ionic liquid electrolyte (4.28%). The cell utilizing this complex system have shown improvement in open-circuit voltage as well as short-circuit current density. The device has also shown enhanced stability and an excellent photoelectric conversion efficiency [45].

10.2.5 Porous Polymer as Energy Conversion Film

10.2.5.1 Polyvinylidene Fluoride-co-Hexafluoropropylene (PVDF-HFP) Membranes

The porous structure of this membrane showed enhanced generation and storage of electricity in addition to energy conversion. This film can generate electricity from simulated solar radiation and can also convert mechanical disturbance to electricity. The piezoelectric voltage and piezoelectric current were found to have maximum values of 8.8 V and 600 nA, respectively. The membrane was prepared by incorporating a chemically modified reduced graphene oxide (rGO) and 1H, 1H, 2H, 2H-perfluorooctyltriethoxysilane (PFOES). Moreover, this can convert solar energy into electricity by rising film temperature. When a film is hit by simulated sunlight, the open-circuit voltage of 1.2 V and a short-circuit current of 25 nA were generated [46].

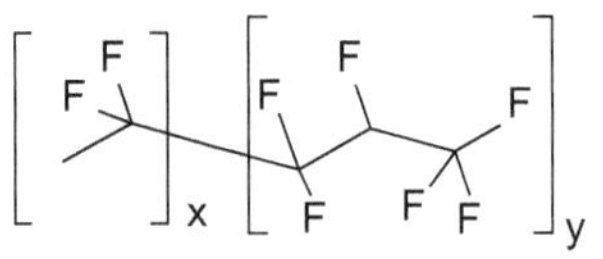

PVDF-HFP

10.2.5.2 Pyridine-based CMPs Aerogels (PCMPAs)

This microporous and nanotubular heterocyclic conjugated polymer is an aerogel-based solar receiver, prepared by one-pot synthesis by reacting 2-amino-3,5-dibromopyridine and 1,3,5-triethynylbenzene, using Sonogashira–Hagihara strategy. The substrate obtained was further sprayed polypyrrole on its surface to get solar steams. This shows a nanotubular structure with 94% of porosity and excellent light absorption capacity (ca. 94%) with low thermal conductivity. These aerogels-based solar systems show high efficiency (84%) of solar energy conversion [47].

10.2.6 Porous Polymers in Coating of Solar Cell

Working with photovoltaics like cleaning, transport results in damage of the cover glass. The strength and durability of surface glass including anti-reflection and anti-soiling coatings show a great impact on the economy. The operational maintenance frequency and methods used for learning also affect the economy directly as well as indirectly. Newkirk et al. used Nylon 6/12 bristles for surface abrasion and determined durability and stability of photovoltaic surfaces which at first come in contact with solar radiation. They examined surfaces, e.g. porous silica-coated-, fluoropolymer-coated-, and ceramic (TiO_2 or $ZrO_2/SiO_2/ZrO_2/SiO_2$)-coated-glass, and compared results with monolithic- poly(methyl methacrylate) and -glass coupons. Results showed that the surface with more durability was the metal oxide layer followed by the porous silica oxide layer [48].

10.2.7 Porous Polymers as Photocatalyst or Electrocatalyst

In 2018, Jeong and co-workers have fabricated cationic porous graphite nanomaterial by sonicating graphite with cationic enzyme media and exfoliating it in active charcoal, which are used as electrocatalyst and photocatalyst in DSSCs. This fabricated cationic porous graphite has a large surface area, large volume of pores, and enhanced conductivity that makes it capable of electrocatalytic reaction and enhanced photoresponse. This enhanced electrocatalytic activity provides DSSCs with small charge transfer resistance and photovoltaic efficiency of 9.59%. Furthermore, it was observed that a large amount of gel electrolyte is included in porous charcoal that increases the rate of interfacial reaction between electrolyte and counter electrode [49].

10.3 PEROVSKITE SOLAR CELLS

10.3.1 Porous Polymers in Electron Transport Layers

Perovskite solar cells with flexible morphology have attracted a lot of researchers, because of their low cost and excellent optoelectronic performance. They have also been seen as potential energy suppliers in soft electronic devices. To make these solar cells efficient and flexible, the designing of charge-transporting layers plays a crucial role. The MOFs NH_2-MIL-101(Fe) and NH_2-MIL-125(Ti) obtained from ligation of 2-aminoterephthalic acid with Fe and Ti respectively show a significant role as an electron transport layer in perovskite solar cells. The optimization of MOF thickness in the perovskite solar cell fabrication resulted in reasonable device efficiencies [50]. It was found that Ti-based MOFs (nTi-MOF) are dispersed in alcoholic solvents in order to generate hole transport layers are also found suitable for injection and transport of charge from perovskite to electrode. On mixing [6,6]-phenyl-C_{61}-butyric acid (PCBM) in this MOF an efficient electron transfer system is formed that surpasses electrode-perovskite intercontact. This combination also results in an increased power conversion efficiency by 18.94% for rigid and by 17.43% for flexible perovskite solar cells [51].

10.3.2 Porous Polymers in Hole Transport Layers

The degradation of perovskite and electron-hole recombination at the interface has hampered power conversion efficiency as well as the stability of perovskite solar cells. This defect is originated from hybrid halide perovskite and because of the low mobility of hole transport materials. To overcome this problem, Zhang et al. gave a strategy to increase hole mobility and passivate perovskite defect. They constructed HTM-FJU-17 (a dual functional layer) by incorporating $(Me_2NH_2)^+$ encapsulated In-based anionic MOF (FJU-17) inside the hole transport material. During the time, when MOF stabilizes positively charged hole transport material (oxidized form), the FJU-17 capsule will release $(Me_2NH_2)^+$ ion to passivate organic cation vacancies. As a result of this phenomenon, perovskite solar cells get suppressed electron-hole recombination leading to improved power conversion efficiency from 18.32% to 20.34%. The stability of the device was also found to be improved, retaining 90% of power conversion efficiency even after 1000 hours in ambient conditions. This demonstrates that dual layer-based anionic MOFs or hole transport layers can be a promising tool to manufacture perovskite solar cells with enhanced performance [52].

Spiro-OMeTAD is one of the hole transport materials used in perovskite solar cells but shows limited stability and less power conversion efficiency. This is because of some intrinsic properties of organic molecules, i.e. low conductivity and non-polar contact interface. To overcome this deficiency hole transport layer (Spiro-OMeTAD) was doped with MOF having rod topology and Lewis base sites ($\{[Zn(Hcbob)]\cdot(solvent)\}_n$ (Zn-CBOB). This doping process oxidized the hole transport layer and results in the hole transport layer with remarkable power conversion efficiency (20.64%). Further increase in porosity passivates the surface traps and hydrophobicity present because Zn-CBOB prevents perovskite surface from water [53]. Geng et al. prepared Co_3O_4@NC composite from ZIF-67 which was found useful in hole extraction and interfacial performance for perovskite solar cells that are mostly carbon-based. A high efficiency of 14.63% was achieved with good stability in carbon-based perovskite solar cells [54].

10.3.3 Porous Polymer as Energy Conversion Film

In 2017, Wu et al. have developed the 3D spirobifluorene-based COFs (SP-3D-COFs) having high thermal stability and permanent porosity. The simple doping of these SP-3D-COFs to perovskite solar cells showed an improvement in the power conversion efficiency by about 18% as compared to undoped perovskite solar cells [55].

10.3.4 Porous Polymers as Interlayers

In 2018, Park et al. have synthesized the two covalent organic nanosheets (CONs), i.e. CON-10 and CON-16 by Stille cross-coupling reaction. These materials have been found useful as interlayers in photovoltaics. These also showed property of hole transport material, which showed improvement after combining with PEDOT: PSS. This combination showed an increase in power conversion efficiency of 4% in CON-16 and 1% in CON-10 [56].

10.3.5 Porous Polymers in Morphology Regulations

One of the studies has shown that supramolecular complexes of Zr(IV), In(III), and Zn(II), formed from 2,6-pyridinedicarboxylic acid and 2,6-pyridinediamine, have found their use as additives in perovskite solar cells. The detailed study showed that a 2 wt% of zinc MOF in the perovskite layer increased the performance of these solar cells. More than 90% of power conversion efficiency was achieved after adding 2 wt% of zinc MOF [57].

10.4 POROUS POLYMERS IN SILICON SOLAR CELL

In one of the works Rajanna and her group reported an approach that was used by them in the fabrication of solar devices based on a single-walled carbon nanotube (SWCNT) film and hydrogenated thin-film amorphous silicon (a-Si:H). The conductivity of these SWCNTs was enhanced by combining with poly(3,4-ethylenedioxythiophene) polystyrene sulfonate and found their use in the window layer and in the front electrode. They have fabricated a series of hybrid solar cells with varying thicknesses of SWCNT film in ambient conditions. The hybrid solar cells fabricated with polymethylmethacrylate layer showed a reduction in reflection by about fourfold. This further enhances short-circuit current, open-circuit current and hence the efficiency was enhanced by about 10% [58]. One more study of hybrid solar devices has shown that the fabrication of [6,6]-phenyl-C71-butyric acid methyl ester ([71]PCBM), and poly-3,4-ethylenedioxythiophene poly styrene sulfonate (PEDOT:PSS) on the hybrid solar device consisting of hydrogenated amorphous silicon (a-Si:H) to get PEDOT:PSS/a-Si:H/[71]PCBM devices. It was observed that both short-circuit current and open-circuit voltage show dependence on the thickness of a-Si:H layer. The device achieved 2.84% of power conversion efficiency at 274 nm a-Si:H thick layer under 1.5 AM solar radiation [59].

In another study, Khatri et al. have investigated silicon hybrid solar cells based on poly(3,4-ethyl enedioxythiophene):poly(stylenesulfonate) (PEDOT:PSS)/n-Si. The study was done with and without embedding green-tea modified multiwalled carbon nanotubes (MWCNTs) at the interface. The devices fabricated with modified MWCNTs have shown better performance as compared to devices without MWCNTs. The better performance can be seen in values of short-circuit current density (30.31 mA cm^{-2}), open-circuit voltage (0.54 V), fill factor (0.66), and power conversion efficiency (10.93%) [60].

10.5 MISCELLANEOUS

10.5.1 Porous Polymers in Solar Evaporators

HCPs are also good paradigms of porous polymers and these have also found their use in solar cells. One of their best use is desalination and purification of water utilizing solar energy (solar evaporation). Li et al. have synthesized one such polymer by Friedel–Crafts reaction from benzene and methylbenzene, inexpensive conventional chemicals. This polymer was modified by coating with polypyrrole to make it superhydrophilic and for light absorption. These modified polymers possess a large surface area of about 773 m^2 g^{-1}, a low apparent density of 0.011 g cm^{-3} but possess a low thermal conductivity of 0.029 W m^{-1} k^{-1}. To improve more further medication was done with sodium alginate and polypyrrole providing HCPs (named as PPy-M-HCPs). This HCP shows superhydrophilicity which can be seen from its water contact angle of about 0°. This polymer also showed better mechanical strength and strong light absorption of about ca. 95%. Further, this polymer showed an outstanding conversion of light to heat with a solar conversion efficiency of 87% under 1 kW m^{-2} irradiation [61].

More porous polymers that have found their applications in solar cell evaporators, desalination, and water purification are nanoporous gold nanoparticle (AuNP)/poly(*p*-phenylene benzobisoxazole) nanofibre (PBONF) with a solar conversion efficiency of 83%[62]; Au@Bi_2MoO_6-CDs having

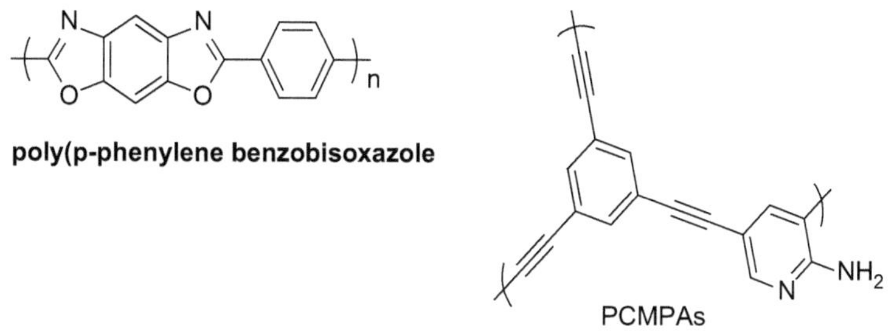

FIGURE 10.6 Structures of poly(p-phenylene benzobisoxazole) and pyridine-based conjugated microporous polymers aerogels (PCMPAs) showing good porous capacity.

coral-like structure with an ion removal rate of 100%; pyridine-based conjugated microporous polymers aerogels (PCMPAs) having high porosity[63] (Figure 10.6), low thermal conductivity, and good mechanical properties with a power conversion efficiency of about 80% [64]; a porphyrin COF (POF) with 80% light to steam conversion efficiency [65]; PPy-coated M-PDVB-PS, a modified form of porous foam based on cross-linked aromatic polymer (PDVB-PS) having contact angle Of 0° light absorption (*ca.* 90%) and solar power conversion efficiency of 87.6% [66]; and PPy-coated VMP foam with 73.81% porosity and 83% of photothermal efficiency [67].

10.5.2 Charge Separation Systems in Solar Cells

Further, Wan et al. have synthesized two COFs (COF-366 and COF-66) that may be used to fabricate heterojunction type cells, to get an impressive charge carrier separation. These porous frameworks (COF-366 and -66) were found to be p-type semiconductors having 1D hole mobilities of 8.1 and 3.0 $cm^2\ V^{-1}\ s^{-1}$, respectively. These multifunctional COFs having good thermal stability, high charge mobility, electrical conductivity, and pore accessibility can be valuable in designing different plastic electronics and optoelectronic materials [68]. In 2015, Jin et al. have synthesized six types of donor-acceptor COFs of Zn, Cu, and Ni with three different diimides, namely DCuPc–APyrDI-COF, DCuPc–ANDI-COF, DCuPc–APDI-COF, DNiPc–APyrDI-COF, DNiPc–ANDI-COF, and DZnPc–APDI-COF. These structures have shown exceptional behaviour as an interface for charge separation in solar cells. Other such examples can be seen somewhere else also [69, 70].

10.5.3 Porous Polymers in ZnO Photoanode

In 2017, Zhang et al. have proposed a novel Fe-based polyoxometalates @ MOF (POM@MOF(Fe)) crystals having leaf-like structure and was considered to show good optical response at 750 nm, an absorption edge of the visible spectrum. Then these crystals were assembled with ZnO photoanode to get a modified form of the solar cell. The photovoltaic performance of the modified solar cell possessing POM@MOF(Fe) modified ZnO as a photoanode was found to be much better compared to bare ZnO. Further results have demonstrated this composite POM@MOF(Fe) as a good electron-hole separation material thus enhancing the photovoltaic property of ZnO. All these properties may be beneficial for enhancing the performance and power conversion efficiency of photoanode [71].

In another work from 2012, the photovoltaic electrode having polyoxometalate (SiW_{12}) and ZnO was fabricated on indium tin oxide glass. The morphology and crystallography of ZnO were strongly influenced by the addition of SiW_{12} as a result of which the power conversion efficiency and photocurrent of SiW_{12}-ZnO film were found to be improved as compared to ZnO film. The enhanced performance of this composite film was attributed to the photoinduced electron transfer between polyoxometalate and ZnO. This electron transfer results in an increase in the dissociation of excitons [72].

10.6 CONCLUSIONS

This chapter describes the brief descriptions of the porous polymers which have been studied by a lot of researchers in the applications of solar cell technology. We have also summarized the applications of porous polymers in various fields such as DSSCs, perovskite solar cells, etc., in solar cell applications. Monocrystalline Si solar cells are easily recognized by colouring; their high purity makes them unique. They also show good power conversion efficiency of about 20% as compared to polycrystalline and thin-film solar cells. MOF (Zn-CBOB) is combined with Spiro-OMeTAD to make these materials work as a porous polymer and this porous material obtained improved the conductivity of Spiro-OMeTAD and hydrophobic Zn-CBOB and was found to prevent perovskite layer, and hence increasing the stability of perovskite solar cells. DSSCs assembled with Bi_2MoO_6/CNFs Ti mesh counter electrode have good power conversion efficiency (9.02%) as compared to that of Pt counter electrode (7.47%) and pure carbon nanofibres (7.48%), respectively. One of the materials [$Co/Co_1P_1N_3$ (MOF)] which was doped as counter electrode in DSSCs, with power conversion efficiency (8.51%), as compared to Co/Co_3N_4 (6.62%) counter electrode and commercial Pt (7.88%) counter electrode.

REFERENCES

[1] Z. Sen, Solar energy in progress and future research trends. Progress in energy and combustion science, *Prog. Energy Combust. Sci.*, 30 (2004) 367–416. https://doi.org/10.1016/j.pecs.2004.02.004

[2] U. Chakravorty, J. Roumasset, K. Tse, Endogenous substitution among energy resources and global warming, *J. Political Econ.*, 105 (1997) 1201–1234. https://doi.org/10.1086/516390

[3] G. Smestad, C. Bignozzi, R. Argazzi, Testing of dye sensitized TiO_2 solar cells I: experimental photocurrent output and conversion efficiencies, *Sol. Energy Mater. Sol. Cells*, 32 (1997) 259–272. https://doi.org/10.1016/0927-0248(94)90263-1

[4] W. Dawidowski, B. Ściana, I. Zborowska-Lindert, M. Mikolasek, J. Kovac, M. Tlaczaia, Tunnel junction limited performance of InGaAsN/GaAs tandem solar cell, *Sol. Energy*, 214 (2021) 632–641. https://doi.org/10.1016/j.solener.2020.11.067

[5] V. Benda, D. A. Grant, J. Gowar, (1999). *Discrete and Integrated Power Semiconductor Devices: Theory and Applications*. John Wiley & Sons.

[6] Y. Ichikawa, T. Yoshida, T. Hama, H. Sakai, K. Harashima, Production technology for amorphous silicon-based flexible solar cells, *Sol. Energy Mater. Sol. Cells*, 66 (2001) 107–115. https://doi.org/10.1016/S0927-0248(00)00163-X

[7] A. S. Najm, P. Chelvanathan, S. K. Tiong, M. T. Ferdaous, S. A. Shahahmadi, Y. Yusof, N. Amin, Numerical insights into the influence of electrical properties of n-CdS buffer layer on the performance of SLG/Mo/p-Absorber/n-CdS/n-ZnO/Ag configured thin film photovoltaic devices, *Coatings*, 11 (2021) 52. https://doi.org/10.3390/coatings11010052

[8] B. P. Jelle, C. Breivik, H. D. Rokenes, Building integrated photovoltaic products: a state-of-the-art review and future research opportunities, *Sol. Energy Mater. Sol. Cells*, 100 (2012) 69–96. https://doi.org/10.1016/j.solmat.2011.12.016

[9] K. L. Chopra, P. D. Paulson, V. Dutta, Thin-film solar cells: an overview. Progress in Photovoltaics, *Research and Applications*, 12 (2004) 69–92. https://doi.org/10.1002/pip.541

[10] A. Chirila, S. Buecheler, F. Pianezzi, P. Bloesch, C. Gretener, A.R. Uh, C. Fella, L. Kranz, J. Perrenoud, S. Seyrling, et al., Highly efficient Cu(In,Ga)Se2 solar cells grown on flexible polymer films, *Nat. Mater.*, 10 (2011) 857–861. https://doi.org/10.1038/nmat3122

[11] W. Hou, Y. Xiao, G. Han, J. Lin, The applications of polymers in solar cells: a review, *Polymers*, 11 (2019) 143–189. https://doi.org/10.3390/polym11010143

[12] Y. Wang, Recent research progress on polymer electrolytes for dye-sensitized solar cells, *Sol. Energy Mater. Sol. Cells*, 9 (2009) 1167–1175. https://doi.org/10.1016/j.solmat.2009.01.009

[13] J. Wu, Z. Lan, S. Hao, P. Li, J. Lin, M. Huang, Y. Huang, Progress on the electrolytes for dye-sensitized solar cells, *Pure Appl. Chem.*, 80 (2008) 2241–2258. https://doi:10.1351/pac200880112241

[14] M. E. Ragoussi, T. Torres, New generation solar cells: concepts, trends and perspectives, *Chem. Commun.*, 51 (2015) 3957–3972. https://doi.org/10.1039/C4CC09888A
[15] A. R. Clavijo, O. C. Calero, M. M. Gonzalez, Revisiting anodic alumina templates: from fabrication to applications, *Nanoscale*, 13 (2021). 2227–2265. https://doi:10.1039/D0NR07582E
[16] L. Calio, S. Kazim, M. Gratzel, S. Ahmad, Hole-transport materials for perovskite solar cells, *Angew. Chem. Int.*, 55 (2016) 14522–14545. https://doi.org/10.1002/anie.201601757
[17] R. Balint, N. J. Cassidy, S. H. Cartmell, Conductive polymers: towards a smart biomaterial for tissue engineering, *Acta Biomater.*, 10 (2014) 2341–2353. https://doi.org/10.1016/j.actbio.2014.02.015
[18] S. A. Khan, L. Li, D. Zhao, S. Chen, Porous counter electrode for dye-sensitized solar cell by simple hydrothermal method, *Polym. Sci. Ser. B*, 61 (2019) 846–855. https://doi.org/10.1134/S1560090420010054
[19] Y. Tian, G. Zhu, Porous aromatic frameworks (PAFs), *Chem. Rev.*, 120 (2020) 8934–8986. https://doi.org/10.1021/acs.chemrev.9b00687
[20] S. Ding, W. Wang, Covalent organic frameworks (COFs): from design to applications, *Chem. Soc. Rev.*, 42 (2013) 548–568. https://doi.org/10.1039/C2CS35072F
[21] K. Wang, K. Amin, Z. An, Z. Cai, H. Chen, B. Z. Tang, Advanced functional polymer materials, *Mater. Chem. Front.*, 4 (2020) 1803–1915. https://doi:10.1039/D0QM00025F
[22] W. Chen, Y. Wu, Y. Yue, J. Liu, W. Zhang, X. Yang, L. Han, Efficient and stable large-area perovskite solar cells with inorganic charge extraction layers, *Science*, 350 (2015) 944–948. https://doi:10.1126/science.aad1015
[23] J. Wang, J. Zhang, Y. Yang, S. Gai, Y. Dong, L. Qiu, R. Fan, New insight into the Lewis basic sites in metal–organic framework-doped hole transport materials for efficient and stable Perovskite solar cells, *ACS Appl. Mater. Interfaces*, 13 (2021), 5235–5244. https://doi.org/10.1021/acsami.0c1996
[24] G. Zhong, D. Liu, J. Zhang, Applications of porous metal–organic framework MIL-100 (M)(M= Cr, Fe, Sc, Al, V), *Crystal Growth & Design*, *18* (2018) 7730–7744. https://doi.org/10.1021/acs.cgd.8b01353
[25] M. I. Khan, M. Saleem, K. A. Bhatti, et al., Comparative study of multilayered $ZnO/TiO_2/ZnO$ and $TiO_2/ZnO/TiO_2$ thin films prepared by sol-gel dip coating method. *J. Mater. Sci: Mater. Electron*, 28 (2017) 17499–17504. https://doi.org/10.1007/s10854-017-7685-9
[26] N. Pavithra, A. M. Asiri, S. Anandan, Fabrication of dye sensitized solar cell using gel polymer electrolytes consisting poly (ethylene oxide)-acetamide composite, *J. Power Sources*, 286 (2015) 346–353. https://doi.org/10.1016/j.jpowsour.2015.03.160
[27] J. Y. Song, Y. Y. Wang, C. C. Wan, Conductivity study of porous plasticized polymer electrolytes based on poly (vinylidene fluoride) a comparison with polypropylene separators, *J. Electrochem. Soc.*, 147 (2000) 3219–3225. https://doi.org/10.1149/1.1393886
[28] M. Saleem, M. Irfan, S. Tabassum, M. D. Albaqami, M. S. Javed, S. Hussain, M. Pervaiz, I. Ahmad, A. Ahmad, M. Zuber, Experimental and theoretical study of highly porous lignocellulose assisted metal oxide photoelectrodes for dye-sensitized solar cells, *Arabian J. Chem.*, 14 (2021) 102937. https://doi.org/10.1016/j.arabjc.2020.102937.
[29] Z. Jin, J. Yan, X. Huang, W. Xu, S. Yang, D. Zhu, J. Wang, Solution-processed transparent coordination polymer electrode for photovoltaic solar cells, *Nano Energy*, 40 (2017) 376–381. https://doi.org/10.1016/j.nanoen.2017.08.028.
[30] T. A. Roberta, W. Xiang, A. S. A. Ahmeda, S. Wua, B. Lia, Q. Liua, C. M. J. Chutia, X. Zhao, Porous rGO/ZnSe/CoSe2 dispersed in PEDOT:PSS as efficient counter electrode for dye-sensitized solar cells. *Mater. Chem. Front.*, (2021). https://doi.org/10.1039/D0QM01078B.
[31] Z. Li, W. Qi, L. Li, Z. Ma, W. Lai, L. Li, X. Jin, Y. Zhang, W. Zhang, Preparation of carbon nanofibers supported Bi_2MoO_6 nanosheets as counter electrode materials on Titanium mesh substrate for dye-sensitized solar cells, *Solar Energy*, 214 (2021) 502–509. https://doi.org/10.1016/j.solener.2020.11.064.
[32] D. Zhou, Z. Xia, H. Shang, H. Xiao, Z. Jiang, H. Li, L. Zheng, J. Dong, W. Chen, Rational design of efficient counter electrode with $Co/Co_1P_1N_3$ atomic interface for promoting catalytic performance, *Mater. Chem. Front.*, (2021). https://doi.org/10.1039/D0QM00806K.
[33] Z. Xie, X. Cui, W. Xu, Y. Wang, Metal-organic framework derived CoNi@CNTs embedded carbon nanocages for efficient dye-sensitized solar cells, *Electrochim. Acta*, 229 (2017) 361–370. https://doi.org/10.1016/j.electacta.2017.01.145.
[34] Y.Y. Wang, S.M. Chen, R. Haldar, C. Wöll, Z.G. Gu, J. Zhang, van der Waals epitaxial growth of 2D Metal–Porphyrin framework derived thin films for dye-sensitized solar cells, *Adv. Mater. Interfac.*, 5 (2018) 1800985. https://doi.org/10.1002/admi.201800985.

[35] W. Moloto, P. Mbule, E. Nxumalo, B. Ntsendwana, Stabilizing effects of zinc(II)-benzene-1,3,5-tricarboxylate metal organic frameworks on the performance of TiO_2 photoanodes for use in dye-sensitized solar cells, *J. Photochem. Photobiol. A: Chem.*, 15 (2021) 113063. https://doi.org/10.1016/j.jphotochem.2020.113063.
[36] S.C. Ramesh, C. C. Columbus, X.S. Shajan, Modeling and simulation of a dye sensitized solar cell with porous aerogel photoanode, *J. Electr. Eng. Technol.* 16, (2021) 509–514. https://doi.org/10.1007/s42835-020-00600-9.
[37] D. Y. Lee, D. V. Shinde, S. J. Yoon, K. N. Cho, W. Lee, N. K. Shrestha, S.-H. Han, Cu-based metal–organic frameworks for photovoltaic application, *J. Phys. Chem. C*, 118 (2014) 16328–16334, https://doi.org/10.1021/jp4079663
[38] J. Dou, Metal–organic framework derived hierarchical porous anatase TiO2 as a photoanode for dye-sensitized solar cell. *Cryst. Growth Des.*, 16, (2015) 121–125. https://doi.org/10.1021/acs.cgd.5b01003
[39] S.M. Wang, Polyoxometalate–anatase TiO_2 composites are introduced into the photoanode of dye-sensitized solar cells to retard the recombination and increase the electron lifetime. *Dalton Trans.*, 42 (2013) 2691–2695. https://doi.org/10.1039/C2DT32403B.
[40] U. H. A. Azeez, A. Gunasekaran, A. Sorrentino, A. Syed, N. Marraiki, S. Anandan, Synthesis and characterization of poly-3-(9H-carbazol-9-yl) propylmethacrylate as a gel electrolyte for dye-sensitized solar cell applications, *Polym. Bull.* (2021). https://doi.org/10.1007/s00289-021-03541-z
[41] K. Mohan, A. Bora, B. Chandra et al., A highly stable and efcient quasi solid-state dye sensitized solar cell based on polymethyl methacrylate (PMMA)/polyaniline nanotube (PANI-NT) gel electrolyte. *Electrochim Acta*, 222 (2016) 1072–1078. https://doi.org/10.1016/j.electacta.2016.11.077
[42] V. Murugadoss, S. Arunachalam, V. Elayappan, S. Angaiah, Development of electrospun PAN/CoS nanocomposite membrane electrolyte for high-performance DSSC. *Ionics (Kiel)* 24 (2018) 4071–4080. https://doi.org/10.1007/s11581-018-2540-4.
[43] R. X. Dong, S. Y. Shen, H. W. Chen et al., A novel polymer gel electrolyte for highly efficient dye sensitized solar cells. *J. Mater. Chem. A*, 1 (2013) 8471–8478. https://doi.org/10.1039/C3TA11331K.
[44] K. Xia, Z. Peng, Z. Hu et al., High efficiency quasi solid-state dye-sensitized solar cells based on a novel mixed-plasticizer modified polymer electrolyte. *Electrochim. Acta*, 153 (2015) 28–32. https://doi.org/10.1016/j.electacta.2014.11.174
[45] M. R. Chirani, E. Kowsari, H. S. Amoli, M. Yousefzadeh, A. Chinnappan, S. Ramakrishna, Covalently functionalized graphene oxide with cobalt-nitrogen-enriched complex containing iodide ligand as charge carrier nanofiller for eco-friendly high performance ionic liquid-based dye-sensitized solar cell, *J. Mol. Liq.*, 325 (2021) 115198. https://doi.org/10.1016/j.molliq.2020.115198.
[46] X. Li, Q. An, H. Li, Y. Zhang, Z. Chen, K. Teng, J. Zhuang, Y. Zhang, W. Tong, A porous piezoelectric-dielectric flexible energy conversion film for electricity generation from multiple sources, *Chem. Phys. Lett.*, 767 (2021) 138357. https://doi.org/10.1016/j.cplett.2021.138357.
[47] Z. Zhu, P. Mu, Y. Fan, W. Bai, Z. Zhang, H. Sun, W. Liang, A. Li, Highly efficient solar steam generation of bilayered ultralight aerogels based on N-rich conjugated microporous polymers nanotubes, *Eur. Polym. J.*, 126 (2020) 109560. https://doi.org/10.1016/j.eurpolymj.2020.109560.
[48] J. M. Newkirk, I. Nayshevsky, A. Sinha, A. M. Law, Q. F. Xu, B. To, P. F. Ndione, L. T. Schelhas, J. M. Walls, A. M. Lyons, D. C. Miller, Artificial linear brush abrasion of coatings for photovoltaic module first-surfaces, *Sol. Energy Mater Sol. Cells*, 219 (2021) 11075. https://doi.org/10.1016/j.solmat.2020.110757.
[49] A. A. Arbab, N. Mengal, I. A. Sahito, A. A. Memon, S. H. Jeong, An organic route for the synthesis of cationic porous graphite nanomaterial used as photocatalyst and electrocatalyst for dye-sensitized solar cell, *Electrochim. Acta.*, 266 (2018) 43–53. https://doi.org/10.1016/j.electacta.2018.02.012.
[50] V. A. G. Andrade, W. O. H. Martínez, F. Redondo, N. B. C. Guerrero, F. Roncaroli, M. D. Perez, Fe and Ti metal-organic frameworks: towards tailored materials for photovoltaic applications, *Appl. Mater. Today*, 22 (2021) 100915. https://doi.org/10.1016/j.apmt.2020.100915.
[51] U. J. Ryu, S. Jee, J.-S. Park, K. Han, J. H. Lee, M. Park, K. M. Choi, Nanocrystalline titanium metal-organic frameworks for highly efficient and flexible Perovskite solar cells, *ACS Nano*, 12 (2018) 4968–4975. https://doi.org/10.1021/acsnano.8b02079.
[52] J. Zhang, S. Guo, M. Zhu, C. Li, J. Chen, L. Liu, S. Xiang, Z. Zhang, Simultaneous defect passivation and hole mobility enhancement of perovskite solar cells by incorporating anionic metal-organic framework into hole transport materials, *Chem. Eng. J.*, 408 (2021) 127328. https://doi.org/10.1016/j.cej.2020.127328.

[53] J. Wang, J. Wang, J. Zhang, Y. Yang, S. Gai, Y. Dong, L. Qi, D. Xia, X. Fan, W. Wang, B. Hu, W. Cao, R. Fan, New insight into the Lewis basic sites in metal-organic framework-doped hole transport materials for efficient and stable Perovskite solar cells, *ACS Appl. Mater. Interfaces*, 13 (2021) 5235–5244. https://doi.org/10.1021/acsami.0c19968.

[54] C. Geng, P. Wei, H. Chen, H. Liu, S. Zheng, H. Wang, Y. Xie, Promoting the hole extraction and interfacial performance with MOFs derived Co_3O_4@NC for efficient carbon-based perovskite solar cells, *Chem. Eng. J.*, 12 (2021) 128878. https://doi.org/10.1016/j.cej.2021.128878.

[55] C. Wu, Y. Liu, H. Liu, C. Duan, Q. Pan, J. Zhu, F. Hu, X. Ma, T. Jiu, Z. Li, Y. Zhao, Highly conjugated three-dimensional covalent organic frameworks based on spirobifluorene for Perovskite solar cell enhancement. *J. Am. Chem. Soc.*, 31 (2018) 10016–10024. https://doi.org/10.1021/jacs.8b06291.

[56] S. Park, M.-S. Kim, W. Jang, J. K. Park, D. H. Wang, Covalent organic nanosheets for effective charge transport layers in planar-type perovskite solar cells, *Nanoscale*, 10 (2018) 4708–4717. https://doi.org/10.1039/C7NR08797G.

[57] M. S. Sowmehesaraee, M. Ranjbar, M. Abedi, S. A. Mozaffari, Fabrication of lead iodide perovskite solar cells by incorporating zirconium, indium and zinc metal-organic frameworks, *Sol. Energy*, 214, (2021) 138–148. https://doi.org/10.1016/j.solener.2020.12.001.

[58] P. M. Rajanna, E. P. Gilshteyn, T. Yagafarov, A. A. Alekseeva, A. S. Anisimov, A. Neumüller, O. Sergeev, S. Bereznev, J. Maricheva, A. G. Nasibulin, Enhanced efficiency of hybrid amorphous silicon solar cells based on single-walled carbon nanotubes and polymer composite thin film, *Nanotechnology*, 29 (2018) 105404. https://doi.org/10.1088/1361-6528/aaa647.

[59] M. H. Yun, J. H. Jang, K. M. Kim, H.-E. Song, J. C. Lee, J. Y. Kim, A hybrid solar cell fabricated using amorphous silicon and a fullerene derivative, *Phys. Chem. Chem. Phys.*, 15 (2013) 19913–19918. https://doi.org/10.1039/C3CP53493F.

[60] I. Khatri, Z. Tang, Q. Liu, R. Ishikawa, K. Ueno, H. Shirai, Green-tea modified multiwalled carbon nanotubes for efficient poly(3,4-ethylenedioxythiophene): poly(stylenesulfonate)/n-silicon hybrid solar cell, *Appl. Phys. Lett.*, 102 (2013), 063508. https://doi.org/10.1063/1.4792691.

[61] F. Liu, W. Liang, C. Wang, J. He, C. Xiao, Z. Zhu, H. Sun, A. Li, Superwetting monolithic hypercross-linked polymers nanotubes with high salt-resistance for efficient solar steam generation, *Sol. Energy Mater. Sol. Cells*, 221 (2021) 110913. https://doi.org/10.1016/j.solmat.2020.110913.

[62] M. Chen, Y. Wu, W. Song, Y. Mo, X. Lin, Q. He, B. Guo, Plasmonic nanoparticle-embedded poly(*p*-phenylene benzobisoxazole) nanofibrous composite films for solar steam generation, *Nanoscale*, 13 (2018) 6186–6193. https://doi.org/10.1039/C8NR01017J.

[63] Z. Zheng, H. Li, X. Zhang, H. Jiang, X. Geng, S. Li, H. Tu, X. Cheng, P. Yang, Y. Wan, High-absorption solar steam device comprising Au@Bi_2MoO_6-CDs: extraordinary desalination and electricity generation, *Nano Energy*, 68 (2020) 104298. https://doi.org/10.1016/j.nanoen.2019.104298.

[64] Z. Zhu, P. Mu, Y. Fan, W. Bai, Z. Zhang, H. Sun, W. Liang, A. Li, Highly efficient solar steam generation of bilayered ultralight aerogels based on N-rich conjugated microporous polymers nanotubes, *Eur. Polym. J.*, 126 (2020) 109560. https://doi.org/10.1016/j.eurpolymj.2020.109560.

[65] Z.J. Xia, H.C. Yang, Z. Chen, R. Z. Waldman, Y. Zhao, C. Zhang, S. N. Patel, S. B. Darling, Porphyrin covalent organic framework (POF)-based interface engineering for solar steam generation, *Adv. Energy Mater.*, 6 (2019) 1900254. https://doi.org/10.1002/admi.201900254.

[66] J. He, G. Zhao, P. Mu, H. Wei, Y. Su, H. Sun, Z. Zhu, W. Liang, A. Li, Scalable fabrication of monolithic porous foam based on cross-linked aromatic polymers for efficient solar steam generation, *Sol. Energy Mater. Sol. Cells*, 201 (2019) 110111, https://doi.org/10.1016/j.solmat.2019.110111.

[67] J. He, Z. Zhang, C. Xiao, F. Liu, H. Sun, Z. Zhu, W. Liang, A. Li, High-performance salt-rejecting and cost-effective superhydrophilic porous monolithic polymer foam for solar steam generation, *ACS Appl. Mater. Interfaces*, 12 (2020) 16308–16318. https://doi.org/10.1021/acsami.9b2283

[68] S. Wan, F. Gandara, A. Asano, H. Furukawa, A. Saeki, S. K. Dey, L. Liao, M. W. Ambrogio, Y. Y. Botros, X. Duan, S. Seki, J. F. Stoddart, O. M. Yaghi, Covalent organic frameworks with high charge carrier mobility, *Chem. Mater.*, 23 (2011) 4094–4097. https://doi.org/10.1021/cm201140r.

[69] S. Jin, M. Supur, M. Addicoat, K. Furukawa, L. Chen, T. Nakamura, S. Fukuzumi, S. Irle, D. Jiang, Creation of superheterojunction polymers via direct polycondensation: segregated and bicontinuous donor-acceptor π-columnar arrays in covalent organic frameworks for long-lived charge separation, *J. Am. Chem. Soc.*, 137 (2015) 7817–7827. https://doi.org/10.1021/jacs.5b03553.

[70] S. Jin, X. Ding, X. Feng, M. Supur, K. Furukawa, S. Takahashi, M. Addicoat, M. E. El-Khouly, T. Nakamura, S. Irle, S. Fukuzumi, A. Nagai, D. Jiang, Charge dynamics in a donor–acceptor covalent organic framework with periodically ordered bicontinuous heterojunctions, *Angew. Chem., Int. Ed.*, 52, (2013) 2017–2021. https://doi.org/10.1002/ange.201209513.

[71] W. Zhang, W. Li, X. He, et al. Dendritic Fe-based polyoxometalates @ metal–organic framework (MOFs) combined with ZnO as a novel photoanode in solar cells, *J. Mater. Sci: Mater Electron*, 29 (2018) 1623–1629. https://doi.org/10.1007/s10854-017-8073-1.

[72] X. Luo, Enhanced photovoltaic response of the first polyoxometalate-modified zinc oxide photoanode for solar cell application, *J. Mater. Chem.*, 22 (2012) 15050–15055. https://doi.org/10.1039/C2JM16018H.

11 Porous Polymers for Hydrogen Production

Ashish Kumar and Abhinay Thakur
Lovely Professional University, Phagwara, India

CONTENTS

DOI: 10.1201/9781003169604-11

11.1 INTRODUCTION

The entire world is facing its "Oil Peak," with oil reserves anticipated to be drained within the next decades. It is important to look for alternatives to fossil fuels. To solve these major environmental challenges and the approaching energy crisis, it is vital to build up sustainable and green energy sources. Hydrogen energy, with its high combustion calorific value, minimum emissions, its immense availability and clean combustion H_2 is recognized as among the foremost potential energy sources and promising alternative fuel for the future. Furthermore, the energy density of H_2 (142 kJ/g) is substantially larger than petroleum oil (47 kJ/g) [1]. Global H_2 production is currently expected to be around 500 billion cubic meters per year (b/m^3) [2, 3]. Hydrogen can be made from a variety of renewable and nonrenewable energy sources, including methane steam reforming, oil/naphtha reforming, coal gasification, biomass, biological sources, and water electrolysis (WE). Fertilizers, petrochemical, petroleum refining operations, fuel cell, and chemical industries all use H_2 generated in this manner. Nonrenewable fossil fuels, primarily steam reformation of methane, currently account for 96% of worldwide H_2 generation. However, the use of fossil fuels results in decreased hydrogen purity and the release of higher concentrations of hazardous greenhouse gases [4]. Many countries are cautiously investigating the positive resultants of shifting to a hydrogen economy. The growing concern in hydrogen derives mostly from its ability to address two fundamental challenges impacting many nations throughout the world: attaining energy sufficiency while minimizing the environmental impact of economic activity. Before a hydrogen economy can be achieved, four crucial technologies need to be developed:

- Low-cost hydrogen production in a carbon-constrained energy grid.
- Hydrogen purification and storage devices are capable of separating, purifying and producing H_2 in an environmentally friendly manner.
- A hydrogen supply and distribution infrastructure that is efficient, readily accessible, and well-managed.
- High-efficiency hydrogen fuel cells and other energy conversion technologies.

Currently, various actions have been focused on environmentally acceptable energy techniques that might potentially replace existing energy production based on fossil fuel consumption. Figure 11.1

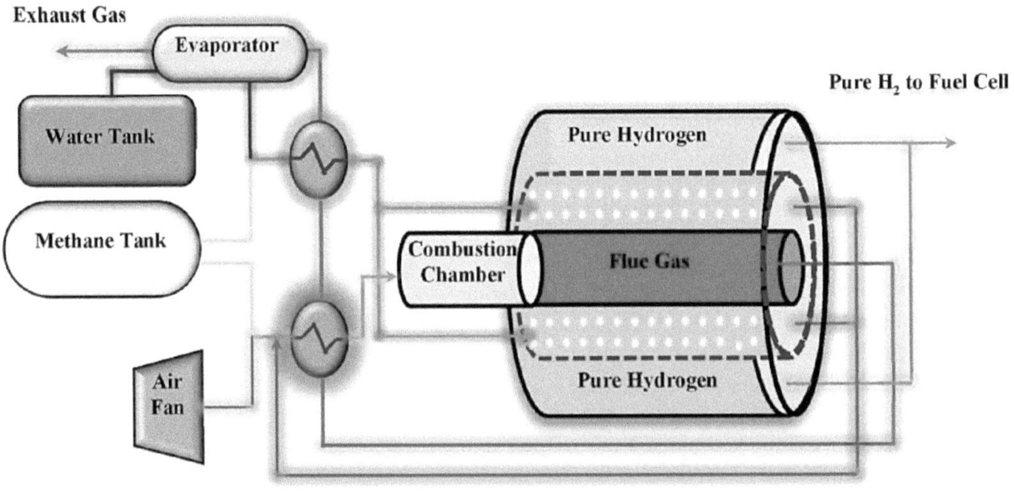

FIGURE 11.1 A potential mechanism for generating pure H_2 in an engine.

(Figure adapted from ref. [1].)

is showing a potential mechanism for the generation of pure H_2 in an engine [5]. This might be accomplished when H_2 is generated using porous polymers. Porous polymers (PPs) comprise pores, or tiny vacant volumes, in them [6]. The challenge with PP is that the vacant volume must be manufactured before it can be set in place for further utilization. The Chemistry of Polymers could be utilized to explain both chemistry of polymerization and the chemistry of macromolecular configurations. These aspects of the Chemistry of Polymers are essential for producing porous polymers efficiently. Both the comprehensive tools of polymerization reactions and the broad tools of macromolecular structures are routinely used in the development and stabilization of pores. Gas evolution after polymer synthesis or gas generation during polymer production can form polymer foams with millimeter-scale pores [7]. Microporous polymers comprising pores as small as a nanometer could be produced by constructing framework architecture throughout polymer synthesis or by constructing hyper-crosslinked networks thereafter. One of the nanometer-scale phases of microphase segregated block copolymers (BCPs) can be eliminated to make mesoporous polymers. A biphasic system is created in a normal colloidal templating scenario, and afterward, the continuous phase is polymerized. After polymerization, the colloidal entities are removed, and the porosity in the resulting polymeric material is increased. Depending on the type of colloidal system being utilized, pore sizes can vary from a couple of nanometers to thousands of micrometers. Figure 11.2 is showing several methods for the production of H_2.

11.1.1 Approaches Utilized for the Generation of Porous Polymers (PPs)

Certain pore production methods use polymerization chemistry, while others use macromolecular structural chemistry, and some use both. Several new approaches to synthesize macromolecules and create unique macromolecular architectures are constantly being discovered due to the ever-expanding toolset of polymerization processes. The macromolecule interactions with its surrounding are influenced by its structural chemistry. The Flory–Huggins interaction parameter has been used to explain these interactions, which is associated with the disparity in the solubility factors assigned to the polymer and solvent. The chemical group attraction constants have been used to link the solubility parameter to the macromolecular structure of the polymer. The final size-scale of the resultant pore is usually intrinsically linked to the pore creation methodology. The most frequent methods for pore formation in polymers are listed here.

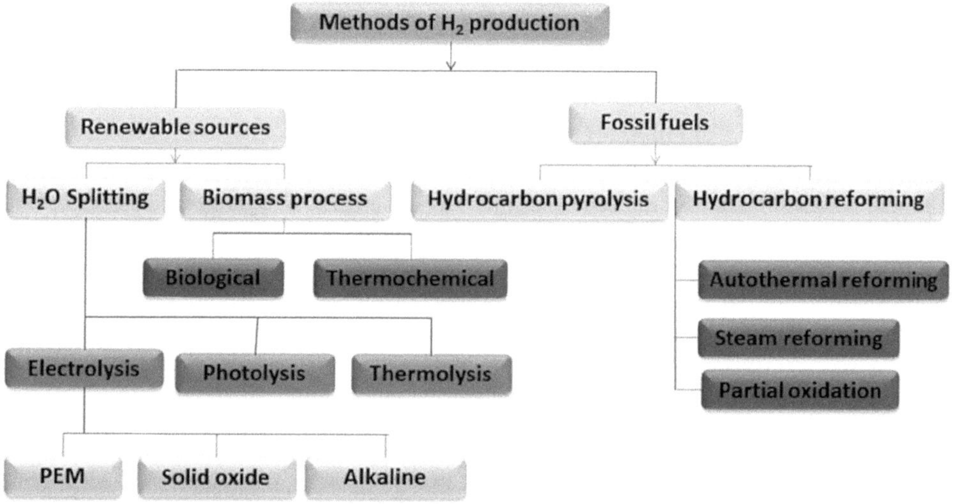

FIGURE 11.2 Various hydrogen production methods.

11.1.1.1 Infiltration

Polymer solutions are frequently inserted into the template holes, where the polymer solidifies. AAO and mesoporous silica are the most commonly used templates. The resultant structure of the recreated PPs is affected by the qualities of the templates, like pore size and surface chemistry, as well as the characteristics of the infiltrating fluids, polymer chain conformation, viscosity and solution concentration. While infiltrating AAO templates, there is usually no requirement for surface modification since infiltrated fluids having low surface energy may be easily distributed on the wide surface inner layers of AAO to produce a dense layer [8–10]. The outer diameter of the produced polymer nanotubes is determined by the AAO pore diameter, while the tube width is mostly determined by the AAO membrane thickness. AAO templating can be used to make solid polymer nanopillars in contrast to hollow polymer nanotubes. The techniques related to the synthesis of nanopillars and polymer nanotubes have received a lot of attention [11].

11.1.1.2 Layer-by-Layer Assembly (LbL)

It was developed in 1991 as a raw material to be commonly used for direct templating with polymers. Polymers can be accumulated onto the surface of the templates using LbL assembly, which involves the alternating adsorption of polymers with oppositely charged moieties. Due to its high tolerance for templates with various shapes, such as AAO, solid spherical nanoparticles, nanowires, and mesoporous silica, it may produce a variety of porous structures. Furthermore, LbL assembly is a low-cost, environmentally friendly method that has the unique ability to organize the sets of layers accumulated on the template to selectively adjust the width of the polymeric shell to be utilized in the porous polymers at a nanoscale level. The development of LbL techniques has also depended on nonionic interactions among uncharged polymers, like hydrogen-bonding interactions, van der Waals interactions and covalent-bonding interactions. This approach broadens the range of polymer sources from polyelectrolytes to uncharged macromolecules, minimizing the need for a further polyelectrolyte in polymeric structures. This method promotes the integration of covalent bonding contacts, which enhance the built-in polymeric wall's durability [12].

11.1.1.3 Conventional Polymerization

Conventional polymerization of monomers in the context of the template is often utilized to create PPs. It is a fundamental process that can be performed in several available monomers. Other processes include oxidative polymerization, polycondensation and others [13]. The templates utilized vary from solid nanoparticles to porous material, allowing a wide range of polymeric porous patterns to be created. Infiltration and consequent polymerization in a porous template, in situ template production and processing polymerization, and polymerization on the interface of nonporous templates are the most common templating techniques.

11.1.1.4 Electrochemical Polymerization

PANI, poly(3-hexylthiophene), PEDOT and poly(carbazole) are commonly implemented in electrochemical polymerization for the creation of porous structures containing conducting polymers. Other one-dimensional particles, like TiO_2 nanotube arrays, track-etched porous polycarbonate membranes and ZnO nanowire have also been utilized as templates to form such porous conducting polymers into structured macroporous and tubular structures. When compared to traditional polymerization, electrochemical polymerization has various advantages. It provides better regulation over the number of accumulated layers, film thickness and the properties of PPs (pore size and pore shape) [14]. Second, it enhances the potency of the invert opal films by enabling control over the polymerization rates and result in a significantly better-confined polymer structure. Finally, porous nanostructures as-built make closer contact with conducting surfaces, making it easier to build an electrical device. Furthermore, it involves choosing which polymerization sites should be used.

11.1.1.5 Controlled/Living Polymerization (CLP)

CLP involving direct templating for the production of PPs regarded as a highly compelling approach, including ring-opening metathesis polymerization, atom transfer radical polymerization (ATRP), and ring-opening polymerization, suitable for the polymerization of varieties of monomers in the existence of various templates including AAO, solid particles and colloidal crystals. In the majority of research, a "grating from" method is used. After that, CLP initiators are mounted to template surfaces, and the polymeric shell is produced through controlled polymerization using those initiating sites [15, 16]. When the shell is formed of a soft linear polymer, copolymerizing using a multifunctional monomer can be used to generate more cross-linking. Surface-initiated CLP has a number of advantages:

(a) By modifying the degree of polymerization (DP), shell thickness may be accurately adjusted at the molecular level,
(b) Additional functionality can be easily added to any section of the chain, particularly while utilizing ATRP,
(c) The shell's "dormant" chain ends could be resurrected, allowing subsequent CLP to build BCP shells that can be used as a barrier to reduce inter-particle aggregation and cross-linking, and then diffuse cross-linked capsules in the desired solvent.

11.1.1.6 Macromolecular Design

Microporous structures could be produced by creating moderately brittle macromolecular structures. Pores could be developed by several ways such as in the case of twisted macromolecules preparation during processing (e.g., covalent organic frameworks (COFs), conjugated microporous polymers (CMPs), covalent triazine frameworks and porous aromatic frameworks) or monomers are engineered to organize during polymerization (e.g., CMPs and COFs) [17].

11.1.1.7 Self-assembly

Mesoporous structures may emerge when enormous molecules with incompatible sections spontaneously self-assemble. Porous structures (hexagonal or cubic) can be produced by self-assembling surfactant molecules with a hydrophobic tail, hydrophilic head and a polymerizable group, which could then be polymerized to generate porous polymers. The self-assembly of BCPs into cylindrical or gyroidal structures could be used to construct PPs by using a block or a block-linker that can be removed afterward.

11.1.1.8 Phase Separation

The phase separation and retention of the phase-separated component are achieved by modulating the polymer-solvent miscibility, which leads to micrometer-scale porous structures typically utilized for membranes.

11.1.1.9 Solid and Liquid Templating

Porous structures with sizes ranging from micrometers to millimeters are formed around a separable solid or liquid templating framework and stabilized via polymerization (for monomers) or solvent extraction (for solvents).

11.1.1.10 Foaming

Gas generation and expansion produce millimeter-scale pores, which are fixed by crosslinking, polymerization and freezing to minimize polymer fluidity.

11.2 VARIOUS POROUS POLYMERS FOR H_2 PRODUCTION

11.2.1 Photocatalysts Based on Conjugated Microporous Polymers

Photocatalytic hydrogen synthesis with solar light irradiation can be considered as an environmentally sustainable advantageous approach to produce sustainable energy as it is easy to perform and non-hazardous to the environment. An electron hole pair is created when photocatalysts are stimulated to sunlight with energy greater or equal toward the photocatalyst's bandgap energy. Photo-excited e^- in a semiconductor move from the valence band to the conduction band, wherein reduction takes place in water to form hydrogen [18–20]. CMPs have large specific surface areas and thermal stabilities, as well as extensive conjugation, tunable molecular topologies and remarkable physical and physicochemical properties, making them appropriate H_2 production photocatalysts. Several researchers have highlighted the importance of rational polymer design in CMP-based photocatalysts for H_2O splitting, resulting in increased photocatalytic efficiency. When exposed to visible light, these polymers displayed high photocatalytic activity due to e^--deficient comonomers that can increase the dipole-polarization effect of the polymers. A group of researchers revealed many nitrogen-containing CMPs (N-CMPs) constituted of D–A complexes with increased H_2 production under visible-light irradiation. On 3D-cored structures, this design method required altering substitution positions, the quantity of e-donors, and the cyano-containing acceptor moieties. These N-CMPs are well-known for their charge carrier separation efficiency and configurable bandgaps in the region of 1.64–2.29 eV. 4-CzPN has a 2,103 mol h^{-1} g^{-1} visible-light-driven HER and a 6.4% AQY at 420 nm. In addition, the efficacy of 4-CzPN photocatalysts is maintained throughout recycling and 25 hours of continuous light irradiation.

Ju Ting Lee et al. [14] used a basic etching SiO_2 approach to create porous poly(vinyl alcohol)/platinum/nitrogen-doped titanium dioxide/strontium titanate composite (PVA/Pt/NT/STO) films with varied pore diameters. With the increased pore size and density of the PVA/Pt/NT/STO film, the photocatalytic hydrogen production efficiency improved as the particle size of the as-prepared SiO_2 expanded. The film's light transmission and energy transference increased as pores appeared on the surface. The ideal photocatalytic H_2 generation rates of PVA/Pt/NT/STO-8S-I-20 film was 33,890 mol, nearly three times larger than that of thick PVA/Pt/NT/STO film, due to porous PVA/Pt/NT/STO film's superior light transmittance and charge separation (18,185 mol h^{-1} g^{-1}).

Figure 11.3 depicts the hydrogen production process, in which the porphyrins' first oxidation potentials are higher than the redox couple of triethanolamine as a sacrificial agent (0.84±0.12 V vs. NHE), allowing dye reconstruction and resisting conjugation between both the injected e- and the dye cation radical [21].

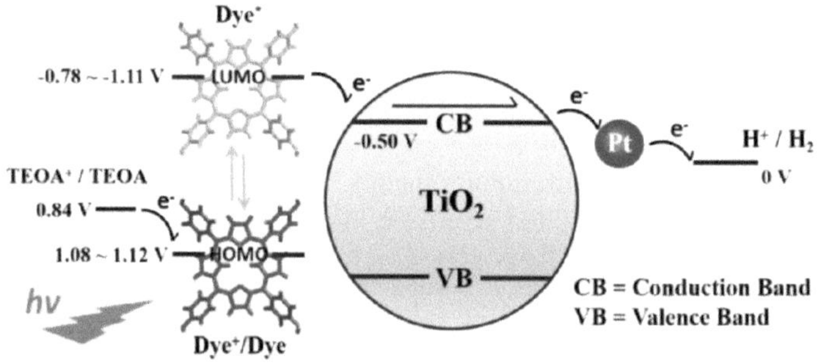

FIGURE 11.3 Photocatalytic H_2 generation over Pt/porphyrin-TiO_2: proposed process.

(Figure adapted from ref. [2].)

11.2.2 Conjugated Microporous Polymers

Extended-conjugation, large specific surface area, outstanding physicochemical, high thermal stability, variable chemical structure, and electronic properties make CMPs suitable photocatalysts for hydrogen evolution. CMPs have a microporous structure that allows for a large surface area, potentially exposing numerous active sites. Further, CMPs may be manufactured in a number of ways, and their band structure could be easily changed to enhance the wavelength of visible light absorption. CMP explorations over the last decade have shown that selecting monomers with different structures can change the structural frame and increase H_2 evolution rate. The Chichibabin reaction involves a metal-free organic synthesis process that generates the pyridine ring by cross-coupling aromatic ketone and aromatic aldehyde, with ammonium acetate as the nitrogen supply. The presence of the pyridine ring enables the formation of N-doped CMPs. There are no metal catalysts used in this synthetic process, and the reaction conditions are low, taking only a short time to complete [22–26]. The initial viability of the Chichibabin reaction improved the production of CMPs, wherein molecular structure could be systematically controlled for hydrogen evolution purposes. The tetra (p-bromophenyl) methane synthesis process and the ferrocene-based CMP porphyrin (FcCMP-1) production pathway (FcCMP-2) has been shown in Figure 11.4 [27].

Qin-ruo et al. [3] developed a set of pyridyl conjugated microporous polymers (PCMPs) for photocatalytic H_2 evolution employing the Chichibabin reaction employing two aromatic ketones and three aromatic aldehydes as monomers. PCMPs with a triazine ring in monomer have a high percentage of H_2 generation (1198.9 mol g^{-1}). In the UV-Vis spectrum, TFPA exhibits extensive

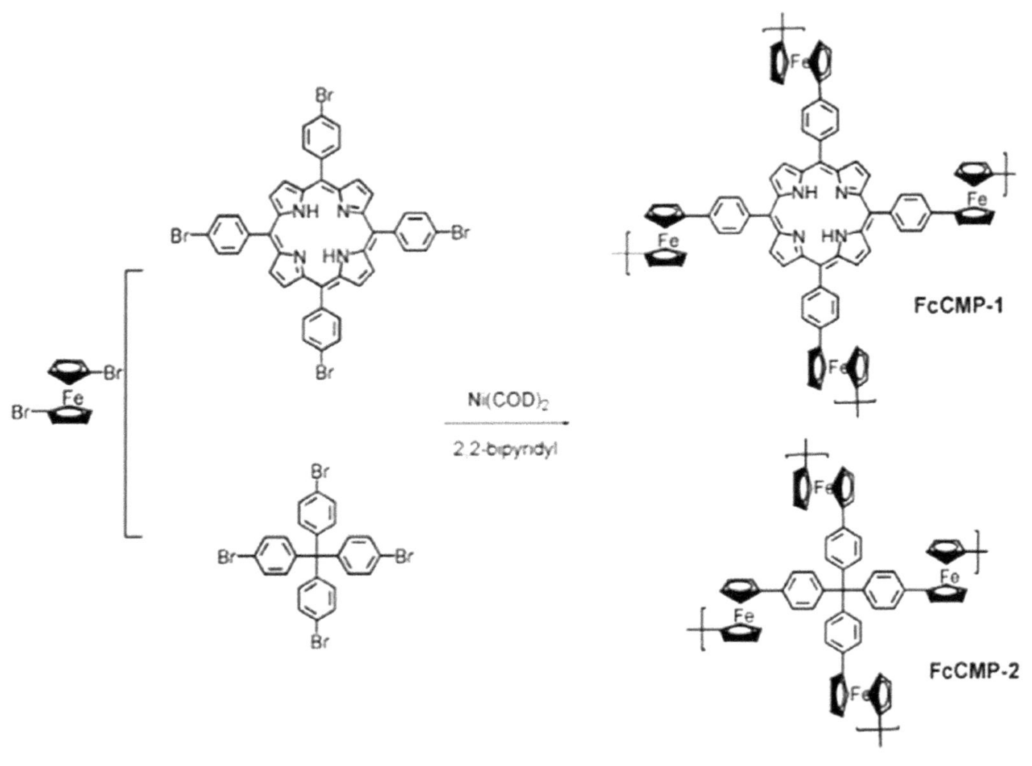

FIGURE 11.4 The tetra (p-bromophenyl) methane synthesis process and the ferrocene-based conjugated microporous polymer porphyrin (FcCMP-1) production pathway (FcCMP-2).

(Figure adapted from ref. [3].)

visible light absorption as a monomer-reacted polymer. In addition, the metal-free synthetic technique opens up new possibilities for creating effective photocatalysts for hydrogen evolution.

Reiner et al. [24] developed an analog with 1,3,5-links (M-CMP) and an extendable analog (ME-CMP). Subsequently, they developed a spirobifluorene CMP (SP-CMP) as well as its extended analog using the planarization method (ESP-CMP). Before operating and using soxhlet extraction to extract low–molecular-weight by-products, all polymerizations were performed using Pd(0)-catalyzed Suzuki–Miyaura polycondensation at a temperature of 150 °C for two days. Although SP-CMP has the maximum porosity and photocatalytic performance, there seems to be no overall relationship between the surface area and photocatalytic efficacy. Instead, photocatalytic performance and light absorption profiles have a strong relationship, and 1,3,5-linkages were substantially harmful to photocatalytic efficiency due to considerably blue-shifted absorption on-sets. As other factors, such as a lack of light absorption, can become prominent as PL lifespan is only partially connected with activity. The SP-CMP combines high surface areas with increased photocatalytic activity over linear, non-porous polymers (p-phenylene). Other applications, such as photocatalytic organic processes, could benefit from this combination of features.

11.2.3 Porous Conjugated Polymer (PCP)

PCPs are a type of entirely crosslinked polymers with a large surface area and porosity in the nanoscale range. They have been used in applications including gas storage, gas production, sensing and (photo)catalysis in the former. The solid-state shape of PCPs is ideal for charge transport. Lianwei et al. [6] determined the reactivity of a 4,8-di(thiophene-2-yl)benzo[1,2-b:4,5-b′]dithiophene (DBD)-based PCP photocatalyst for PHP application, emphasizing the significance of the electronic characteristics of the acceptor comonomer. This study emphasizes the importance of comonomer electric matching in influencing the photocatalytic efficiency of PCP for photocatalytic hydrogen production (PHP) [28–31]. It was revealed that monomeric ligand monomers having N_2 atoms, such as pyridine, could increase PHP activities in the PCP network, in contrast to bidentate ligand monomers. A higher internal dipole polarization impact is mostly responsible for increased improvement. However, the introduction of diazine monomer resulted in a synergistic improvement in light absorption, solid-state form, interfacial wettability and crystalline properties.

Minhua et al. [11] reported MnOx-decorated 3D porous C_3N_4 with internal donor–acceptor motifs for effective photocatalytic hydrogen generation. Carbon nitride (C_3N_4) is an enticing visible-light photocatalyst because it is a conjugated polymeric semiconductor with substantial sources, excellent thermodynamic stability, an adequate energy band configuration and a customizable molecular framework. C_3N_4 with a flexible porous architecture generated by soft or hard templating could generate multiple active sites and a shorter charge diffusion distance for increased photocatalytic activity. Using NaCl as a template and in-situ MnOx growth, they alter the electrical structure and carrier behaviors of C_3N_4, resulting in increased surface area, improved visible light absorption, and interfacial charge [32, 33]. MnOx, as a hole-trapping site, effectively facilitates charge carrier separation. The developed photocatalyst has a large surface area, which allows for more effective charge transfer and visible light absorption. As a result, it has an 18-fold higher photocatalytic H_2 generation rate than bulk C_3N_4 under visible light.

11.2.4 Membrane Reactor

A computer simulation experiment was used to first document the improved efficiency of steam reforming comprising the actual membrane catalytic framework. Membrane catalysis centered on the solution-diffusion transport principle already had achieved 100% specificity of Pd-based membranes for H_2 separation using spirals with large surface area. Membranes must have both high permeability and strong separation selectivity, i.e., the frequency of permeation must be proportional to the frequency of catalytic reaction, to be effective membrane reactors. E kikuchi [1] formulated a

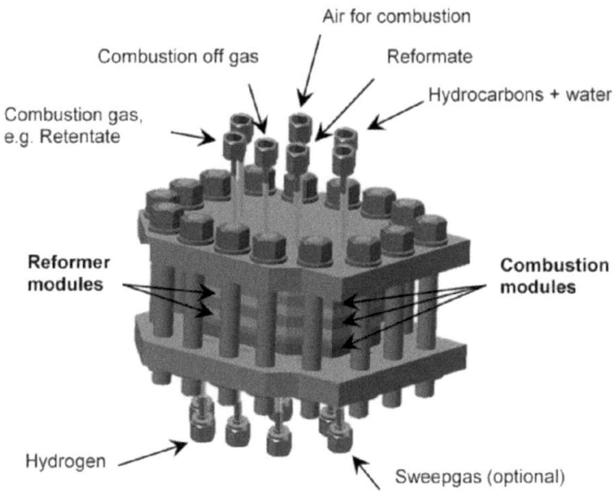

FIGURE 11.5 The versatile membrane reactor system includes reformers and combust modules and inlets and outputs for reac. gas.

(Figure adapted from ref. [4].)

composite membrane composed of a fine Pd layer formed on the exterior surface of porous ceramics by electroless-plating. It was experimentally proved that a Pd layer with a width greater than 4.5 mm was adequate for obtaining 100% hydrogen permeation selectivity. He constructed a membrane reactor out of a single membrane tube and investigated the effects on the stimulation of many processes, including CO_2 reforming, steam reforming, propane dehydrocyclization and the water–gas shift reaction. He operated the membrane reformer at temperatures between 500 and 600 degrees Celsius and pressures between 270 and 610 kPa. Owning to the increased driving power of H_2 penetration across the membrane, the thermodynamically favorable state of increasing reaction pressure enhanced town gas conversion and yield of H_2. Figure 11.5 depicts the new ultra-compact reactor system for on-site production, in which the modules described above are combined/connected to supply tubes, etc. [34]. These reactor units can easily be linked into decentralized processes that only require a minimal amount of hydrogen.

11.2.5 Paper-Structured Catalyst with Porous Fiber-Network Microstructure

Structural catalysts comprising microns porosity have attracted a lot of attention in recent years as they allow efficient heat and reactant dispersion within the catalytic component directly after catalytic activities. Several structural catalysts have been developed, including string catalysts, honeycomb catalysts, metal foam-based catalysts and egg-shell catalysts, but further research and development are needed to put them into practice. Many researchers discovered that utilizing paper-structured catalysts can also stabilize hydrogen production behavior. It was proposed that the paper composite's particular porous microstructure must play a significant influence on ATR performance.

Through continuous load testing in the ATR procedure, Hirotaka et al. [7] compared a paper-structured catalyst's catalytic persistence to that of the conventional catalyst powder. The sintering behavior of Cu–ZnO catalyst coatings on powder-shaped and paper-structured catalysts was investigated using X-ray diffractometry. Here's a summary of the papermaking process: Each stage took three minutes and consisted of a ceramic fiber/water suspension (1.0%, w/v), PDADMAC (0.5 wt% per solids), alumina sol (Cu–ZnO catalyst powder) and A-PAM (0.5 wt% per total solids). According to TAPPI Test Methods T205, the combination was added to a pulp fiber suspension (0.125%, w/v) and reinforced by drying with a 200-mesh wire. After being pressed at 350 kPa for three minutes, the hand sheets were dried in the oven at 105 °C for 1 hour. Cu–ZnO catalyst powders (1.5 g), ceramic fibers (5.0 g), alumina sol (0.50 g) and pulp fibers were used to make the paper-structured catalyst (0.00 or 0.25 g). To remove pulp fibers, the paper-structured catalyst was thermally heated at 350 °C for 12 hours. Then the paper-structured catalyst outperformed both commercial catalyst pellets and the original catalyst powder in terms of methanol conversion and carbon monoxide concentration. Furthermore, as it's simple to manufacture using a defined high-speed and low-cost papermaking methodology, the paper-structured catalyst is projected to have a high practical value and substantial effectiveness in the catalytic gas-reforming mechanism.

11.2.6 Porous Organic Polymers (POPs)

Due to their simple synthesis process and high porosity, porous organic polymers (POP) have sparked a lot of interest for their vast applications in separation, catalysis and gas adsorption. Gargi et al. [16] developed the TpTph porphyrin organic polymer, which is chemically stable and has a large surface area. The photocatalytic efficacy of the TpTph–TiO_2 hybrid was investigated by stimulating hydrogen generation from H_2O using visible light, and a broadband TiO_2 was efficiently integrated into the polymer. The composite demonstrated a considerable improvement in hydrogen evolution when compared to pure TpTph and bulk TiO_2 systems. TpTph's porphyrin units make it an excellent supportive component for semiconductor nanoparticles, as well as a successful photogenerated e^- hole pair separator for H_2 production.

11.2.7 PEM Water Electrolysis

In the mid-1950s, Grubb invented the first PEM water electrolysis, and General Electric Co. established it in 1966 to solve the issues of alkaline water electrolysis. PEM water electrolysis primarily uses polysulfonated membranes as an electrolyte, which is connected to PEM fuel cell technology (proton conductor). Low gas permeability, strong proton conductivity (~0.02–0.1 S cm^{-1}), texture (radius around 20–300 mm), and high-pressure operations are all advantages of these proton exchange membranes. PEM water electrolysis is among the finest environmentally benign methods for transforming renewable energy to highly pure hydrogen in terms of both rapid and long-term impacts. In PEM water electrolysis, H_2O is electrochemically separated into H_2 and O_2 at their corresponding electrodes, featuring O_2 at the anode and H_2 at the cathode. Pumping water to the anode, where it is split into oxygen (O_2), protons (H^+), and electrons, is the working principle of PEM water electrolysis. A proton-conducting membrane transferred the protons from the anode to the cathode. The electrons leave the anode via the reaction's driving force, which is generated by an external power circuit (cell voltage). On the cathode side, protons and electrons recombine to produce hydrogen. The overall reaction as:

Anode: $H_2O \rightarrow 2H^+ + ½O_2 + 2e^-$
Cathode: $2H^+ + 2e^- \rightarrow H_2$
Overall cell: $2H_2O \rightarrow H_2 + ½O_2$

11.2.8 Microporous Inorganic Membranes

A few of the promising alternatives for H_2 separation and purification are microporous inorganic membranes, which have shown prominence in membrane reactors in H_2-producing operations. These microporous membranes are chemically and thermally stable, with high permeability and moderate to high selectivity. As a result, they are appealing for use in hydrogen generation operations. For example, silica and silica functionalized ceramic membranes have a lot of potential in terms of hydrogen separation and production. In the recent decade, there has been a lot of progress in silica membrane research, with various groups in the United States, Germany, Australia and Japan heading the way.

11.2.9 Hybrid Porous Solids for Hydrogen Evolution

Inorganic semiconductors represent a type of photocatalyst in the subject of hydrogen evolution, particularly at the nanoscale. However, significant obstacles persist, including carrier recombination and harsh preparation conditions, which limit the efficiency of inorganic semiconductors. Hybridization of inorganic catalysts with PPs having comparable bandgaps enhances catalyst dispersibility, thereby theoretically permitting photo-induced charge carriers to flow between various materials and improving photocatalytic activity. The composite of triptycene-based polymer and $Cd_{0.5}Zn_{0.5}S$ demonstrates an ultrahigh hydrogen production rate of 50,670 mol h^{-1} g^{-1} with Na_2S/Na_2SO_3 as a sacrificial agent, which is 11 times greater than that of $Cd_{0.5}Zn_{0.5}S$.

11.3 OTHER ALTERNATIVES FOR HYDROGEN PRODUCTION

11.3.1 Metal–Organic Frameworks (MOFs)

MOFs are crystalline porous organic–inorganic hybrid compounds composed of a homogeneous grid of positively charged metal ions enclosed by organic linker molecules. The metal ions connect the arms of the linkers to build a cage-like structure by forming nodes. MOFs have a very large internal surface area due to their hollow structure. In respect to other porous materials, MOFs have a unique structural diversity: homogeneous pore structures, atomic-level structural uniformity, variable porosity, a wide range of variations, and flexibility in network topology, geometry, size and chemical functionality [35–37]. This enables researchers to successfully maintain the topology, porosity and functionality of frameworks.

Using MV^{2+}, $Ru(bpy)_3{}^{2+}$ and $EDTA^{-2}$ Na, Yusuke et al. [4] constructed a unique open porous MOF model that functions as an active site for conversion of H_2O into H_2 molecules when exposed to visible light. To test the material's porosity and gas absorption capacity, they tested the gaseous N_2 sorption isotherm upon dehydrated R1 specimens at a temperature of approx. 77.4 K. The N_2 sorption isotherm exhibited type I behavior, as well as the Langmuir and Brunauer–Emmett–Teller equations, were used to fit the data, yielding effective active sites of 516.5 and 803.9 m^2, respectively. To improve light-harvesting efficiency, MV^{2+} (N,N0-dimethyl-4,40-bipyridinium) as an e^- relay and $Ru(bpy)_3{}^{2+}$ as a photosensitizer (ps) were introduced to an aqueous solvent of EDTA–2Na. R_1,$Ru_2(CH_3COO)_4BF$ and $[RuII,III\ 2(p\text{-}BDC)_2BF_4]_n$ were used as photocatalyst powders. When visible light was shined on R1–R3, H_2 evolution was observed. The Langmuir MOFs ($[M_2(p\text{-}BDC)_2]$ n;M 14Cu,10) have a greater surface area than the MOFs ($[M_2(p\text{-}BDC)_2]$n;M 14Cu,10). For all photocatalytic reactions, measurements were conducted in an oxygen-free environment. The apparent quantum yield was determined using a photodiode and a bandpass and cut-off filter (AQY) (PM_3; Coherent).

11.3.2 Covalent Organic Frameworks

COFs are a new category of porous polymers that combine molecular functioning with the structural definition and robustness of crystalline solids. COFs, a closely related family of organic polymers, have the potential to overcome some of the inherent shortcomings of CN by the application of chemical adaptability and modularity (high porosity and crystallinity) [38–40]. COFs have yet to be investigated as photosensitizers in photocatalytic applications for solar fuel production. By the addition of Pt as p^+ reduction catalyst, Linus et al. [26] reported the initial COF that participates in hydrogen evolution triggered by visible light. The hydrazone-based COF (TFPT-COF) was made up of the building blocks 2,5-diethoxy-terephthalohydrazide and 1,3,5-tris-(4-formylphenyl) triazine (TFPT) with mesopores measuring 4 nm in diameter and the surface area of hydrazone-based COF. The amount of H_2 produced in the initial five hours reached 1970 mol h^{-1} g^{-1}, equivalent to a quantum effectiveness of 2.2%, whereas individual batches attained maximum QEs of up to 3.9% by utilizing a 10 vol% aq. triethanolamine soln. as a sacrificial donor. The photocatalytic system deactivates at a higher rate as a result of the high rate. By lowering the pH of the solution to pH = 7 and reducing the quantity of triethanolamine (1 vol%), continuous H_2 emission was measured over a longer period (24 hours). The significant quantity of H_2 generated (1970 mol h^{-1} g^{-1}) under common basic circumstances reveals that TFPT-COF/Pt outperforms Pt-modified amorphous melon, g-C_3N_4.

11.3.3 Photochemical Device

Photocatalytic substances and related light-gathering properties under a suitable solar spectrum are required for effective solar to chemical conversion. The three main types of sensitizers utilized in photocatalytic H_2 generation are natural or bioinspired dyes, organometallic complexes, and metal-free organic dyes. In photochemical devices for hydrogen production, various derivatives of coumarin, indoline, triarylamine, carbazole and thioalkyl derivatives are employed. Tetrathiafulvalene and its derivatives demonstrate excellent light-harvesting capabilities, as well as high charge separations and prolonged periods. TTF is a chemical having a high electron-donor capability that is available commercially. By constructing hierarchical porous TiO_2 and metal-free thioalkyl substituted tetrathiafulvalene sensitizers, Amritanjali et al. [9] proposed an innovative strategy to generate active photocatalysts. These composites were employed for the first time in the availability of a Pt cocatalyst and a sacrificial electron donor to achieve efficient photocatalytic water reduction and H_2 production. A simple hydrothermal approach was used to synthesize the hierarchical porous TiO_2. TTF's extended pie conjugation (ex-TTF) and strong donor property imply that H_2 evolution has evolved significantly. They achieved 1,22,998 mol g^{-1} cat H_2 yields, a high AQY of 41% G3T3 utilizing TTF as an H_2 evolution catalyst after five hours of light irradiation.

11.3.4 Conjugated Polymer Dots (Pdots)

Several organic semiconductors and high purity graphene-based polymers are H_2O insoluble, necessitating the use of an organic solvent during the reaction period to improve solubility. Polymer dots (Pdots) can be a feasible substitute to pure and metal-oxide-based photocatalysts for H_2 generation due to the factors like solubility and customizable band structures. As a result, various researchers have created conjugated Pdots, which are new organic semiconducting substances having diameters ranging from 1 to 100 nm. It offers a lot of potential in applications like bioimaging, detecting and theranostics, also in drug delivery because of its biocompatibility and easily modifiable surfaces. Pdots have been the focus of numerous researches. However, no comprehensive study on the manufacture of conjugated Pdots for H_2 production has yet been reported [34].

11.4 PREPARATION TECHNOLOGY AND POST-PROCESSING

Almost each PPs synthesis approach is challenging in some way. Long reaction periods, severe reaction conditions, and purification approaches (e.g., Soxhlet extraction) are all restricting issues in enhancing laboratory research effectiveness and manufacturing PPs in industry. The remaining catalyst could harm photocatalysis. Furthermore, the high polymerization degree, the rigid backbone and aggregation of PPs make subsequent processes challenging. As a result, developing mild and green preparation methods, as well as improving processability, are challenges and hotspots for PPs research. A group of researchers discovered a reversible imine condensation process assisted by simulated sunlight irradiation (200–2500 nm) and acetic acid that produces moderate crystalline COF in a short period of time (3 h). In addition, the water-promoted hydrothermal reaction among hydrazine and aldehyde eliminates the requirement for an acid catalyst, reduces reaction duration (8–12 h), and produces COFs to a vast extent. Processability is higher for porous organic membranes and PPs with a specific solubility than for PPs that have been efficiently generated in the form of particles. However, the technology is still not widely developed and maintained, it is indeed conceivable that it will be confined with polymers that have identical linking patterns and components. More research toward more efficient technology is still needed [5, 27].

11.5 MATERIAL COST AND ENERGY SOURCE

Whenever noble metals are often used in photocatalytic operations, the cost of catalysts raises. Combining PPs with a readily available metal (e.g., Ni) can reduce photocatalysis costs and improve H_2 production. Metal-free PPs do not include any metal and have a high photostability, however, the cost of designing and synthesizing various functionalized monomers for photocatalytic PPs is not insignificant. As a result, it's critical to select out PPs with high performance and a large supply of raw materials [21, 41]. When it comes to light sources, this is widely recognized that solar energy is practically limitless and inexpensive. Furthermore, the most often used light source of the research is driven by electricity and limited to a small area, requiring additional energy and preventing solar energy from being completely utilized. Several photocatalytic PPs that work well in natural sunlight have been reported, however, they still fall short of meeting the various production criteria. It is critical to develop low-cost PPs with strong photocatalytic performance under natural sunlight to increase the practicality of materials.

11.6 APPLICATION OF POROUS POLYMERS

- **Health**. Blood purification requires porous polymer dialysis membranes. Porous polymers with controlled discharge and tissue engineering functions are currently being explored.
- **Water**. The reverse osmosis process, which produces drinkable water, relies heavily on porous polymer membranes. Porous polymers are being developed for eliminating organic and metal ion pollutants from water, as well as cleaning up oil spills.
- **Air**. Gas purification and recovery approaches rely on porous polymer membranes, which are being explored for CO_2 capture purposes.
- **Energy**. Environmental priorities include reducing fossil fuel consumption and exploring alternative energy sources. Either in the automobile or aerospace industries, reducing vehicle weight will result in significant fuel savings. Polymers, whether as polymer composites, plastics or lightweight polymer foams, have been at the forefront of such initiatives. Porous polymers are important in the development of lithium-polymer storage batteries, the design of hydrogen storage frameworks, as well as the fabrication of polymer electrolyte membrane fuel cells.

11.7 ADVANTAGE AND LIMITATIONS

11.7.1 Advantage

Porous polymers have many advantages over several materials. As an example, extremely porous polymers are more stable in ambient environments where metal ions in MOFs are hypersensitive to moisture. Different organic synthesis pathways and starting materials can be used to regulate the polymeric structure. PEM water electrolysis also provides some advantages, including a concise design, a higher current density (over 2 A cm^{-2}), high efficiency, quick reaction and the potential to function at moderate temperatures (20–80 °C) while producing ultrapure H_2 and O_2 as byproducts. PPs are lightweight and can be compressed, pelletized and other engineering processes can be used to transform them to a high volumetric density. With current infrastructure, PPs may be scaled up for industrial production at a low cost. The PP–H_2 interaction is based on the physical-adsorption/desorption principle that uses the least amount of parasitic energy. Under the temperature and humidity conditions required for hydrogen storage, PPs remain stable.

11.7.2 Limitations

Sacrificial templates are required for the synthesis of PPs, resulting in difficulties scaling up production and materials with limited value; products lack micropores. While utilizing lab-made and expensive amphiphilic BCPs, scaling up is observed to be often difficult due to a deficiency of micropores. Raw materials are generally costly, precise design and controls over the mesoporous structure are unattainable. Surface area is limited and lacks mechanical strength frequently.

11.8 CHALLENGES AND FUTURE OUTLOOKS

Catalytic membrane reactors may often be easily incorporated into conventional plants and no major alterations to the PPs for the H_2 production process are required. There are several studies on PPs and their productive applications, such as membrane reactors, which could be due to the fact that the technology is still in its infancy as well as several shortcomings must be addressed before it can be implemented at a larger scale. Future research should focus on developing defect-free, homogenous membranes that can perform for long periods at elevated pressures and temperatures and are even more impervious in adverse conditions; better permeability and selectivity will also be required. As PPs have advanced to the point where they can be employed for photocatalytic H_2 evolution, a universal bottom-up strategy to developing customized photocatalysts and photosensitizers with customizable electrical and optical properties has become possible. PPs as photocatalysts have the potential to open up new paths for precision heterogeneous photocatalysts, as well as broaden the existing PP development for H_2 production applications.

11.9 CONCLUSIONS

Hydrogen has received a lot of interest as a source of environmentally beneficial energy, and efficient manufacturing processes like polymer electrolyte fuel cells (PEFCs) have encouraged individuals to use it. In this chapter, we've covered a brief overview of the various hydrogen production strategies, as well as their advantages and limitations. Inorganic membrane reactors, which can be separated into metallic and ceramic types, have advantages including high temperature and high flux operation. The most difficult but crucial goal for all researchers is to scale up hydrogen membrane technologies. In research, forming an extensive quality of membranes saves money since unsuccessful units may be disregarded. Because the costs of producing membrane-based PPs are considerable, the most serious issue is to prepare high-quality membranes on wide surfaces which could be incorporated in the module's processing with extreme temperature sealing. More PP experimental investigations are needed to design such membrane reactor units. Although noble metal cocatalysts from the polymerization process, such as Pt and Pd, are used in numerous photocatalytic reactions, it is vital to investigate noble metal-free photocatalytic interactions for environmental and economic considerations. Instead of Pt, affordable and ubiquitous Ni, Co and Fe are being used as cocatalysts in several photocatalytic systems, resulting in low photocatalytic efficiency. PPs have superior efficiency, selectivity, stability and reusability when compared to equivalent homogeneous catalysts. It will be possible to manufacture PPs with superior performance and attain large practical applications as chemical science and intelligent automation technologies progress. However, for these systems to be efficient and competitive with other well-developed materials, significant advancements must be made. All of these initiatives to develop polymers with excellent porosity configurations and specific functionality are being guided by the quest for enhanced porous materials which could be employed in a variety of high-value applications. Despite the diversity of hierarchical structures occurring naturally, the development of technologies for the direct synthesis of stratified porous polymers with well micro-, meso- and macropores is highly needed. These porous structures can be retro-designed using the knowledge gained by examining the relationships between porous structure, functionality and performance, but new strategies are still needed. Constant efforts are still required to overcome the problems and challenges in obtaining porous polymers with excellent post abilities for efficient H_2 production.

REFERENCES

1. Kikuchi E, (2000) Membrane reactor application to hydrogen production. *Catal Today* 56: 97–101
2. Langton R, Clark C, Hewitt M, Richards L, (2009) Fuel storage. *Aircr Fuel Syst* 9: 31–51.
3. Zeng Q, Cheng Z, Yang C, He Y, Meng N, Faul CFJ, Liao Y, (2021) Metal-free synthesis of pyridyl conjugated microporous polymers for photocatalytic hydrogen evolution. *Chinese J Polym Sci* 25: 1–9.
4. Kataoka Y, Sato K, Miyazaki Y, Masuda K, Tanaka H, Naito S, Mori W, (2009) Photocatalytic hydrogen production from water using porous material $[Ru_2(p\text{-}BDC)2]_n$. *Energy Environ Sci* 2: 397–400.
5. Parvasi P, Jokar SM, Basile A, Iulianelli A, (2020) An on-board pure H2 supply system based on a membrane reactor for a fuel cell vehicle: A theoretical study. *Membranes (Basel)* 10: 1–17.
6. Li L, Lo WY, Cai Z, Zhang N, Yu L, (2016) Donor–acceptor porous conjugated polymers for photocatalytic hydrogen production: The importance of acceptor comonomer. *Macromolecules* 49: 6903–6909.
7. Koga H, Fukahori S, Kitaoka T, Nakamura M, Wariishi H, (2008) Paper-structured catalyst with porous fiber-network microstructure for autothermal hydrogen production. *Chem Eng J* 139: 408–415.
8. Lu GQ, Diniz da Costa JC, Duke M, Giessler S, Socolow R, Williams RH, Kreutz T, (2007) Inorganic membranes for hydrogen production and purification: A critical review and perspective. *J Colloid Interface Sci* 314: 589–603.
9. Tiwari A, Duvva N, Rao VN, Venkatakrishnan SM, Giribabu L, Pal U, (2019) Tetrathiafulvalene Scaffold-based sensitizer on hierarchical porous TiO_2: Efficient light-harvesting material for hydrogen production. *J Phys Chem C* 123: 70–81.
10. Lu GQ, Diniz JC, Duke M, Giessler S, Socolow R, Williams RH, Kreutz T, (2007) Inorganic membranes for hydrogen production and purification: A critical review and perspective. *J Colloid Interface Sci* 314: 589–603.
11. Ai M, Zhang JW, Gao R, Pan L, Zhang X, Zou JJ, (2019) MnOx-decorated 3D porous C3N4 with internal donor–acceptor motifs for efficient photocatalytic hydrogen production. *Appl Catal B Environ* 256: 117805.
12. Jayakumar J, Chou HH, (2020) Recent advances in visible-light-driven hydrogen evolution from water using polymer photocatalysts. *ChemCatChem* 12: 689–704.
13. López-Calixto CG, Barawi M, Gomez-Mendoza M, Oropeza FE, Fresno F, Liras M, De La Penã O'Shea VA, (2020) Hybrids based on BOPHY-conjugated porous polymers as photocatalysts for hydrogen production: Insight into the charge transfer pathway. *ACS Catal* 10: 9804–9812.
14. Lee JT, Wey MY, (2020) PVA/Pt/N-TiO_2/$SrTiO_3$ porous films with adjustable pore size for hydrogen production under simulated sunlight. *J Colloid Interface Sci* 573: 158–164.
15. Thomas A, (2020) Much ado about nothing – A decade of porous materials research. *Nat Commun* 11: 11–13.
16. Mukherjee G, Thote J, Aiyappa HB, Kandambeth S, Banerjee S, Vanka K, Banerjee R, (2017) A porous porphyrin organic polymer (PPOP) for visible light triggered hydrogen production. *Chem Commun* 53: 4461–4464.
17. Song C, Liu B, Nie J, Ma C, Lu C, Wang F, Yang G, Chen Z, (2020) Photocatalytically active conjugated porous polymers via click chemistry for heterogeneous dehydrogenation of hydrazo aromatics. *ACS Sustain Chem Eng* 8: 14377–14385.
18. Sachs M, Sprick RS, Pearce D, Hillman SAJ, Monti A, Guilbert AAY, Brownbill NJ, Dimitrov S, Shi X, Blanc F, Zwijnenburg MA, Nelson J, Durrant JR, Cooper AI, (2018) Understanding structure-activity relationships in linear polymer photocatalysts for hydrogen evolution. *Nat Commun* 9: 1–11.
19. Yu J, Chang S, Xu X, He X, Zhang C, (2020) The effect of single atom substitution (O, S or Se) on photocatalytic hydrogen evolution for triazine-based conjugated porous polymers. *J Mater Chem C* 8: 8887–8895.
20. Silverstein MS, (2014) PolyHIPEs: Recent advances in emulsion-templated porous polymers. *Prog Polym Sci* 39: 199–234.

21. Porphyrins MH, (2020) porous hybrid materials based on mesotetrakis(hydroxyphenyl) porphyrins and TiO_2 for efficient visible-light-driven hydrogen production. *Catalysts* 10: 656.
22. Wang TX, Liang HP, Anito DA, Ding X, Han BH, (2020) Emerging applications of porous organic polymers in visible-light photocatalysis. *J Mater Chem A* 8: 7003–7034.
23. Shiva Kumar S, Himabindu V, (2019) Hydrogen production by PEM water electrolysis – A review. *Mater Sci Energy Technol* 2: 442–454.
24. Sprick RS, Bonillo B, Sachs M, Clowes R, Durrant JR, Adams DJ, Cooper AI, (2016) Extended conjugated microporous polymers for photocatalytic hydrogen evolution from water. *Chem Commun* 52: 10008–10011.
25. Sprick RS, Bonillo B, Sachs M, Clowes R, Durrant JR, Adams DJ, Cooper AI, (2016) Extended conjugated microporous polymers for photocatalytic hydrogen evolution from water. *Chem Commun* 52: 10008–10011.
26. Stegbauer L, Schwinghammer K, Lotsch BV, (2014) A hydrazone-based covalent organic framework for photocatalytic hydrogen production. *Chem Sci* 5: 2789–2793.
27. Tan Z, Su H, Guo Y, Liu H, Liao B, Amin AM, Liu Q, (2020) Ferrocene-based conjugated microporous polymers derived from yamamoto coupling for gas storage and dye removal. *Polymers (Basel).* doi: 10.3390/polym12030719.
28. Vilela F, Zhang K, Antonietti M, (2012) Conjugated porous polymers for energy applications. *Energy Environ Sci* 5: 7819–7832.
29. Li B, Huang X, Liang L, Tan B, (2010) Synthesis of uniform microporous polymer nanoparticles and their applications for hydrogen storage. *J Mater Chem* 20: 7444–7450.
30. Wong YL, Tobin JM, Xu Z, Vilela F, (2016) Conjugated porous polymers for photocatalytic applications. *J Mater Chem A* 4: 18677–18686.
31. Wu D, Xu F, Sun B, Fu R, He H, Matyjaszewski K, (2012) Design and preparation of porous polymers. *Chem Rev* 112: 3959–4015.
32. Xu Y, Mao N, Zhang C, Wang X, Zeng J, Chen Y, Wang F, Jiang JX, (2018) Rational design of donor-Π-acceptor conjugated microporous polymers for photocatalytic hydrogen production. *Appl Catal B Environ* 228: 1–9.
33. Kang Z, Yang G, Mo J, Yu S, Bender G, Pivovar BS, Green JB Jr., Zhang F, (2018) Developing titanium micro/nano porous layers on planar thin/tunable LGDLs for high-efficiency hydrogen production highlights. *Int. Jr. Hydro Energy* 43: 1–30.
34. Wunsch A, Kant P, Mohr M, Haas-Santo K, Pfeifer P, Dittmeyer R, (2018) Recent developments in compact membrane reactors with hydrogen separation. *Membranes (Basel).* doi: 10.3390/membranes8040107.
35. Zhang Y, Liu J, Wu G, Chen W, (2012) Nanoscale for efficient sunlight-driven photocatalytic hydrogen production. *Nanoscale* 4: 1–8.
36. Xu C, Zhang W, Tang J, Pan C, Yu G, (2018) Porous organic polymers: An emerged platform for photocatalytic water splitting. *Front Chem.* doi: 10.3389/fchem.2018.00592.
37. Deng P, Hong W, Cheng Z, Zhang L, Hou Y, (2020) Facile fabrication of nickel/porous g-C3N4 by using carbon dot as template for enhanced photocatalytic hydrogen production. *Int J Hydrogen Energy* 45: 33543–33551.
38. Hentze HP, Antonietti M, (2002) Porous polymers and resins for biotechnological and biomedical applications. *Rev Mol Biotechnol* 90: 27–53.
39. Cousins K, Zhang R, (2019) Highly porous organic polymers for hydrogen fuel storage. *Polymers (Basel).* doi: 10.3390/polym11040690.
40. Gondolini A, Fasolini A, Mercadelli E, Basile F, Sanson A, (2021) Freeze cast porous membrane catalyst for hydrogen production via oxy-reforming. *Fuel Process Technol* 213: 106658.
41. Yu K, Bi S, Ming W, Wei W, Zhang Y, Xu J, Qiang P, Qiu F, Wu D, Zhang F, (2019) Side-chain-tuned π-extended porous polymers for visible light-activated hydrogen evolution. *Polym Chem* 10: 3758–3763.

12 Porous Polymers in Photocatalysis

Mehdi Mogharabi-Manzari and Shahla Rezaei
Mazandaran University of Medical Sciences, Sari, Iran

Zahra Pakdin-Parizi
Razavi Hospital, Mashhad, Iran

Masoud Salehipour
Parand Branch of Islamic Azad University, Parand, Iran

Ali Motaharian
Birjand University of Medical Sciences, Birjand, Iran

CONTENTS

12.1 INTRODUCTION

Since the beginning of the industrial revolution, increasing demand for clean and sustainable chemical processes has led to the development of renewable energy sources. Photosynthesis shows the ability of using sunlight as the most available renewable energy source to drive chemical processes (Zhang et al., 2019; Zhang and Sun, 2019). Therefore, nature-inspired efficient synthetic catalysts that would ultimately mimic photosynthesis have been developed to facilitate chemical reactions using UV-Vis light (Cui et al., 2020). Nowadays, photocatalysts are successfully applied in various fields such as water purification (Bedia et al., 2019), water splitting (Fajrina & Tahir, 2019), chemo- and regioselective syntheses (Jiang et al., 2019).

Although titanium oxide is known as the most successful photocatalyst due to its relatively high availability, high efficiency, and low cost, some cations with the d^0 and d^{10} configuration (Nb^{5+}, Ta^{5+}, Ga^{3+}, In^{3+}, and Bi^{3+}) are interesting novel photocatalysts (Nasalevich et al., 2014). Despite the performance, challenging problems still remain with the metallic photocatalysts such as high bandgap energy and toxicity of heavy metals. However, porous organic polymers attract attention in the recent decade for solving the drawbacks of conventional photocatalytic systems. In comparison with solid polymers, porous organic polymers usually possess high surface area to volume ratio that improves the absorption of light irradiation (You et al., 2021). Conjugated porous polymers

DOI: 10.1201/9781003169604-12

combining the electronic properties with tunable pore features are ideal photocatalyst candidates for light-driven reactions (Dai & Liu, 2020). In addition, porous organic polymers are known as competitive candidates in several applications including energy storage, separation, CO_2 reduction, optical devices, biomedical applications, pollutant degradation, and organic catalysis (Wang et al., 2020). The development of green and sustainable organic synthesis approaches is a driving force for the replacement of conventional methods with photocatalytic organic synthesis techniques which exploit clean solar energy to achieve final products under environmentally friendly and non-toxic conditions (Xiao et al., 2020).

This chapter describes the molecular structures and physical and chemical properties of conjugated porous polymers associated with their photocatalytic characteristics. Moreover, this perspective aims to highlight the unique fundamental and practical importance of porous photocatalysts.

12.2 PHOTOCATALYSIS

Photocatalysis is inspired by natural photosynthesis and represents the acceleration of a photoreaction in the presence of a catalyst in which light energy is used to drive pairs of chemical reactions (Cui et al., 2020). The oxidation and reduction reactions during the photosynthesis are described by Z-scheme which was described firstly by Bard in 1979 (Bard, 1979). The Z-scheme photocatalytic model employs photosystem I (PSI) and photosystem II (PSII) reaction center proteins which can be identified by the adsorbed photons from the sunlight with the wavelength of 700 and 680 nm for PSI and PSII, respectively (Figure 12.1).

The energy of light is trapped by PSII and causes the production of photogenerated electrons that can be transported to the photogenerated holes in PSI via a shuttling channel (Low et al., 2017). The excited electron from chlorophyll P680 is transferred quickly to pheophytin as a primary electron–acceptor molecule in PSII. In the next step, the electron is transferred to chlorophyll P700 by several electron carriers and the energy of the excited electron is used to transfer protons across the thylakoid membrane into the lumen (Huang et al., 2019). The chlorophyll P680 in PSII which is

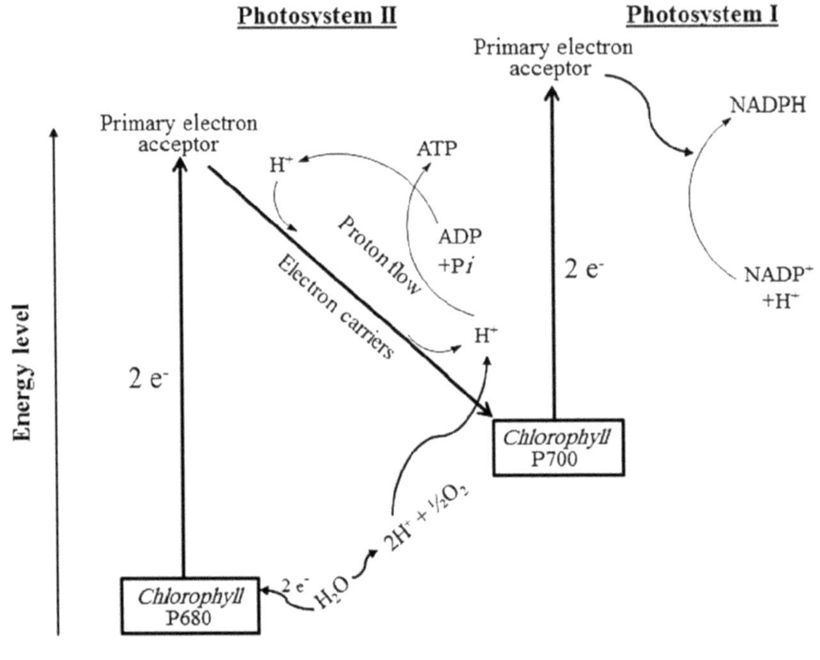

FIGURE 12.1 Schematic representation of Z-scheme model.

depleted of electrons acts as an oxidizing agent with the capability of splitting the water molecule. The hydrogen ions generated by the oxidation of water lead to the formation of a proton gradient that is consumed by adenosine triphosphate synthase to generate adenosine triphosphate (Xu et al., 2018). On the other hand, PSI is a chlorophyll–protein complex that uses light energy to transfer electrons from the lumenal electron carriers (cytochrome c6 or plastocyanin) across the thylakoid membrane to the cytosolic electron carriers (flavodoxin or ferredoxin). However, PSI transfers electrons to $NADP^+$ resulting in the formation of a powerful reducing agent NADPH which is required for the carbon fixation process (Medlyn et al., 2002).

The ability of natural photosynthesis to drive chemical reactions and production of organic molecules using sustainable solar energy encourages researchers to develop photocatalysts that can drive thermodynamically uphill reactions (Zhang et al., 2019). Water splitting and carbon dioxide reduction are widely studied photocatalytic reactions. Using photocatalysts for efficient splitting of water to oxygen and hydrogen molecules is known as a promising and sustainable photocatalytic process for converting sunlight or solar energy into clean and renewable hydrogen fuel. Moreover, photocatalysts are used for the reduction of carbon dioxide to hydrocarbons via a photocatalytic process under light irradiation in an aqueous solution (Ding et al., 2021). Although many photocatalysts with varying mechanisms are used in chemical reactions, most of them can be described by four distinguished steps including: (1) generating electron–hole pairs by light absorption, (2) excited charge separation, (3) immigration of electrons and holes to the surface of the catalyst, and (4) redox reaction (Ryu et al., 2019).

12.3 MECHANISM OF ACTION

Light-harvesting ability is the most important property of the organic porous polymer photocatalysts arising from their conjugated chemical structure and delocalized π-system. The backbone of conjugated porous polymers is composed of sp^2 hybrid orbitals and p_z orbitals which lead to the π-delocalization and energy bandgaps of the highest occupied molecular orbital–lowest unoccupied molecular orbital (HOMO–LUMO) (Byun & Zhang, 2020). Light absorption and generation of electron–hole pairs is the first step of organic porous polymer that could be migrated to the surface and transferred to substrates (Banerjee et al., 2014). The migrated electron and hole are transferred to an electron acceptor and an electron donor via reduction and oxidation processes, respectively (Banerjee et al., 2014). Therefore, an efficient photocatalytic process includes three steps including electron–hole pair generation, charge separation, and transferring of electron or hole to the target substrates (Zhang et al., 2018). The portion of the absorbed solar spectrum which is related to the length of the conjugation system and porous polymers with a narrow bandgap (less than 3.18 eV) can absorb the energy of visible light with wavelengths ranging from 390 to 700 nm (Park, 2017). After absorption of photons the electron–hole pairs are formed and the efficient charge separation is achieved by donor–acceptor interfaces within the structure of the polymer. Fast charge recombination is prevented when the polymer consists of electron-poor acceptor and electron-rich donor groups (Günes et al., 2007).

Electrons and holes generated in the photocatalyst are able to interact with target substrates and cause redox reactions. Difference between the highest occupied molecular orbital (HOMO) and lowest occupied molecular orbital (LUMO) energy levels of conjugated porous polymers is a key parameter for their application as photocatalysts. For example, the energy of HOMO/LUMO levels of conjugated porous polymer can be tuned by the change in length of conjugation, the ratio of various monomers, and substitution positions of linkers (Byun & Zhang, 2020). In addition, using various compositions of donor and acceptor moieties into the conjugated porous polymer is a promising technique for the engineering of HOMO and LOMO energy levels (Taylor et al., 2020).

Several studies have been focused to optimize the chemical and physical properties of conjugated porous polymer for catalysis of photocatalytic reactions (Vilela et al., 2012). Fine control of the molecular structure of conjugated porous polymer has been achieved by controlling the type and amount of monomers. Moreover, the porosity, morphology, and hydrophilicity of conjugated porous polymers could be effectively controlled using various template molecules with suitable functional groups (Zhang et al., 2020).

12.4 POROUS POLYMER CATALYZED LIGHT INDUCED ORGANIC TRANSFORMATIONS

12.4.1 POLYCARBAZOLE

In recent years, polycarbazoles with electron-rich conjugated systems and rigid skeletons are widely used to form permanent porous materials with intrinsic photoelectric properties. Most of the reported porous polycarbazoles polymers have novel photoelectric properties and are prepared via Sonogashira–Hagihara coupling (Figure 12.2), Suzuki reaction (Figure 12.3), Yamamoto reaction (Figure 12.4), electrochemical coupling, and oxidative coupling reactions (Figure 12.5).

The conjugated porous polymers have been used for catalysis of various redox reactions as effective visible light photocatalysts (Figure 12.6) (Wang et al., 2014; Dadashi-Silab et al., 2014). Visible light-stimulated photocatalytic activity of carbazole-based catalyst was used for hydroxylation and α-alkylation of arylboronic acids with natural aldehydes. The lengthy π-conjugation in carbazole

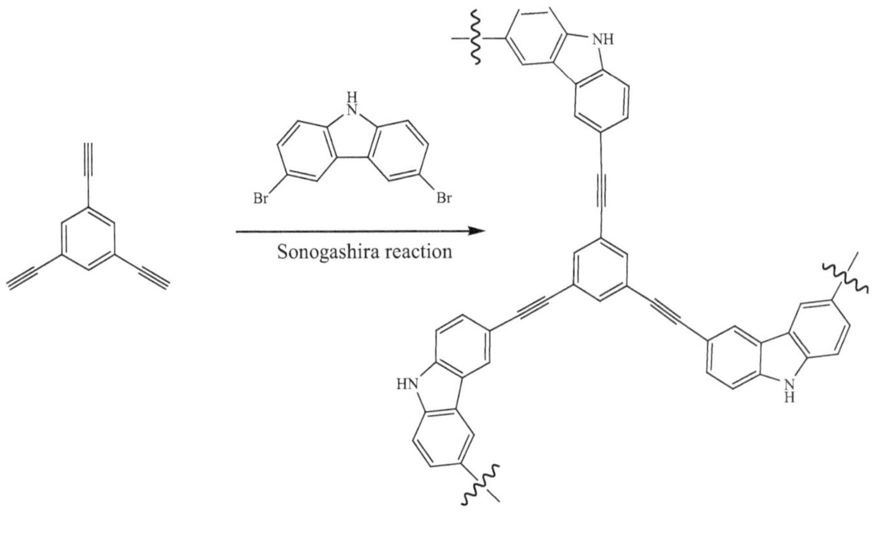

FIGURE 12.2 Synthesis of polycarbazoles polymers via Sonogashira–Hagihara coupling.

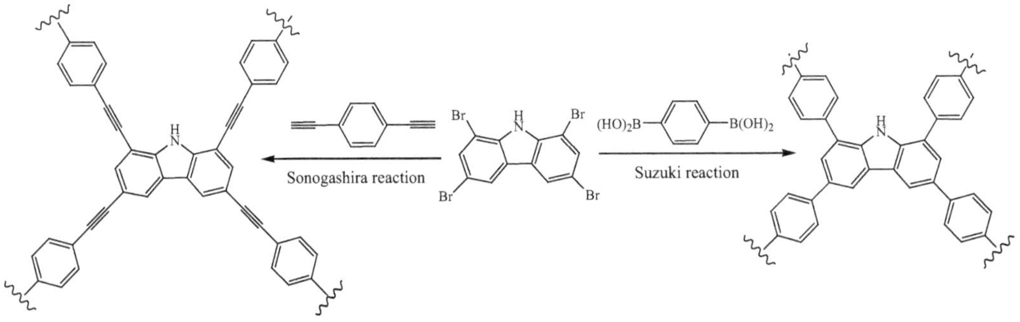

FIGURE 12.3 Synthesis of polycarbazoles polymers via Suzuki reaction.

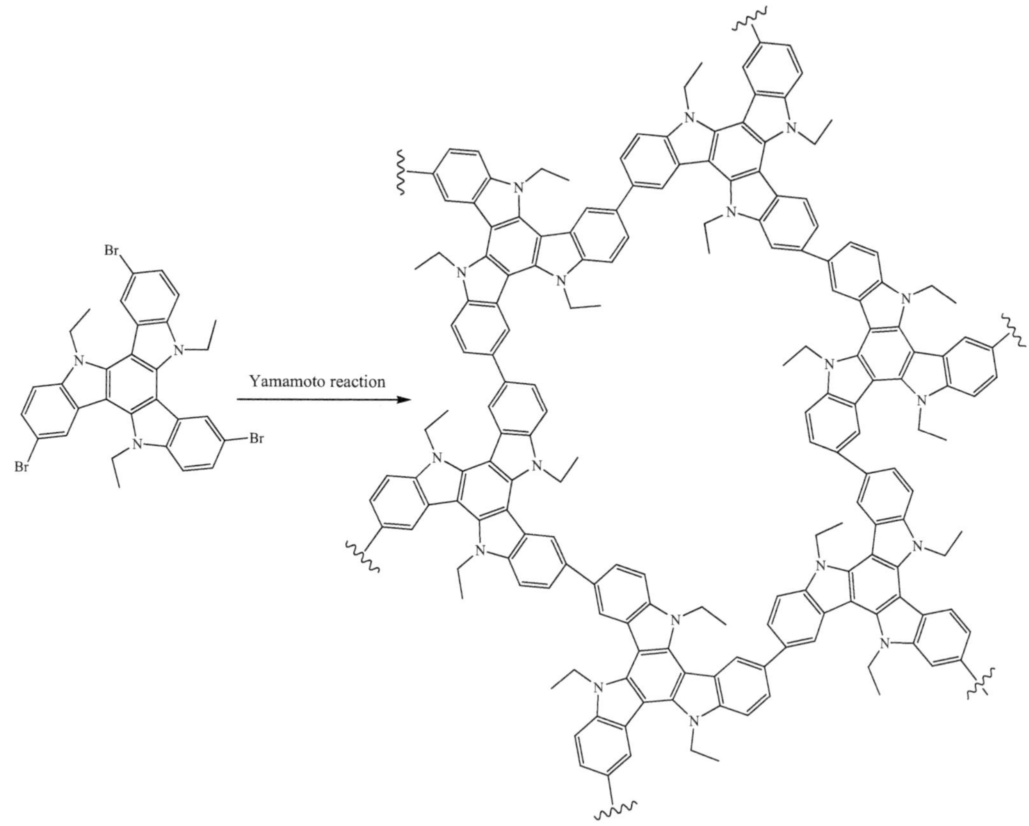

FIGURE 12.4 Synthesis of polycarbazoles polymers via Yamamoto reaction.

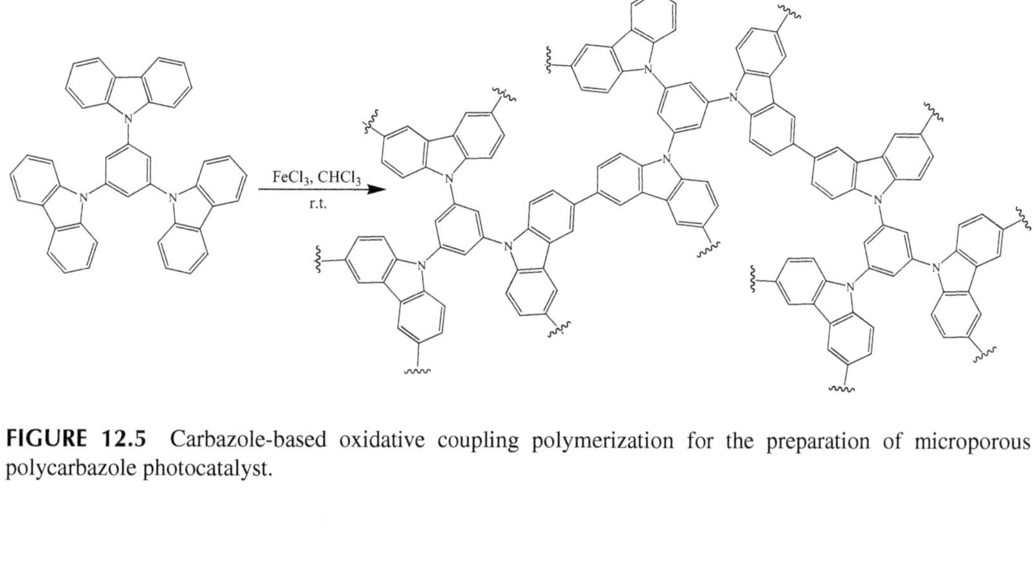

FIGURE 12.5 Carbazole-based oxidative coupling polymerization for the preparation of microporous polycarbazole photocatalyst.

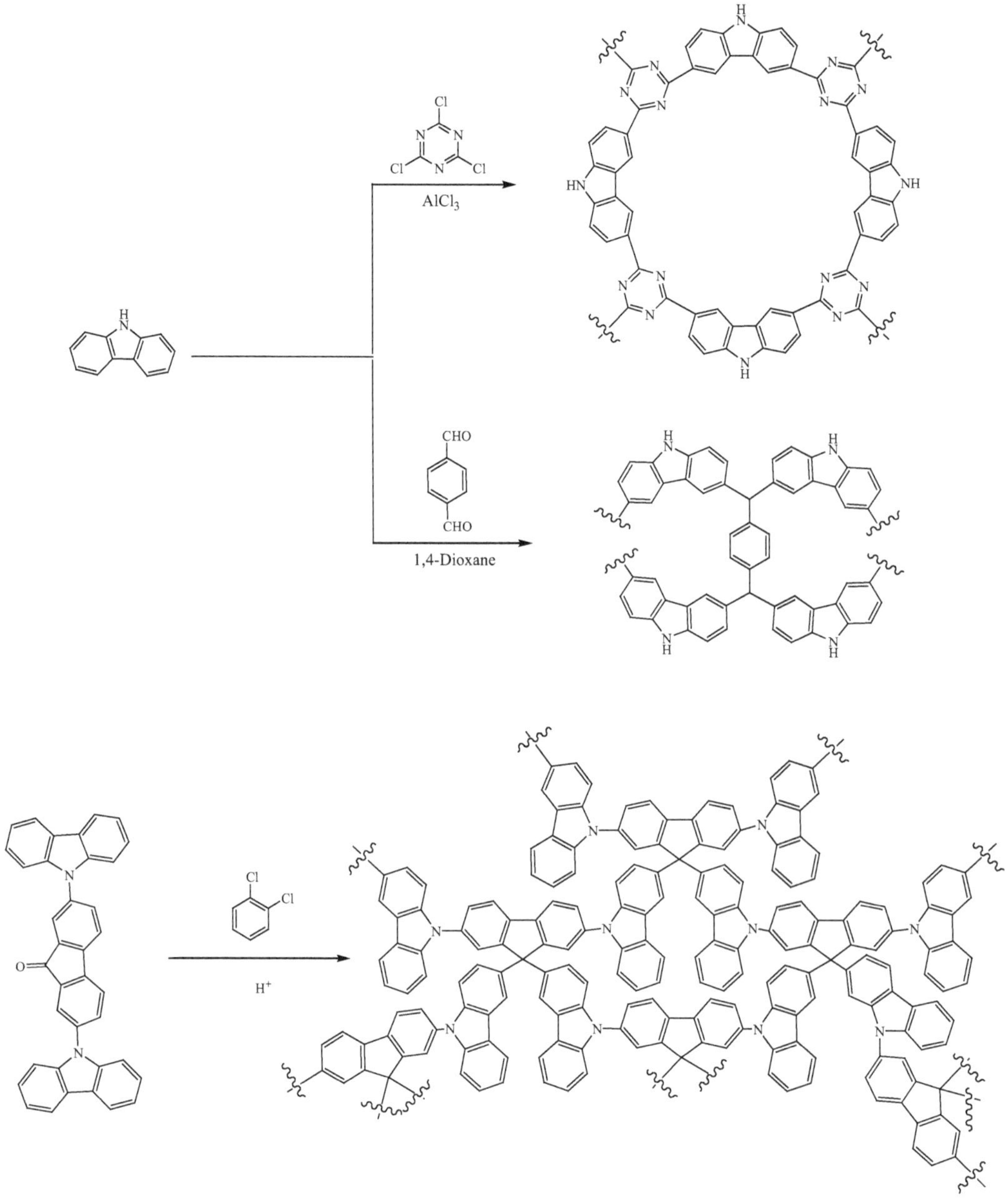

FIGURE 12.6 Synthesis of carbazole-based polymers by the Friedel–Crafts reaction.

porous organic framework increases its visible light absorption, and the large porosity accelerates the reaction rate (Luo et al., 2015). The porous polymer was produced via $FeCl_3$-induced polymerization of 3,3′,5,5′-tetra(9H-carbazol-9yl)-1,1′-biphenyl (Luo et al., 2015).

Dehalogenation of phenacyl bromide derivatives was performed to investigate the potential application of the catalyst in redox reactions, in which diethyl-1,4-dihydro-2,6-dimethyl-3,5-pyridinedicarboxylate acts as a stoichiometric reductant reagent to reduce C–Br bond to C–H bond in the presence of overnight fluorescent irradiation (14 W) (Figure 12.7). In addition, the capability of the photocatalyst was investigated via photo-induced oxidative hydroxylation of arylboronic acids (Luo et al., 2015).

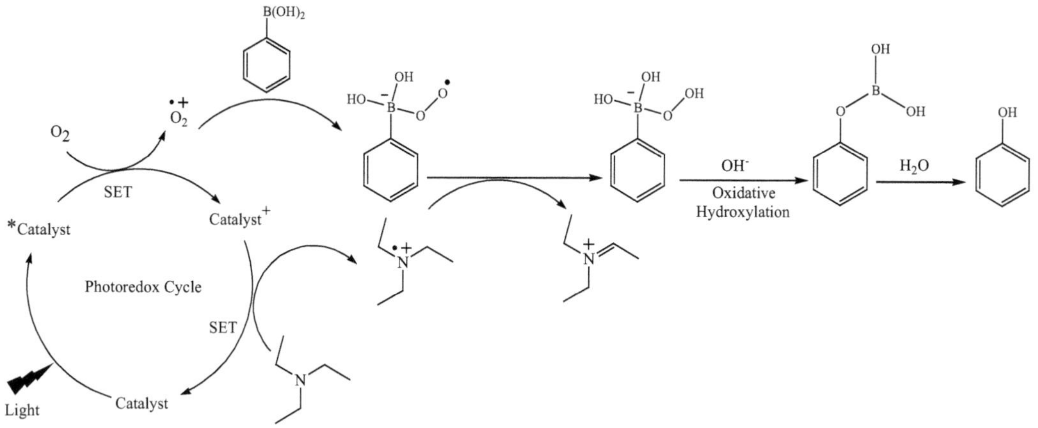

FIGURE 12.7 Mechanistic representation of the oxidative hydroxylation of arylboronic acids.

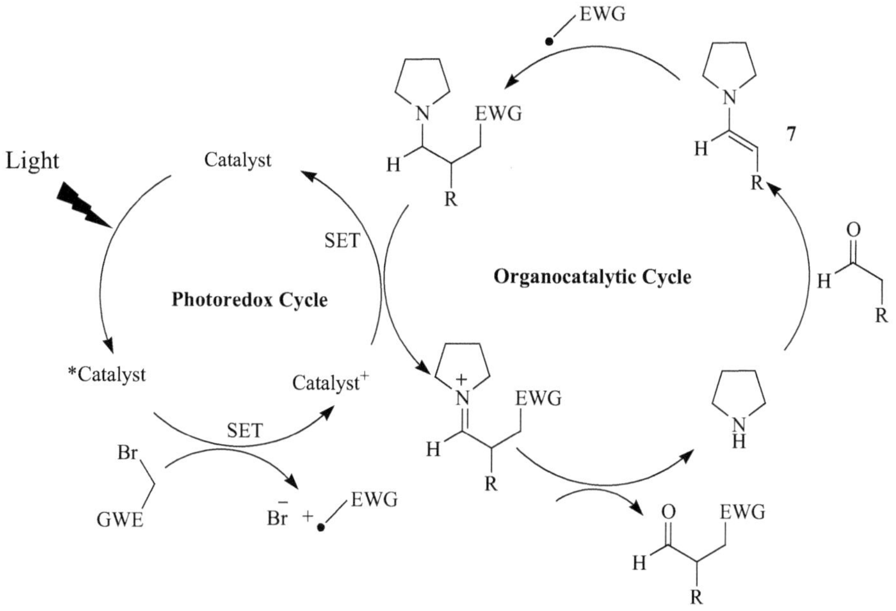

FIGURE 12.8 Mechanistic representation of α-alkylation of aldehydes.

A combination of a carbazole-based photocatalyst and an enamine-based organocatalyst was used for α-alkylation of aldehydes via a single electron redox process (Figure 12.8) (Luo et al., 2015).

Aerobic oxidation of various compounds was studied using porous polymers as photocatalysts such as alcohols, amines, sulfides, and arylboronic acids. Luo et al. (2018) reported a series of carbazole–triazine-based donor–acceptor porous polymers with the ability to catalyze various aerobic oxidation reactions under visible light irradiation (Figure 12.9).

The photocatalytic activity of the prepared catalysts was investigated via Mannich reaction, sulfide oxidation, and oxidative amine coupling (Figure 12.10). Thioanisole was converted to methyl phenyl sulfide (yield 95%) using 0.5 mol% catalysts under fluorescent light at 25 °C for 12 h. The photocatalytic Mannich reaction involved α-amino C–H bond oxidative activation to produce

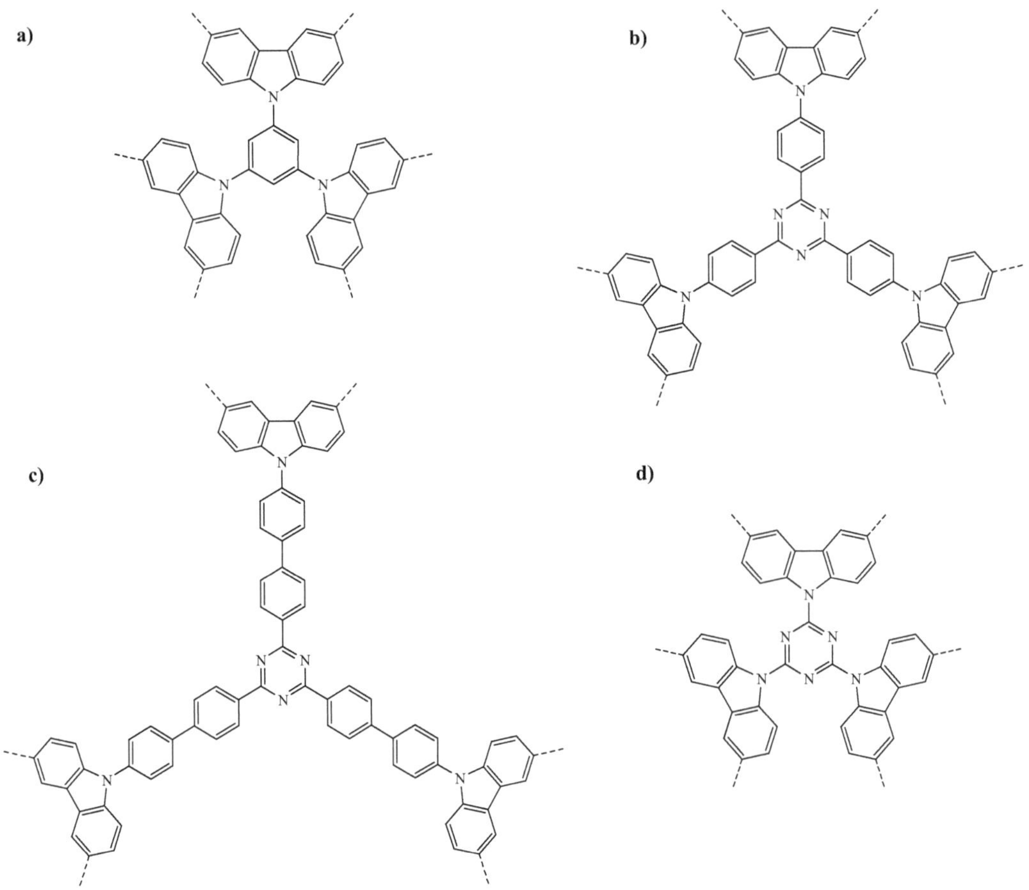

FIGURE 12.9 Chemical structure of some carbazole-based conjugated organic photocatalysts used in aerobic oxidations.

reactive iminium ions leading to C(sp^3)–C(sp^3) cross-coupling reaction in the presence of 0.5 mol% porous polymer catalyst and L-proline as co-catalyst (Luo et al., 2018).

Su et al. (2016) reported carbazole-based polymeric catalyst achieved via straightforward oxidative polymerization that exhibited highly photocatalytic activity for an extensive range of organic reactions such as oxidative coupling of primary amines, aerobic dehydrogenation of non-active secondary amines, and selective oxidation of sulfide. A donor–acceptor carbazole-based photocatalyst was used successfully for oxidation of 2-chloroethyl ethyl sulfide (a warfare agent) in the presence of molecular oxygen and under visible light irradiation (Zhi et al., 2019).

Carbazole-based photocatalysts are able to catalyze various oxidation reactions without using metals and toxic oxidizing agents.

12.4.2 Benzimidazole, Benzoxazole, and Benzothiazole

Synthesis of various carbonyl-containing compounds directly from alkenes through cleavage of C=C bond is a very key reaction in organic synthesis. Photocatalysts are promising alternatives to traditional metal catalysts that suggest sustainable and environmentally friendly organic synthesis approaches. Ayed et al. (2018) report a visible light-promoted C=C bond cleavage procedure using

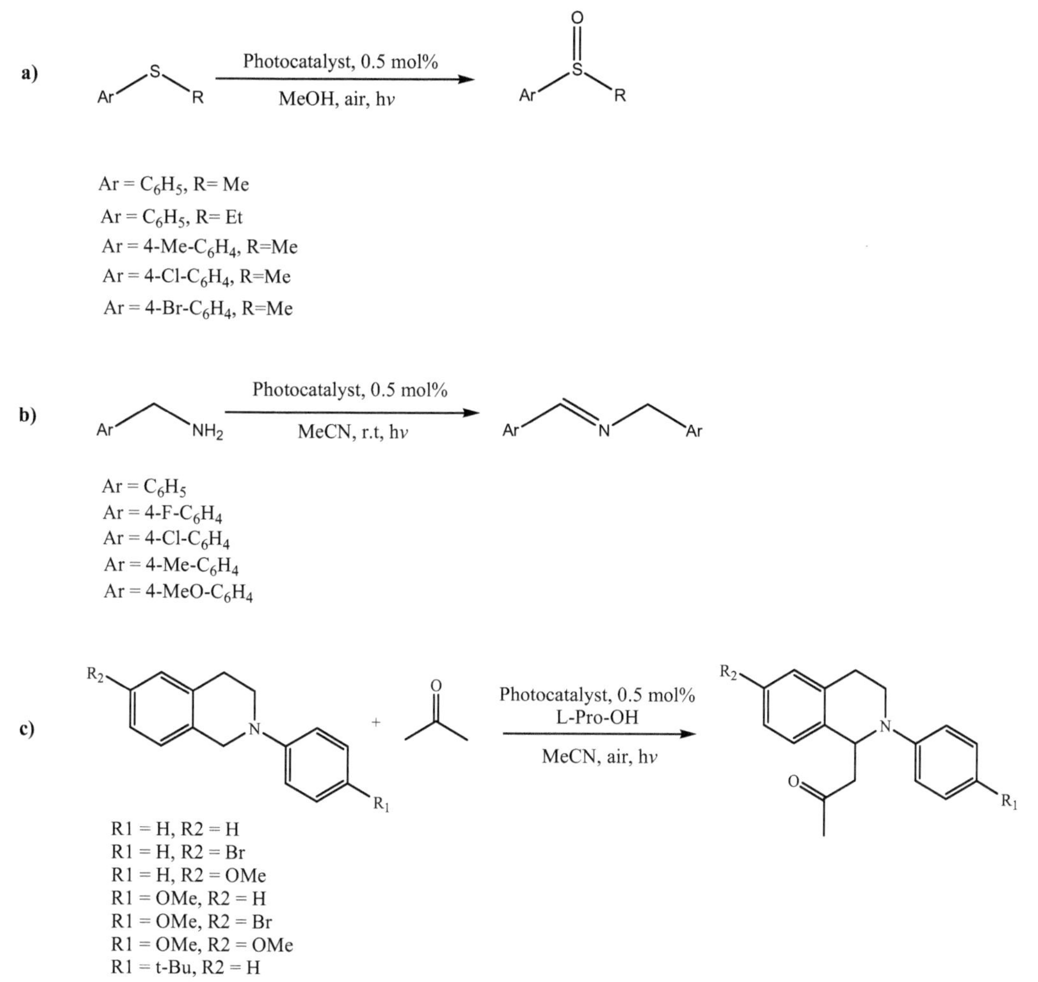

FIGURE 12.10 Porous conjugated polymer catalyzed (a) oxidation of sulfides (b) oxidative amine coupling (c) Mannich reaction.

benzothiazole-containing conjugated microporous polymers as heterogeneous metal-free and reusable photocatalysts (Figure 12.11).

To explore the potential catalytic activity of the designed photocatalyst, the C=C bond cleavage was performed applying styrene as a model substrate using a blue LED lamp (460 nm) and at room temperature. The mechanistic investigations of the reaction showed that the charge separation occurred within the catalyst under visible light irradiation. Styrene is oxidized into the corresponding cationic radical intermediate by the photogenerated hole of the catalyst (Ayed et al., 2018).

Recently, Yang et al. (2020) developed highly functionalized metal-organic frameworks (MOFs) by acid-catalyzed Aldol reaction of 1,3,5-triazine derivatives catalyzed by trifluoroacetic acid (TFA) in a solvent mixture at 150 °C for 72 h (Figure 12.12).

The prepared nonmetallic and heterogeneous photocatalyst was applied in a trifluoromethylation reaction via a mild photocatalytic approach (Figure 12.13). The experiments were conducted using 1,3,5-trimethoxybenzene (as a model compound) with 300 W Xe lamp as the light source. The obtained results revealed a 98% yield after 16 h incubation at 25 °C.

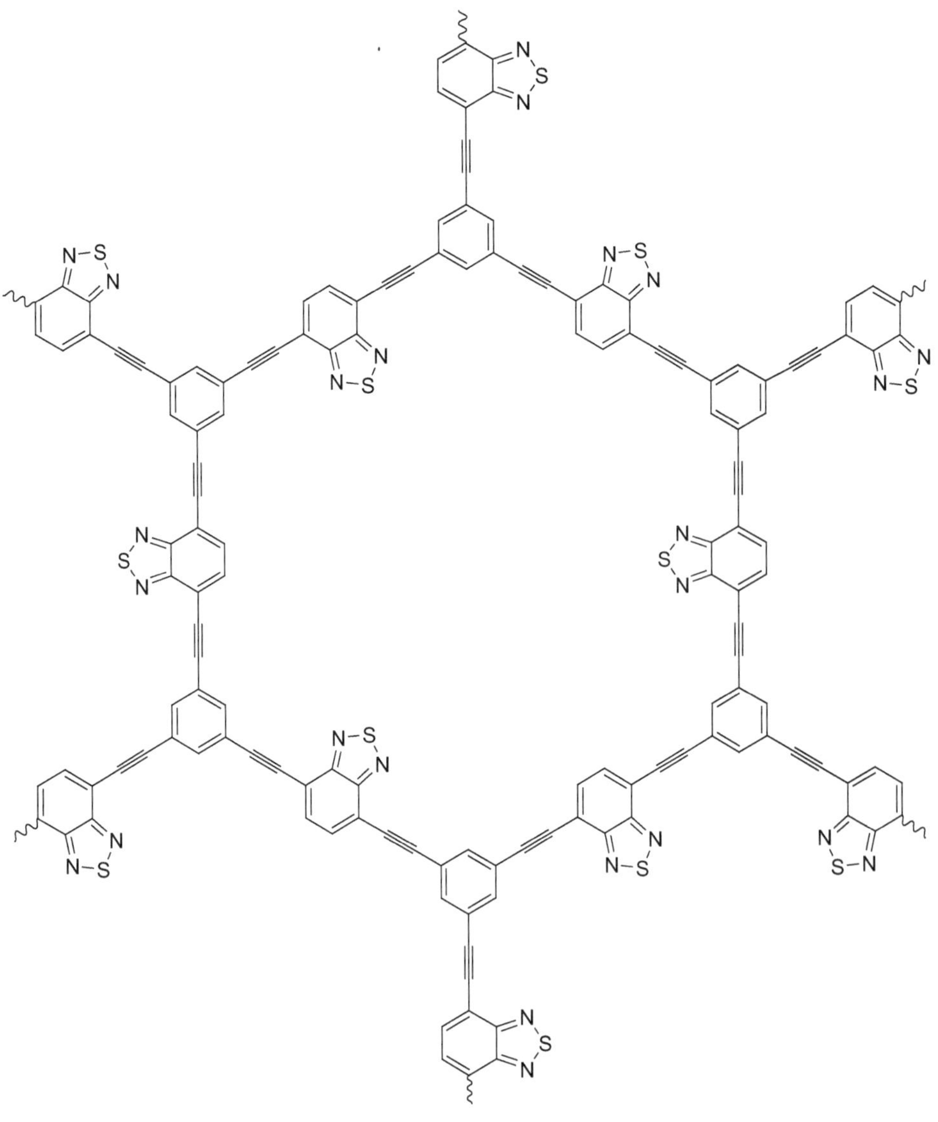

FIGURE 12.11 Benzothiazole-containing conjugated microporous polymers as a photocatalyst.

12.4.3 Xanthene

Xanthene dyes (such as Rose Bengal, Fluorescein, and Eosin Y) with large π-conjugated systems have been applied as suitable alternatives to transition metal complexes in photoredox catalysis due to their unique properties such as less expense and toxicity compared to conventional catalysts and good visible light absorption capacity. Therefore, xanthene-containing conjugated porous polymers might be applied to improve the visible light absorption of the photocatalysts (Liu et al., 2020).

Recently, Li et al. (2020) synthesized 1,4-diethynylbenzene-linked xanthene dye porous polymer via direct palladium-catalyzed Sonogashira–Hagihara cross-coupling polycondensation as potential heterogeneous photocatalysts of photoinduced polymerization (Figure 12.14). Mechanical and morphological characteristics of 2D planar networks of the catalyst enable the reusability of the prepared catalyst by centrifugation. In addition, no significant loss was observed in the catalytic performance of the catalyst after eight independent runs.

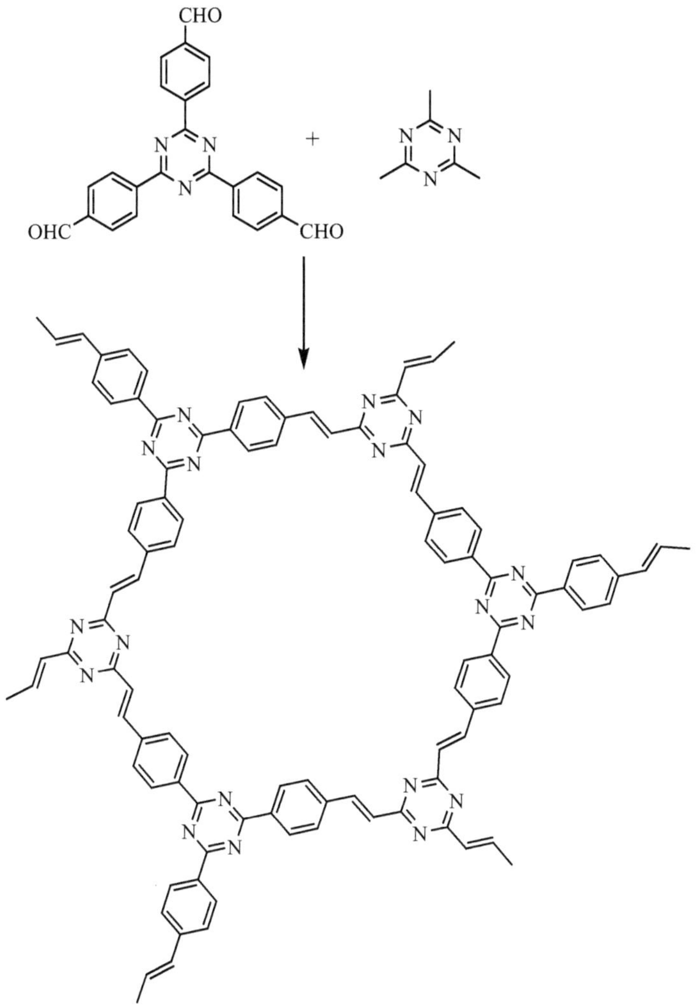

FIGURE 12.12 Conjugated covalent organic frameworks via acid-catalyzed Aldol reaction.

Rose Bengal is known as an effective metal-free organocatalyst that has been studied in the area of homogeneous catalysis because of its low cost and low toxicity. Jiang et al. (2013) reported Rose Bengal functionalized conjugated porous polymer via a bottom-up strategy. Aza-Henry reaction (nitro-Mannich reaction) is known as an efficient synthetic tool for the formation of a new C–C bond between nitroalkanes and imines. They explored the catalytic activity of the prepared photocatalyst using aza-Henry (Figure 12.15). In this reaction, N-substituted tetrahydroisoquinoline and nitromethane were applied as substrates and the reaction was carried out under air and fluorescent lamp (60 W) irradiation at room temperature.

12.4.4 Porphyrin

A porphyrin-based conjugated porous polymer is porphyrin subunits that are linked by strong covalent bonds acting as building blocks. Polyphenylporphyrin benzobisoxazole, a porphyrin-based polymer, was produced by polycondensation between 5, 15-bis(4-carboxyphenyl)-10,20-diphenylp orphyrin and 4,6-diaminoresor-cinol dihydrochloride (Figure 12.16). Extended fully π-conjugated

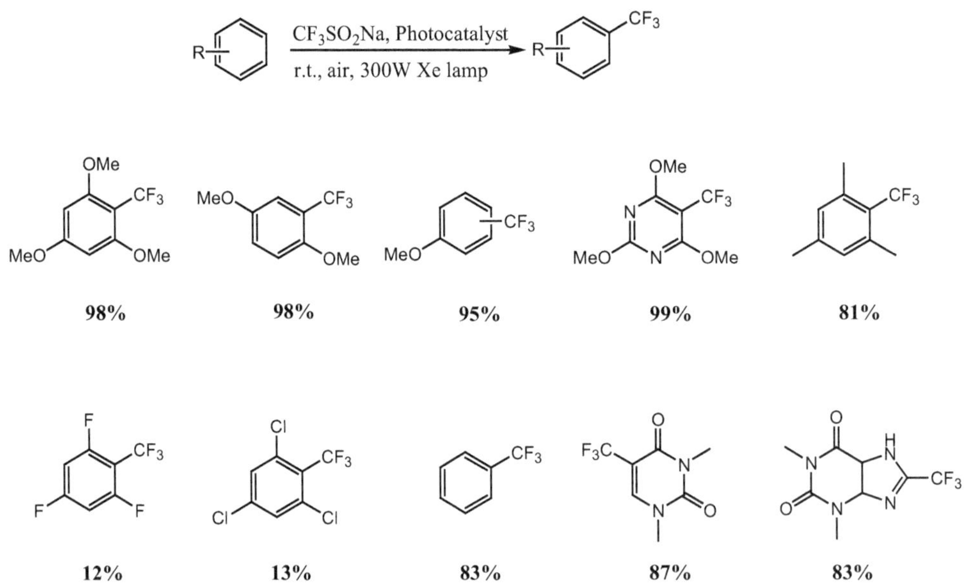

FIGURE 12.13 Trifluoromethylation reaction using azole-based photocatalyst.

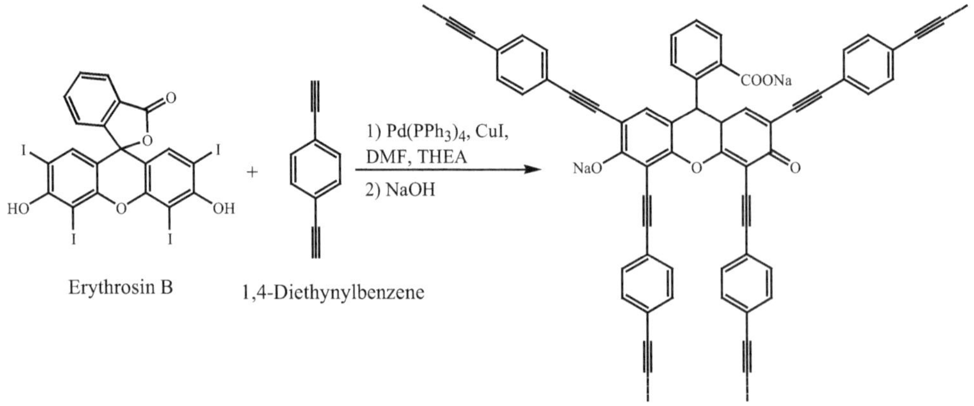

FIGURE 12.14 Synthesis of diethynylbenzene-linked xanthene dye porous polymer as a photocatalyst.

systems of the polymer enhanced its chemical stability, photocatalytic activity, and the ability of visible light absorption (Yu et al., 2019).

Chen et al. (2019) developed a novel porphyrin-based conjugated polymer with permanent porosity and high chemical stability under harsh conditions (Figure 12.17). The conjugated heterogeneous polymer was applied for aerobic oxidation of amines under visible light irradiation.

N-benzylidenebenzylamine was synthesized using a porphyrin-based photocatalyst under irradiation of a 3 W white LED. The electron paramagnetic resonance (EPR) spectroscopy technique was used to understand the photocatalytic reaction mechanism. The obtained results showed that the main mechanistic steps include the light-induced generation and separation of electron–hole pairs. In the next step, the superoxide radical anion is produced by the reduction of oxygen and leading to oxidation of the substrates to the products (Chen et al., 2019).

Benzobisoxazole-containing porous polymer was produced via a condensation reaction between tetra(4-carboxyphenyl)porphyrin and resorcinol derivatives in acidic solution (Li et al., 2018).

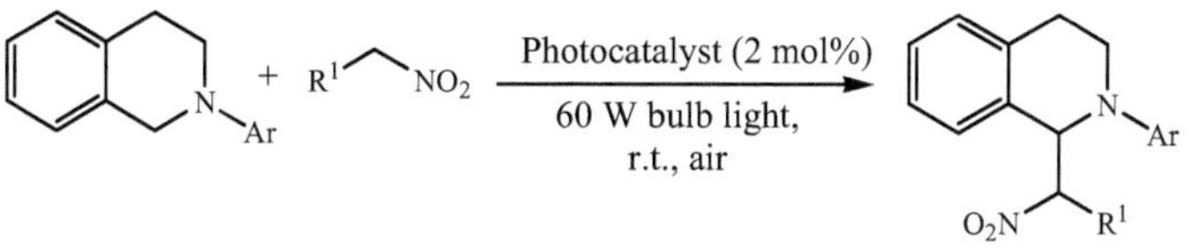

Ar	R^1	Time (h)	Yield (%)
Ph	H	15	95
4-OMe-C_6H_4	H	15	97
3-Me-C_6H_4	H	15	95
4-Br-C_6H_4	Me	18	93
3,5-Me_2-C_6H_3	H	18	96
1-naphthyl	H	15	90

FIGURE 12.15 Aza-Henry reaction as a model reaction to evaluate xanthene-based photocatalyst.

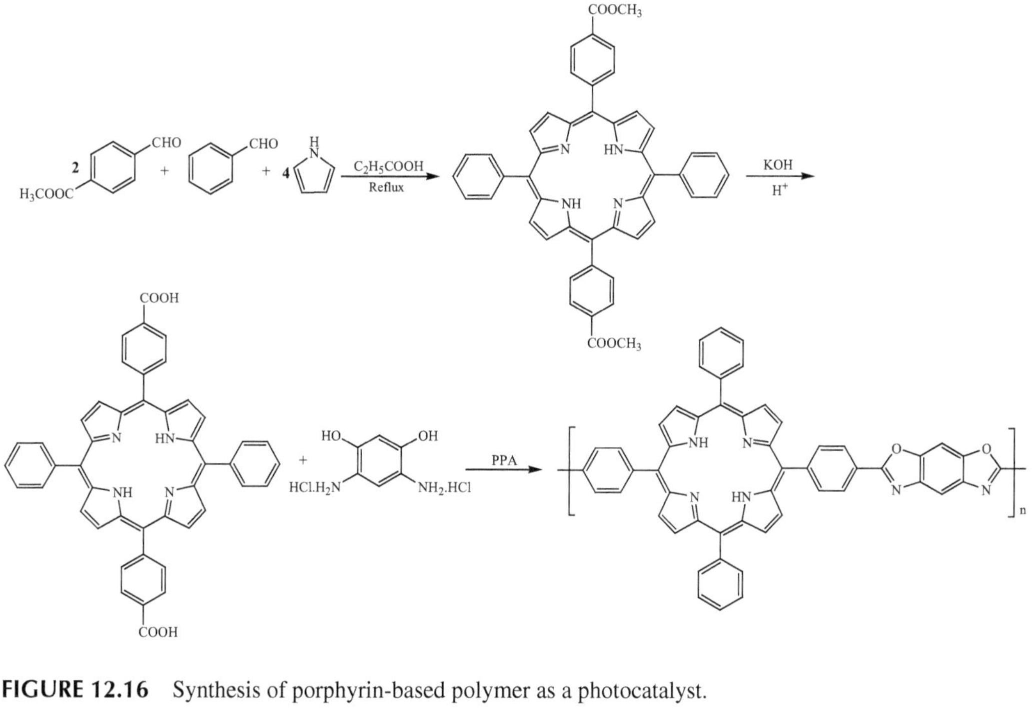

FIGURE 12.16 Synthesis of porphyrin-based polymer as a photocatalyst.

The prepared photocatalyst with an improved light-harvesting capability prevents the aggregation of porphyrins (Li et al., 2018). A porous organic polymer exhibiting both electron donor and electron acceptor properties was synthesized to obtain an efficient catalytic system using Zinc porphyrin-based conjugated microporous polymer (Figure 12.18). Nitrogen gas adsorption investigations on pretreated porous polymer at 77 K to investigate the permanent porosity revealed type I isotherms in the nitrogen adsorption–desorption isotherm. The obtained results from Brunauer–Emmett–Teller (BET) surface area analysis showed 465 m^2 g^{-1} and a total pore volume of 0.20 cm^3 g^{-1} indicating a high surface area and large pore volume and diameters of 0.9–1.9 nm according to the nonlocal density functional theory method (Guo et al., 2020).

The catalyst was able to catalyze selective oxidation of amines under the illumination of a white LED lamp and molecular oxygen. The single excited state containing electron–hole pairs is

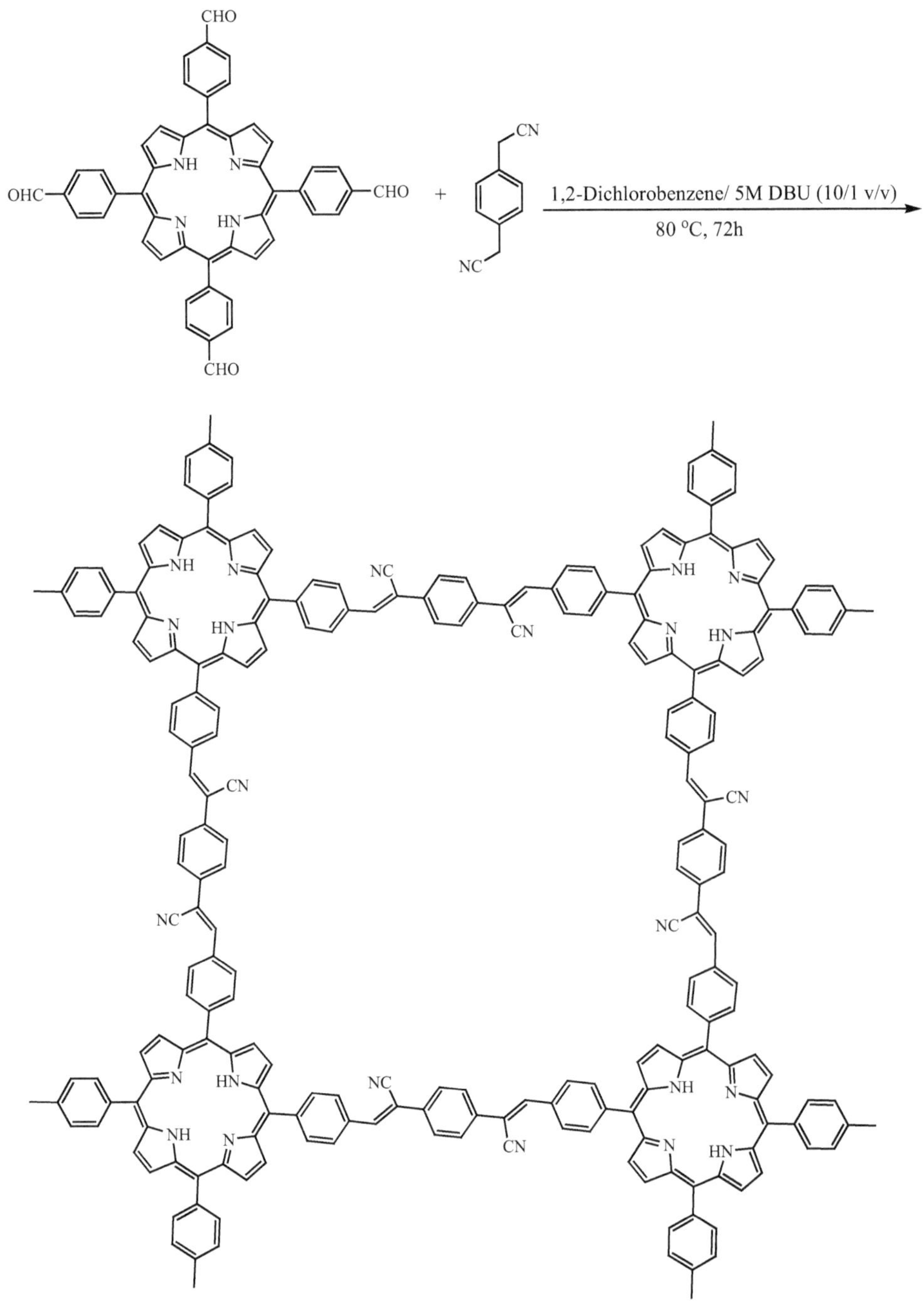

FIGURE 12.17 Synthesis of a novel porphyrin-based conjugated polymer with large-size pores.

generated by absorption light energy. Benzylamine free radical cations are generated by oxidation of benzylamine and the produced electrons to oxygen molecule. The oxygen molecule can be activated by the triplet excited states via energy transfer to obtain $^1O_2$. In the next step, O_2^- and $^1O_2$ react with benzylamine free radical cations and benzylamine, respectively. The two reactions produce H_2O_2 and imide intermediates. Then, H_2O_2 will form OH, which reacts with benzylamine to form an imine intermediate. Finally, the imine intermediate reacts with the benzylamine leading to imine production (Guo et al., 2020).

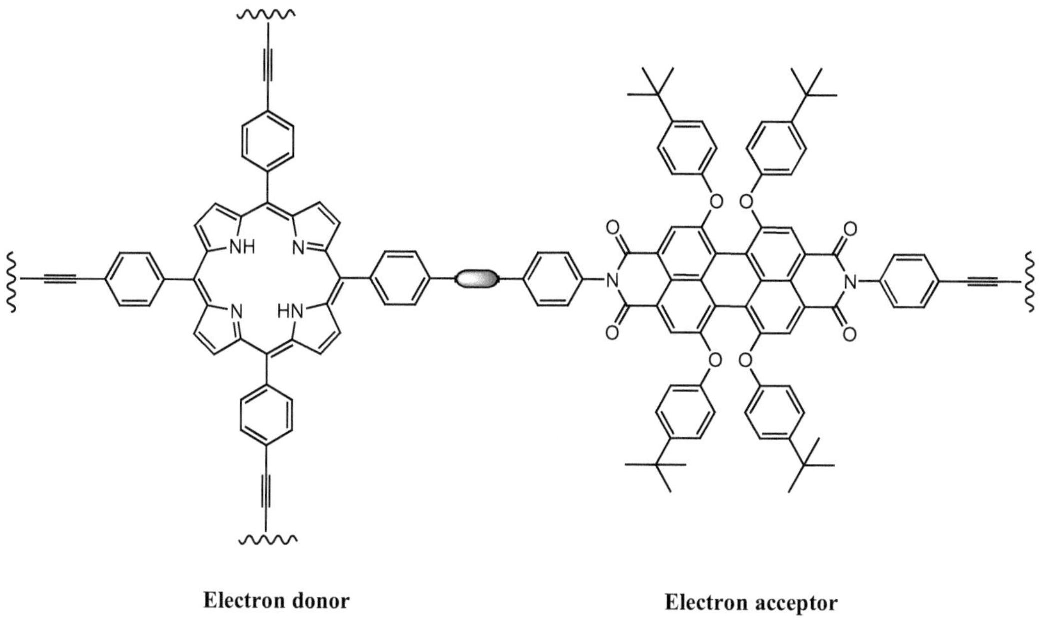

FIGURE 12.18 Porous organic polymer exhibiting both electron acceptor and donor properties.

12.5 CONCLUSION

Porous polymers are known as green and environmentally friendly platforms for applying in various photocatalytic applications including organic synthesis, energy storage, carbon dioxide reduction, and degradation of pollutants. Porous polymer photocatalysts present good sustainability and high activity through easy recycling and recovery processes which are key factors for achieving industrial applications. The progress of microporous and nanoporous polymers as photoredox catalysts offers some benefits, most noticeably by decreasing the usage of costly metals and simplifying catalyst recovery processes and reusability. Although porous polymer-based photocatalysts have been progressed in the last decade, more effort is required to design novel photocatalysts, particularly on larger scales. The interdisciplinary nature of photocatalysts suggests the connotation of various scientific fields to accelerate the development of this research area such as chemistry, physics, and materials.

REFERENCES

Ayed, C., L. C. da Silva, D. Wang, and K. A. Zhang. 2018. Designing conjugated microporous polymers for visible light-promoted photocatalytic carbon–carbon double bond cleavage in aqueous medium. *J. Mater. Chem. A.* 6 (44): 22145–22151.

Banerjee, S., S. C. Pillai, P. Falaras, K. E. O'Shea, J. A. Byrne, and D. D. Dionysiou. 2014. New insights into the mechanism of visible light photocatalysis. *J. Phys. Chem. Lett.* 5 (15): 2543–2554.

Bard, A. J. 1979. Photoelectrochemistry and heterogeneous photo-catalysis at semiconductors. *J. Photochem.* 10 (1): 59–75.

Bedia, J., V. Muelas-Ramos, M. Peñas-Garzón, A. Gómez-Avilés, J. J. Rodríguez, and C. Belver. 2019. A review on the synthesis and characterization of metal organic frameworks for photocatalytic water purification. *Catalysts* 9 (1): 52.

Byun, J., and K. A. Zhang. 2020. Designing conjugated porous polymers for visible light-driven photocatalytic chemical transformations. *Mater. Horiz.* 7 (1): 15–31.

Chen, R., J. L. Shi, Y. Ma, G. Lin, X. Lang, and C. Wang. 2019. Designed synthesis of a 2D Porphyrin-based sp^2 carbon-conjugated covalent organic framework for heterogeneous photocatalysis. *Angew. Chem. Int. Ed.* 58 (19): 6430–6434.

Cui, W., J. Li, L. Chen, H. Wang, J. Sheng, Y. Sun, Y. Zhou, and F. Dong. 2020. Nature-inspired $CaCO_3$ loading TiO_2 composites for efficient and durable photocatalytic mineralization of gaseous toluene. *Sci. Bull.* 65 (19): 1626–1634.

Dadashi-Silab, S., H. Bildirir, R. Dawson, A. Thomas, and Y. Yagci. 2014. Microporous thioxanthone polymers as heterogeneous photoinitiators for visible light induced free radical and cationic polymerizations. *Macromolecules* 47 (14): 4607–4614.

Dai, C., and B. Liu. 2020. Conjugated polymers for visible-light-driven photocatalysis. *Energy Environ. Sci.* 13 (1): 24–52.

Ding, Y. M., X. Nie, H. Dong, N. Rujisamphan, and Y. Li. 2021. Predicting a new graphene derivative C_3H as potential photocatalyst for water splitting and CO_2 reduction. *Phys. E.* 127: 114562.

Fajrina, N., and M. Tahir. 2019. A critical review in strategies to improve photocatalytic water splitting towards hydrogen production. *Int. J. Hydrog. Energy* 44 (2): 540–577.

Günes, S., H. Neugebauer, and N. S. Sariciftci. 2007. Conjugated polymer-based organic solar cells. *Chem. Rev.* 107 (4): 1324–1338.

Guo, X. X., J. Jiang, Q. Han, X. H. Liu, X. T. Zhou, and H. B. Ji. 2020. Zinc porphyrin-based electron donor–acceptor-conjugated microporous polymer for the efficient photocatalytic oxidative coupling of amines under visible light. *Appl. Catal. A* 590, 117352.

Huang, D., S. Chen, G. Zeng, X. Gong, C. Zhou, M. Cheng, W. Xue, X. Yan, and J. Li. 2019. Artificial Z-scheme photocatalytic system: What have been done and where to go?. *Coord. Chem. Rev.* 385: 44–80.

Jiang, J. X., Y. Li, X. Wu, J. Xiao, D. J. Adams, and A. I. Cooper. 2013. Conjugated microporous polymers with Rose Bengal dye for highly efficient heterogeneous organo-photocatalysis. *Macromolecules* 46 (22): 8779–8783.

Jiang, C., W. Chen, W. H. Zheng, and H. Lu. 2019. Advances in asymmetric visible-light photocatalysis, 2015–2019. *Org. Biomol. Chem.* 17 (38): 8673–8689.

Li, Y. W., Q. Duan, H. Wang, B. Gao, N. Qiu, and Y. Li. 2018. Construction of two-dimensional porphyrin-based fully conjugated microporous polymers as highly efficient photocatalysts. *J. Photochem. Photobiol. A.* 356: 370–378.

Li, X., Y. C. Zhang, Y. Zhao, H. P. Zhao, B. Zhang, and T. Cai. 2020. Xanthene dye-functionalized conjugated porous polymers as robust and reusable photocatalysts for controlled radical polymerization. *Macromolecules* 53 (5): 1550–1556.

Liu, Y., B. Li, H. S. Li, P. Wu, and J. Wang. 2020. Metal–organic frameworks containing xanthene dyes for photocatalytic applications. *Dalton Trans.* 49 (48): 17520–17526.

Low, J., C. Jiang, B. Cheng, S. Wageh, A. A. Al-Ghamdi, and J. Yu. 2017. A review of direct Z-scheme photocatalysts. *Small Methods* 1 (5): 1700080.

Luo, J., J. Lu, and J. Zhang. 2018. Carbazole–triazine based donor–acceptor porous organic frameworks for efficient visible-light photocatalytic aerobic oxidation reactions. *J. Mater. Chem. A.* 6 (31): 15154–15161.

Luo, J., X. Zhang, and J. Zhang. 2015. Carbazolic porous organic framework as an efficient, metal-free visible-light photocatalyst for organic synthesis. *ACS Catal.* 5 (4): 2250–2254.

Medlyn, B. E., E. Dreyer, D. Ellsworth, M. Forstreuter, P. C. Harley, M. U. F. Kirschbaum, X. Le Roux, P. Montpied, J. Strassemeyer, A. Walcroft, K. Wang, and Loustau, D. 2002. Temperature response of parameters of a biochemically based model of photosynthesis. II. A review of experimental data. *Plant Cell Environ.* 25 (9): 1167–1179.

Nasalevich, M. A., M. Van der Veen, F. Kapteijn, and J. Gascon. 2014. Metal–organic frameworks as heterogeneous photocatalysts: advantages and challenges. *CrystEngComm* 16 (23): 4919–4926.

Park, J. 2017. Visible and near infrared light active photocatalysis based on conjugated polymers. *J. Ind. Eng. Chem.* 51: 27–43.

Ryu, Y., J. A. Berry, and D. D. Baldocchi. 2019. What is global photosynthesis? History, uncertainties and opportunities. *Remote Sens. Environ.* 223: 95–114.

Su, C., R. Tandiana, B. Tian, A. Sengupta, W. Tang, J. Su, and K. P. Loh. 2016. Visible-light photocatalysis of aerobic oxidation reactions using carbazolic conjugated microporous polymers. *ACS Catal.* 6 (6): 3594–3599.

Taylor, D., S. J. Dalgarno, Z. Xu, and F. Vilela. 2020. Conjugated porous polymers: incredibly versatile materials with far-reaching applications. *Chem. Soc. Rev.* 49 (12): 3981–4042.

Vilela, F., K. Zhang, and M. Antonietti. 2012. Conjugated porous polymers for energy applications. *Energy Environ. Sci.* 5 (7): 7819–7832.

Wang, Z. J., S. Ghasimi, K. Landfester, and K. A. Zhang. 2014. Highly porous conjugated polymers for selective oxidation of organic sulfides under visible light. *Chem. Comm.* 50 (60): 8177–8180.

Wang, T. X., H. P. Liang, D. A. Anito, X. Ding, and B. H. Han. 2020. Emerging applications of porous organic polymers in visible-light photocatalysis. *J. Mater. Chem. A*. 8 (15): 7003–7034.

Xiao, J., X. Liu, L. Pan, C. Shi, X. Zhang, and J. J. Zou. 2020. Heterogeneous photocatalytic organic transformation reactions using conjugated polymers-based materials. *ACS Catal.* 10 (20): 12256–12283.

Xu, Q., L. Zhang, J. Yu, S. Wageh, A. A. Al-Ghamdi, and M. Jaroniec. 2018. Direct Z-scheme photocatalysts: Principles, synthesis, and applications. *Mater. Today* 21 (10): 1042–1063.

Yang, Y., H. Niu, L. Xu, H. Zhang, and Y. Cai. 2020. Triazine functionalized fully conjugated covalent organic framework for efficient photocatalysis. *Appl. Catal. B Environ.* 269: 118799.

You, J., Y. Zhao, L. Wang, and W. Bao. 2021. Recent developments in the photocatalytic applications of covalent organic frameworks: A review. *J. Clean. Prod.* 291: 125822.

Yu, Y., Y. Li, Y. Li, H. Wang, Q. Zuo, and Q. Duan. 2019. A 1D porphyrin-based rigid conjugated polymer as efficient and recyclable visible-light driven photocatalyst. *React. Funct. Polym.* 143: 104340.

Zhang, S., G. Cheng, L. Guo, N. Wang, B. Tan, and S. Jin. 2020. Strong-base-assisted synthesis of a crystalline covalent triazine framework with high hydrophilicity via benzylamine monomer for photocatalytic water splitting. *Ang. Chem.* 132 (15): 6063–6070.

Zhang, J., B. Tian, L. Wang, M. Xing, and J. Lei. 2018. *Mechanism of Photocatalysis*. In *Photocatalysis* (pp. 1–15). Singapore: Springer.

Zhang, F., X. Wang, H. Liu, C. Liu, Y. Wan, Y. Long, and Z. Cai. 2019. Recent advances and applications of semiconductor photocatalytic technology. *Appl. Sci.* 9 (12): 2489.

Zhang, B., and L. Sun. 2019. Artificial photosynthesis: opportunities and challenges of molecular catalysts. *Chem. Soc. Rev.* 48 (7): 2216–2264.

Zhi, Y., Z. Yao, W. Jiang, H. Xia, Z. Shi, Y. Mu, and X. Liu. 2019. Conjugated microporous polymers as heterogeneous photocatalysts for efficient degradation of a mustard-gas simulant. *ACS Appl. Mater. Interfaces* 11 (41): 37578–37585.

13 Porous Polymers for CO_2 Reduction

Sughra Gulzar, Zeeshan Abid, Ammara Mushtaq, Syed Arfan Haider, Raja Shahid Ashraf, and Muhammad Altaf
Government College University Lahore, Lahore, Pakistan

Munazza Shahid
University of Education, Lahore, Pakistan

CONTENTS

DOI: 10.1201/9781003169604-13

13.1 INTRODUCTION

A moderate and suitable temperature is required on this planet Earth for survival and living. The average temperature of the Earth is between the freezing and boiling point of water which is appropriate for our kind of life to survive on this planet and it cannot be justified by assuming that our planet Earth is orbiting around the Sun at the accurate space to absorb the right amount of solar radiation. The precise kind of atmosphere is required to flourish on this planet. The Earth's atmosphere acts as a shielding blanket to receive an appropriate amount of solar energy to keep the temperature of this planet in an average range. This shield of the atmosphere is known as greenhouse and is made up of gases, mostly water vapour, methane, carbon dioxide and nitrous oxide. These gases are named greenhouse gases (GHG) and they act as an effective global insulator. Ultraviolet (UV) radiation having a shorter wavelength than infrared (IR) radiation passes through the glass wall of the greenhouse gets absorbed by the plants; however, the IR radiation has difficulty in passing through the walls and is trapped inside the atmosphere thereby warming the greenhouse. The absorbing of IR radiation and trapping the heat in the atmosphere and increasing the natural greenhouse effect is termed global warming (Socolow 2005). This led to the temperature of the atmosphere suitable for survival for all kinds of living things on this planet Earth.

The level of GHG has been raised because of anthropogenic activities. The use of fossil fuels as a source of energy without appropriate regulation leads to the enhancement of GHG. Human activities are altering the atmospheric concentrations and distributions of GHG resulting from climate change (Kweku et al. 2017). One of the major challenges of the 21st century is climate change. In the late 19th century, GISS (Goddard Institute for Space Studies) studied that the atmospheric temperature is globally changing and there is an increase of at least 0.8 °C in global warming since 1851 and it is continually increasing as the emissions of GHG is raising day-by-day (Hansen et al. 2010). According to the IPCC "*The Intergovernmental Panel on Climate Change*" report the excessive CO_2 emissions are altering the climate globally and unfavourably affecting the temperature globally (Livingston, Lövbrand, and Olsson 2018). To meet the energy demand, there are multiple ways that can satisfy the demand for energy globally such as fossil fuel switching, renewable energy sources, energy conversation and efficiency, nuclear and carbon capture and storage (CCS).

Carbon dioxide capture and storage (CCS) is an efficient and quick way these days to diminish the level of carbon to achieve the low carbon content in the atmosphere. Concerning the processes of CCS, various methodologies have been adopted like physical adsorption, absorption and membrane separation. The respective has been stored in the adsorbent and buried deep underground. Traditionally, some liquid amines like monoethanolamine and diethanolamine, being exploited by CCS processes to store CO_2. Besides the extensive usage of these methodologies, there are some limitations, for example, volatility of amines, corrosive nature, the disintegration of the produced salts and costly regeneration expense (Vega et al. 2018). Therefore, new technologies were required. Porous materials dominate the previous methods, especially the porous organic polymers (POPs) recently developed synthesized from lightweight elements like carbon, hydrogen, oxygen and nitrogen. These species link together with strong covalent bonds and are potential candidates for gas storage and separation to cleanse the environment.

13.2 ROLE OF CO_2 IN CLIMATE CHANGE

CO_2 is a major GHG for heat-trapping. From the time of global usage of fossil fuels which include coal, oil and gas, approximately 1850, fulfilling the energy demands led to the rapid increase in CO_2 emissions. The emissions of GHG are continually growing and the level of CO_2 is increasing in the atmosphere mainly through the burning of fossil fuels, deforestation, solid waste and other biological materials, and also as a consequence of certain chemical reactions, for example, the manufacture of cement. The researchers have found CO_2 is responsible for the major portion of GHG. According to NASA's official website, the level of CO_2 has reached nearly 416 ppm, in 2021 and still rising. It shows a 48% increase since the start of the Industrial Revolution.

13.3 MITIGATION STRATEGIES OF CO_2

As anthropogenic activities increase the concentration of CO_2 in the atmosphere and as a result, living beings are facing the problems like global warming and climate change. To reduce CO_2 emissions, various countries adopted various methodologies; which includes:

- Encourage energy conservation and increase energy efficiency.
- More utilization of low-carbon fuels, like natural gas, nuclear or hydrogen power.
- Install renewable energy sources, for example, wind, solar, bioenergy and hydropower.
- Enforce geoengineering approaches, such as afforestation and reforestation.
- Carbon dioxide capture and storage (CCS)

Every approach has certain advantages, applications and limitations. All these are summarized in Table 13.1 below. Some approaches concern with the reduction in pollution emissions like clean coal and clean fuels technologies and some with preservation as energy conservation. Each technology has its pros and cons. Implementation of a single approach cannot meet the demand of IPCC regarding CO_2 reduction, which is 50–85% from 2000 to 2050. The last approach, carbon dioxide capture and storage (CCS) reduces CO_2 emissions by 85–90%. In this method, the climate-changing gas CO_2 is first captured, separated by using sorbent and moved either towards storing permanently or reutilizing industrially.

TABLE 13.1
Pros and Cons of Different CO_2 Mitigation Technologies

Methodology	Application	Advantages	Disadvantages
Encourage energy conservation and increase energy efficiency	Implement majorly in commercial and industrial structures	Saves energy easily up to 10% to 20%	May need a heavy budgeted investment for device installation
More utilization of low-carbon fuels, like natural gas, nuclear or hydrogen power	Utilization of natural gas instead of coal as an energy-driving source	Natural gas has higher combustion efficiency, reduce CO_2 emissions as compared to coal by almost 40–50%, release fewer particulates and sulphur content	Cost-intensive for conventional natural gas but comparable for shale gas
Install clean coal technology	Substitute conventional combustion technologies by implementing IGCC, PFBC, etc.	Usage of coal with lower emitted pollution	Needed extensive investment to spread these technologies extensively
Renewable energy utilization	Solar/thermal, wind power, hydro and biofuel broadly establish	Use the native natural resources and emit less GHG as well as toxic emissions	These technologies are not mature enough to implement with efficient efficiencies
Afforestation and deforestation	Easily applicable in all countries	Easier natural way to sink CO_2	Restraint the land for other applications
Carbon capture and storage	Valid to extensive emission sources	Reduce CO_2 effectively >80%	Not proved to apply commercial scale

13.4 TECHNOLOGIES FOR CO_2 CAPTURE

The other proposed technologies as pollution-free renewable energy sources need sufficient time for their implementation and efficient working. The alternative way to reduce CO_2 emissions, until the widely accessible alternative clean energy sources, is carbon dioxide capture and sequestration (CCS) (Muhammad et al. 2019). CO_2 needs to be captured from gas mixtures before sequestration.

Carbon dioxide capture and storage (CCS) consists of three steps:

- Strips out, purges, and compresses CO_2 emissions before entering into the atmosphere from fossil fuel burning.
- The seized CO_2 is then transported to the storage site by forming a liquid under high pressure.
- The fluid is instilled deeper into the rock pores (800 m below the surface)

(Rubin 2008).

Currently, three strategies are used for carbon dioxide capture which include post-combustion, pre-combustion and oxy-fuel combustion processes.

13.4.1 Post-combustion Process

In this process, CO_2 is separated after the combustion of fuel from flue gas. The high nitrogen content is used for the burning of the fuel so in this separation process selective carbon dioxide is captured over nitrogen. This technique is easily employed in existing power plants that is why different sorbents for CO_2 have been explored (Wang et al. 2017).

13.4.2 Pre-combustion Process

On the other hand, in the pre-combustion process, CO_2 is separated by first converting the fossil fuel into a mixture of H_2 and CO_2 which is followed by the combustion of H_2. Separation of CO_2 from a mixture of CO_2/H_2 is much easier than from the mixture of CO_2/N_2 because of the low polarizability of H_2 (Jansen et al. 2015).

13.4.3 Oxy-fuel Combustion

In this process, N_2 is removed from the air to get a denitrified air which is then utilized for the combustion of fossil fuels leads to the production of CO_2 and H_2O and the separation of this mixture can be easily achieved by existing technologies (Stanger et al. 2015).

13.4.4 Chemical Looping Combustion

Unlike oxy-fuel combustion, the chemical looping combustion used a metal oxide as an oxygen carrier as compared to pure oxygen. Redox reaction happens and metal oxide gets reduced to metal and fuel is oxidized to CO_2 and H_2O. Metal is recycled in this process by getting oxidized at some other stage. Water is removed simply by condensation and pure CO_2 is obtained without the utilization of energy. Many metal oxides can be used for this process and also economical, for example, Fe_2O_3, NiO, Mn_2O_3 and CuO (Lyngfelt 2020).

The phenomena of physical adsorption possessed by porous materials are effective for capturing CO_2. Several kinds of sorbents have been explored for CO_2 capturing such as amine solutions, porous carbons, zeolites, metal–organic frameworks (MOFs) and POPs. The sorbents like porous carbons and zeolites are the potential candidates as having high chemical and thermal stability, also they are cost-effective. But there are some limitations for these sorbents as the strong hydrophilic character of zeolites (Krishna and van Baten 2012) and low selectivity towards CO_2/N_2 of porous carbons make them less suitable for capturing CO_2 from flue gas. MOFs increase the selectivity towards CO_2/N_2 mixture by exposing the metal sites but the low thermal and chemical stability hinders MOFs application under extreme conditions (Zhou, Long, and Yaghi 2012). Another important class for CO_2 capture is POPs, which are built from elements having light weight and linked by a strong covalent bond. POPs show a variety of applications in various fields such as catalysis, optoelectronics, gas storage and separation, chemical sensing and energy storage (Das et al. 2018).

13.5 POROUS ORGANIC POLYMERS

POPs are considered as a potential candidate for the application of gas capture materials because of their quite excellent features like high thermal stability, low density, plentiful surface functionalization and the major one is the diversity of synthetic approaches. The status of POPs being organic exhibits versatility in their structural motifs for capturing and selecting CO_2 and offering additional scope for the improvement of gas adsorption in CCS applications. The plentiful existence of CO_2-philic sites in the sorbent enhances the CO_2 capturing and selecting an ability from the mixture of gases such as from flue gas (CO_2/N_2) or natural gas (CO_2/CH_4).

POPs can be divided into two types; amorphous or crystalline. Hyper-crosslinked polymers (HCPs), conjugated microporous polymers (CMPs), polymers of intrinsic microporosity (PIMs) and porous aromatic frameworks (PAFs) are categorized under amorphous POPs. On the other hand, covalent organic frameworks (COFs) and covalent triazine frameworks (CTFs) are represented as crystalline POPs (Zou et al. 2017). These several classes of POPs are described below

13.5.1 Covalent Organic Frameworks (COFs)

COFs were first discovered in 2005 by Yaghi's group and it is the first classification of porous organic materials that were discovered (Cote et al. 2005) and having the structure ordered of the nanoporous materials that are only made up of light non-metal elements as hydrogen, boron, carbon and oxygen and are linked by a strong covalent bond (B–O, C–C and B–C). The structure of COFs is periodic, ordered, rigid and plain due to B–O five- or six-member rings which give them an ordered nanoporous form. Different precursors can be used for the synthesis of COFs and giving various pore sizes and controlling space structure in two or three dimensions (Zeng, Zou, and Zhao 2016). Figure 13.1 shows two representative structures of COFs.

Some of the COFs materials with their pore size and BET surface area have been shown in Table 13.2.

13.5.2 Covalent Triazine Frameworks

CTFs were first discovered in 2008 by Thomas' group (Kuhn, Antonietti, and Thomas 2008). The structural features of CTFs are the triazine bonding centre which is formed by the condensation of aromatic nitriles catalysed by $ZnCl_2$.

Many factors can affect the properties of CTFs' surface areas, like the monomers, the ratios of monomers, heating conditions and $ZnCl_2$ catalyst. The synthetic conditions should be selected carefully to obtain the uniform pores for crystalline structure. The strong acid catalyst CF_3SO_3H can be used for the synthesis of CTFs at room temperature but not in crystalline form (Ren et al. 2012). Figure 13.2 shows two representative structures of CTFs (Table 13.3).

13.5.3 Polymers of Intrinsic Microporosity

This class of POPs was discovered in 2004 by McKeown's group (Budd 2004). PIMs hold special importance among the classes of POPs because of their special property of solubility in certain organic solvents like chloroform and THF. PIMs are widely used for the application of separation of gas (Table 13.4).

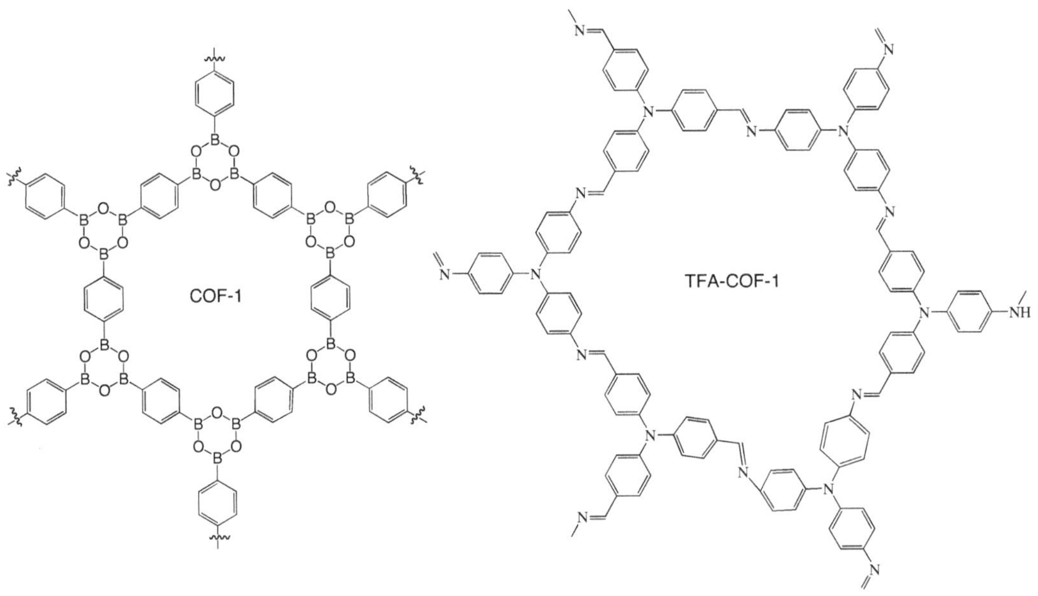

FIGURE 13.1 Some representative structures of COFs.

TABLE 13.2
Precursor, Surface Properties and CO_2 Adsorption Capacity of a Few Crystalline COFs

				CO_2 adsorption (mmol g^{-1})		
Monomer	**Covalent bond category**	**Pore size (nm)**	**SA_{BET} ($m^2\ g^{-1}$)**	**273 K, 1 bar**	**298 K, 1 bar**	**Ref.**
PyTTA-BFBImiCOF	Imine	2.3	1532	177	93	(Huang et al. 2017a)
TPA-COF-1	Imine	1.93	714	68.68	38.2368	(El-Mahdy et al. 2018)
COF-JLU6	Imine	3.1	1450	129		(Zhi et al. 2018)
COF-TpAzo	Azo	2.58	1552	112	68	(Ge et al. 2016)
TPE-COF-I	Imine	1.65 ($cm^3\ g^{-1}$)	1535	134	73.8	(Gao et al. 2018)

13.5.4 Porous Aromatic Frameworks

PAFs were first discovered in 2009 by Zhu's group (Ben et al. 2009). It has a 3-D porous structure. They are well-known for their peculiar properties like high specific surface areas, chemical stability and thermal stability. PAF-1 has the highest BET surface area which is 5640 $m^2\ g^{-1}$ among all the POPs.

Owing to their property of large surface area PAFs show a very good CO_2 storage capability at high pressure but this property is quite poor at low pressure because of its non-polar structures and weak affinity towards CO_2 (Ben et al. 2012) (Table 13.5).

13.5.5 Hyper-Crosslinked Polymers

HCPs are synthesized through Friedel–Crafts alkylation using $FeCl_3$ as a Lewis acid catalyst from poly(chloromethyl)styrene. HCPs form a cross-linked structure using various precursors to a profound degree thus making the networks highly compact and rigid making them unable to collapse (Dawson, Cooper, and Adams 2013).

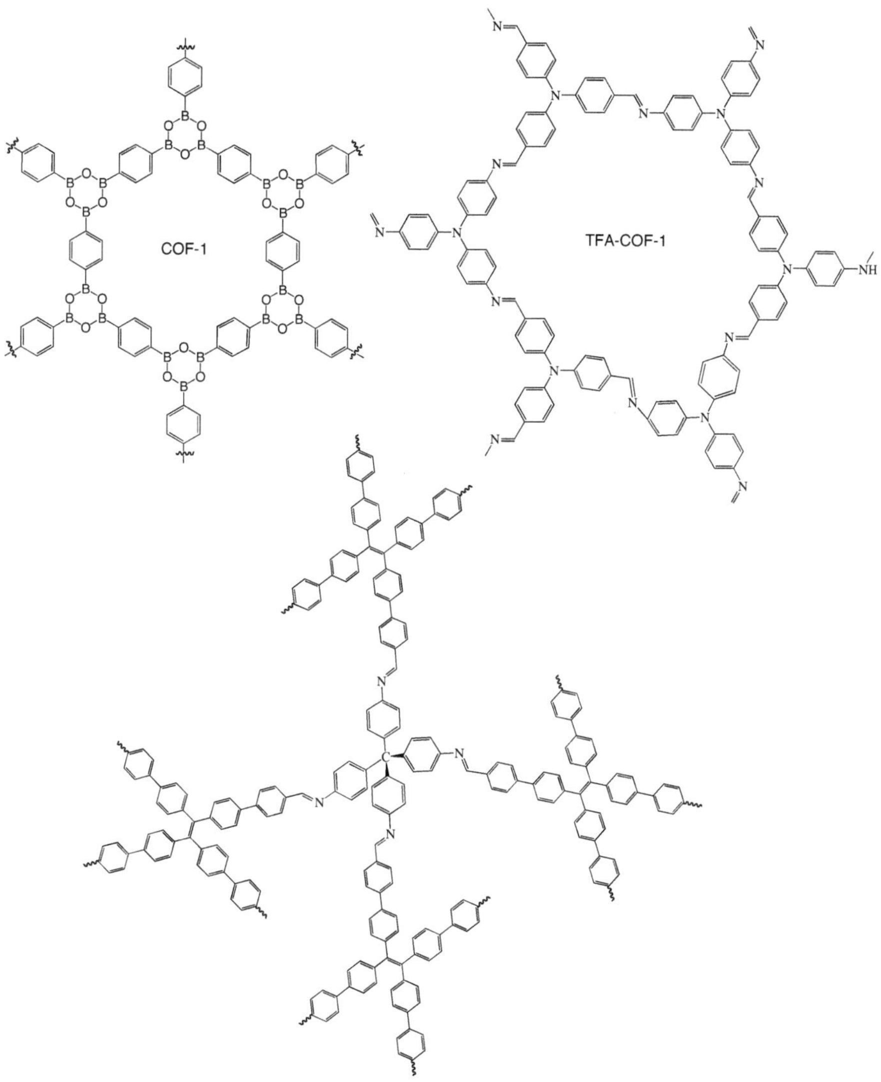

FIGURE 13.2 Two representative structures of CTFs.

TABLE 13.3
Precursor, Surface Properties and CO_2 Adsorption Capacity of a Few CTFs

Monomer	Covalent Triazine Framework	SA_{BET} ($m^2\ g^{-1}$)	CO_2 Adsorption ($mmol\ g^{-1}$) 273 K, 1 bar	298 K, 1 bar	Ref.
7,7,8,8-Tetracyanoquinodimethane	TCNQ-CTF-900	4000	5.99	3.41	(Deng et al. 2020)
4,4′-Malonyldibenzonitrile	acac-CTFs-5-400	1131	2.87	1.89	(Jena et al. 2018)
9H-Fluorene-2,7-dicarbonitrile	Fl-CTF-350	1235	18.4 wt%	10.1 wt%	(Hug et al. 2014)
2,2′,3,3′,5,5′,6,6′-Octafluoro-4,4′-biphenyldicarbonitrile	F-DCBP-CTF-1	1574	26.31 wt%	16.81 wt%	(Wang et al. 2018)
1,3-Bis(4-benzocynide) imidazoline	CTF-Cl-1	516	40.0 wt%	26.3 wt%	(Xu et al. 2019)

TABLE 13.4
Examples of a Few PIMs with Surface Properties and CO_2 Adsorption Capacity

PIMs	Pore Size (nm)	SA_{BET} ($m^2 g^{-1}$)	CO_2 Adsorption (mmol g^{-1})		Ref.
			293 K, 1 bar	333 K, 1 bar	
Trip-TB ladder	0.237	899	3.812	2.857	(Del Regno et al. 2013)
Trip-TB network	0.241	1035	4.337	3.059	
TB-cat-A	0.184	525	2.568	1.828	
Indan-TB	0.187	535	1.933	1.293	
Trip(H)-PIM	0.310	1032	6.309	6.309	

TABLE 13.5
Examples of a Few PAFs with Surface Properties and CO_2 Adsorption Capacity

PAFs	Pore size (nm)	SA_{BET} ($m^2 g^{-1}$)	CO_2 adsorption Mmol g^{-1}		Ref.
			293 K, 1 bar	298 K, 1 bar	
PAF-1	1.5	5640	2.07	1.09	(Ben et al. 2009)
PAF-3	—	2932	3.48	1.82	
PAF-4	—	2246	2.43	1.16	
PAF-1-CH_2NH_2	1.17	1363	—	4.38	(Garibay et al. 2013)
PAF-1-SO_3H	0.6,1.5	885	—	2.64	(Zhang et al. 2018)
PAF-1-SO_3Co	0.6,1.5	696	—	2.52	

HCPs have diversity in their structural materials and are also cost-effective. This property makes them a special class of porous organic materials and this is owing to their synthetic ability from a variety of small different aromatic monomers which includes benzene, pyrrole, phenol, aniline and many more.

They are potential candidates to be considered for CO_2 capture property as many of their precursors are cheap and commercially available. HCP structure is rich in micropores and even the micropores survive in cracks of polymers because of the deep degree of cross-linking and that property increases the ability of a material to absorb the gas even at low pressure (Table 13.6).

Unlike COFs, the new designs of HCPs cannot be synthesized easily as the size of the pores is not distributed uniformly in the entire structure. Thus, for the new designs, the pore sizes are difficult to control. On the other hand, the Lewis acid catalyst $FeCl_3$ makes the synthesizing reaction of HCPs complex because $FeCl_3$ plays a dual role in the reaction; as an oxidant and Lewis acid as well and, as a result, the aromatic polymers may get oxidized to some extent in the polymerization process (Li et al. 2011).

13.5.6 Conjugated Microporous Polymers

Several kinds of organic reactions, for example, Suzuki coupling, Sonogashira coupling, Yamamoto coupling, click coupling and many others have been already applied for the synthesis of this class of porous organic materials. CMPs are considered as the subclass of HCPs (Dawson, Cooper, and Adams 2013).

The CMPs can be functionalized with different functional groups to compare the CO_2 capturing abilities of various CMPs having the same structure but possessing different functional groups

TABLE 13.6
The CO_2/N_2 Selectivity of Some HCPs with Respective CO_2 Adsorption Values

HCPs	CO_2/N_2 Selectivity	CO_2 Adsorption Mmol g^{-1}	Ref.
Aniline-HCP	49.2	0.35 (300 K)	(Dawson et al. 2012)
HC-PCz-8	17	3.5	(Zhu et al. 2017)
Py-1	117	2.7 (273 K)	(Luo et al. 2012)
Gal-2	34	1.38	(Li et al. 2015)
TNB	19.4	2.35	(Hou and Tan 2018)
BOLP-5	65	1.8	(Rabbani, Islamoglu, and El-Kaderi 2017)
PMOP	28.7	3.17	(Luo et al. 2018)

TABLE 13.7
Surface Properties, Adsorption Heat and CO_2 Uptake Capacity of Various CMPs

CMPs	SA_{BET} (m^2 g^{-1})	Adsorption Heat (kJ mol^{-1})	CO_2 Adsorption (mmol g^{-1})	Ref.
CMP-1	837	26.9	2.05	(Huang et al. 2017b)
CMP-1-$(OH)_2$	1043	27.8	1.80	(Dawson, Adams, and Cooper 2011)
CMP-1-$(CH_3)_2$	899	1.64	27.0	(Dawson, Adams, and Cooper 2011)
CMP-1-COOH	522	32.6	1.60	(Zulfiqar, Sarwar, and Mecerreyes 2015)
CMP-1-NH_2	710	29.5	1.64	(Wang et al. 2015)
BFCMP-2	1316	25	2.77	(Zhang et al. 2015)

(Dawson, Adams, and Cooper 2011). There is another fact that simply introducing different functional groups to enhance the CO_2 capturing by enhancing the surface gas affinity under ambient conditions is not sufficient (Dawson et al. 2009). Some other factors like the surface also need to be considered in order to improve the performance of the material (Table 13.7).

13.6 FACTORS AFFECTING THE CO_2 UPTAKE

As the pollution in the environment has been increasing and consequently increasing the concentration of CO_2 in the environment and enhancing the global warming and results in climate change. So, to control the concentration of CO_2, carbon dioxide capture and storage is one of the strategies employed these days. POPs, in this matter, have attracted much attention because of their ability to capture CO_2. Now the prior concern is how the CO_2 capturing property can be enhanced by POPs. Luckily, many strategies have been devised already for this purpose. There are some of the factors that affect CO_2 uptake. Some main factors are described below:

13.6.1 Surface Area

The first more prominent feature that affects the CO_2 capturing property of POPs is the surface area. POPs having high surface areas are considered as the most optimistic attribute for capturing CO_2. Under high pressure, the CO_2 capture capacity gets increased when the POPs possess a high surface area. This as a result condenses the CO_2 packing inside the pores. On the other hand, at low pressure, the POPs show poor performance because of relatively large pores and little CO_2 interaction with the sorbent materials (Wang, Zhou, and Yuan 2017b).

The sorbent PAF-1 synthesized from a tetrahedral monomer of tetrakis(4-bromophenyl)methane via Yamamoto coupling possessed a high surface area having BET 5600 $m^2 g^{-1}$, diamondoid framework topology having the broadly exposed and interrelated pores to efficiently avoid the foundation of dead space (Yuan and Zhu 2019). Some other sorbents like PPN-3 (possess adamantane), PPN-4 (possess silicon) and PPN-5 (possess germanium) (Abdelnaby et al. 2020) were synthesized by replacing the central carbon of tetrahedral monomers with quadricovalent centres. These three POPs have an exceptionally high surface area, especially PPN-4 having the BET 6461 $m^2 g^{-1}$, which gives them a tremendous CO_2 storage capacity.

The factor of a high surface is essential for capturing CO_2 but other factors like affinity of gas for adsorbent material and pore size of the network are also crucial for CO_2 uptake capacity and selectivity.

13.6.2 Functionalization of Pore Size

Pore structure like the surface area is also an important factor that has a prominent impact on the gas selectivity and storage capacity. Thus, the functionalization of the pore surface specifically the pore size, shape and environment is a valuable approach to increase the CO_2 uptake by sorbents. Multiple interactions between adsorbent and adsorbate can be initiated by introducing functional groups and guest molecules in the network to tune the pore size according to the kinetic diameter of CO_2. Many POPs materials such as CMPs and COFs are successfully modified (Bhanja, Modak, and Bhaumik 2019).

As for the storage of CO_2 high pressure is required for most of the sorbents, if the pore size gets accurately controlled at almost 3.3 Å near the kinetic diameter of CO_2, the sorbents would be able to capture the gas even at low pressure (Xu et al. 2020). POPs of the required pore size are still a challenge because most of the synthesized POPs are amorphous by nature. However, the crystalline POPs as COFs are tuned with different functional units with a decrease in pore size.

13.6.3 Swellable Polymers

HCPs are an amorphous kind of POPs (POPs) and are not rigid like the other polymers. They have the predominant feature of gas adsorption. This sorbent does not reach the point of thermodynamic equilibrium and it collapses when the solvent is removed from it. The polymer chains can be prevented from collapsing by thoroughly cross-linking reactions. As a consequence, permanent pores can be created. Although the surface area of these POPs is low, they exhibit the unique property that by improving the pressure, their gas capturing capacity gets increased (Saleh et al. 2014). The polymer is hydrophobic by nature due to which it is hard to get affected by the adsorption of water vapours. Even they sustain their chemistry and remain stable in boiling concentrated acid. It is affordable as well because they do not need precious metal catalysts for their preparation. Through HCPs, 22.5 (wet) and 29.9 (dry) at 40 bar were achieved in the post-combustion method. All these factors reveal that the polymer is a promising candidate for CO_2 capture and quantity production (Tan and Tan 2017).

These polymers are suitable for practical applications and can be synthesized by an economic reaction condition. These are the special "soft" polymers. By applying high pressure, the CO_2 capture capacity can be much improved as on polymers with high surface areas. This property makes them a potentially good candidate for the application of CO_2 capture.

13.6.4 Heteroatomic Skeleton

At low pressure, CO_2 uptake is not affected by BET surface area. The CO_2 has stronger interactions for adsorbents as compared to the other gases as CO_2 has a much higher quadrupole moment and this feature can be utilized as an effective way of attaining high selectivity and storage capacity for CO_2.

The isosteric heat of adsorption between adsorbent and adsorbate can be improved by introducing heteroatoms in the sorbents. In order to increase the efficiency of POPs, oxygen, nitrogen, phosphorus and sulphur atoms are mainly infused as a heteroatom.

The phenyl group can be replaced with pyridyl or thiophenyl groups having different functional groups to enhance the activity of POPs. Although the surface area of pyridyl based PON-2 (189 $m^2 g^{-1}$) and thiophenyl-based PON-3 (44 $m^2 g^{-1}$) are considerably lower as compared to phenyl-based PON-1 (1447 $m^2 g^{-1}$), selectivity for CO_2 is much greater for nitrogen- and sulphur-containing POPs despite of those containing only carbon atoms (Choi et al. 2015). The carbazole group is used as a nitrogen-rich skeleton. CPOP-1 (microporous polycarbazole) is prepared by carbazole-based oxidative coupling polymerization. The CPOP-1 possessing high specific BET (2220 $m^2 g^{-1}$) and narrow pore size (0.62 nm) interact strongly with small gas molecules. CPOP-1 exhibits tremendously high CO_2 uptake capacity, i.e. 4.8 mmol g^{-1} at 1 bar and 273 K. The azo (–N=N–) groups are introduced into the POPs to form covalent organic polymers, azo-COPs (Wang and Han 2020). They possess a nitrogen-rich skeleton of nanoporous polymers. The CO_2 uptake capacity becomes 3.4 mmol g^{-1} at 263 K and 1 bar due to the high affinity between azo groups and polarizable CO_2 by dipole–quadrupole interactions. Besides, the nitrogen in the framework makes N_2-phobic. These azo-COPs that are N_2-phobic can be used as a potential candidate for extremely selective gas separations because at a high temperature they show excellent CO_2/N_2 selectivity (Yang et al. 2015).

CTFs, benzimidazole and tetraphenyladamantane-based microporous polybenzimidazoles are also the major classes that constitute nitrogen in a functionalized skeleton and possess high selectivity for CO_2 uptake (Wang et al. 2019). Introducing heteroatoms in the skeleton of POPs is an efficient methodology to enhance CO_2 capture capability even at low pressure.

Another method to improve the CO_2 capture is the codoping of different heteroatoms in POPs. The codoping of nitrogen and oxygen shows good selectivity towards CO_2/CH_4 at low pressure (Wu et al. 2020). Additionally, by a simple polycondensation reaction between 3,3′-diaminobenzidine and hexachlorocyclotriphosphazene, the heteroatoms like nitrogen and phosphorus can impinge into PECNOFs. The nitrogen and phosphorus groups make the skeleton of sorbent electron-rich and as CO_2 is an electron-poor moiety, therefore, Lewis acid-base interactions happened among both of them and subsequently exhibit high selectivity towards CO_2/N_2 and enhance the CO_2 uptake capacity (Mohanty, Kull, and Landskron 2011). The carbon dioxide capture capacity of PECNOF-1 is 1.86 mmol g^{-1}, PECNOF-2 is 2.85 mmol g^{-1} and PECNOF-3 is 3.49 mmol g^{-1} at 373 K and 1 bar (Wu et al. 2020).

13.6.5 Surface Functionalized POPs

The surface modification of POPs with various functional groups is an effective methodology to increase the CO_2 capture capability rely on chemisorption and physisorption. The organic functional groups and inorganic ions are usually used to modify the polarity of POPs to increase the interactions between sorbent and sorbate. These modifications play a crucial role in CO_2 uptake. These alterations in POPs can be achieved by two methods, either by post-synthetic method or by functional monomers in-situ for task-specific POPs.

13.6.5.1 Organic Functional Groups

The CO_2 adsorption and selectivity are one of the major concerns, organic functional groups impinged in the framework of POPs enhance the CO_2 capturing capability majorly due to improved interaction between adsorbent and adsorbate. The drawback of introducing a functional group is that the surface area is reduced but the CO_2 capture capacity is either improved or remains the same due to interaction between CO_2 and organic functional groups.

The amino (–NH_2) groups are considered as potential candidates for CO_2 capture. The introduction of an amino group within PPN-6 forms a series of amine-tethered POPs as PPN-6-CH_2EDA,

PPN-6-CH_2TETA and PPN-6-CH_2DETA (Lu et al. 2012). Although surface area decreased, the performance increased effectively. The trend towards the improvement of CO_2 adsorption is DETA > TAEA > TETA > EDA. The hydrophobicity of certain POPs is also an essential feature to capture CO_2. It allows the uptake of CO_2 in moist conditions as well. Triazine-phenol polymer HPF-1 possesses these properties. Triazine units and the phenol groups cooperatively capture CO_2 even in humid conditions (Kong et al. 2019).

The nitro (–NO_2) groups functionalized the networks of POPs and increase the adsorption of CO_2 even at low pressure. NO_2 groups in PAFs form NO_2-PAF-1 and its surface area 610 $m^2\ g^{-1}$ is lower than non-modified PAF-1 but the uptake capacity of CO_2 is 2.2 mmol g^{-1} which is higher than PAF-1.

Some POPs can be modified by the post-synthesis modification process to functionalize the framework for capturing CO_2. The NPOF-4 can be modified by –NO_2 and –NH_4 to obtain NPOF-4-NO_2 and NPOF-4-NH_4, and CO_2 uptake capacity increases 20% and 16%, respectively (Satar et al. 2019). The –NO_2 and –NH_4 have large dipole moments that cause dipole–quadrupole interactions with the gas CO_2 and consequently enhance CO_2 capture uptake.

Different kinds of functional groups can be incorporated into the POPs to enhance efficiency like –NO_2, –NH_2, –CH_3, –SO_2Cl and –OH. These successive POPs are achieved by the Yamamoto-type Ullmann cross-coupling reaction. Bifunctionalized (COP-21 (–OH and –NH_2)) and multi-functionalized (COP-22 (–OH, –NH_2 and SO_2Cl)) POPs can also be prepared and they are useful in order to improve CO_2 uptake capacity and selectivity (Balibar 2017).

In order to consider the isosteric heats of sorption for carbon dioxide, the calculations indicate that the incorporation of the carboxylic acid (–COOH) group into the POPs shows the highest isosteric heat. The order of functional groups on the isosteric heats of CO_2: –COOH > $(OH)_2$ > NH_2 > H > $(CH_3)_2$ (Bhanja, Modak, and Bhaumik 2019).

13.6.5.2 Inorganic Ions

Inorganic ions are used to alter the POPs structure so that isosteric heats and adsorption selectivity gets enhanced for capturing CO_2. Like organic functional moieties, the incorporation of inorganic ions reduces the surface area which affects its application but pore size usually comes in a good range. The CO_2 selectivity gets enhanced due to electrostatic interactions held between CO_2 and the respective inorganic ions.

Porous polymer network (PPN) embedded with sulfonic acid (PPN-6-SO_3H) and its lithium salt (PPN-6-SO_3Li) shows a significant decrease in BET surface area and pore size. The BET surface area for PPN-6 is 4023 $m^2\ g^{-1}$, it gets reduced from 1254 $m^2\ g^{-1}$ for PPN-6-SO_3H and 1186 $m^2\ g^{-1}$ for the lithium salt of sulfonic acid. The pore size get decreases in a range of 5–10 Å which is good for capturing CO_2. The grafted POPs have increased in CO_2 uptake as 3.6 mmol g^{-1} for PPN-6-SO_3H and 3.7 mmol g^{-1} for PPN-6-SO_3Li as compared to simple nongrafted PPN-6 as 1.4 mmol g^{-1} at 295 K and 1 bar (Li et al. 2017). Due to strong electrostatic interactions between CO_2 and polar moieties, the adsorption selectivity gets increases for CO_2/N_2. The salt of sulfonic acid shows greater performance as it possesses three open coordination sites once it is activated completely and, as a result, the electrostatic interactions increase between CO_2 and Li-cations (Wang et al. 2021). Sulfonate ammonium is also grafted into PPN-6 to get PPN-6-SO_3NH_4. The NH_4 group leaves the surface area only at 593 $m^2\ g^{-1}$. It exhibits a high CO_2 capture capacity at 295 K and 0.15 bar. The CO_2 and –SO_3NH_4 hold strong interactions which result in high adsorption selectivities of 796 mmol g^{-1} for CO_2/N_2 and 180 mmol g^{-1} for CO_2/CH_4 at 313 K and 1 bar.

The ions of transition metals such as Co, Ni and Cu replaced the H_2 in non-metal porphyrin (H_2P PON) to form isostructural metalloporphyrin porous organic networks (MP PONs). The BET surface area gets decreased in the order of CoP PON (908 $m^2\ g^{-1}$) > NiP PON (847 $m^2\ g^{-1}$) > CuP PON (757 $m^2\ g^{-1}$) as compared to H_2P PON (1477 $m^2\ g^{-1}$). Although there is a significant decrease in BET surface area, the CO_2 capture capacity is almost the same because of the electrostatic interactions between CO_2 molecules and positively charged metal ions.

13.7 SELECTIVITY OF CO_2

The framework of POPs is not only modified according to the uptake capacity of CO_2 but also the selectivity of CO_2 should be considered. The adsorbent should be more vulnerable to CO_2 as compared to other gas species. The flue gas from power plants usually consists of nitrogen, carbon dioxide, water, excess oxygen and small quantities of sulphur and nitrogen oxides. Particulate matters are also present in coal-fired plants. In order to achieve effective separation of CO_2 from other gases and particulate matters, the adsorbent should be selective towards carbon dioxide.

The interaction between the adsorbent and adsorbate must be strong, for the selectivity of CO_2, the greater quadrupole moment of carbon dioxide molecule can be the effective route for attaining high selectivity towards CO_2 as compared to the other gases. Usually, the incorporation of heteroatoms in the framework of POPs is an influential pathway to enhance the isosteric heat of adsorption between adsorbent and adsorbate. Heteroatoms like nitrogen, oxygen, sulphur and phosphorus are usually merged into the POPs skeleton to obtain a high yield of CO_2. CPOP-1, for example, possesses high selectivity towards CO_2/N_2 that can be described by the electron-rich nitrogen atoms in the framework and greater charge density in the nitrogen sites. These things simultaneously strengthen the dipole–quadrupole interactions among the polarizable CO_2 molecules (Huang, Krishna, and Jiang 2015).

Ideally, there must be a balance between CO_2 uptake capacity and selectivity or both should be in a direct relationship with each other. But that is not the case, most of the POPs materials show the inverse trend towards the CO_2 uptake capacity and selectivity. POPs with the highest CO_2 uptake capacity generally possess lower selectivity. For example, BILP-2 exhibits high selectivity towards CO_2/N_2 at 298 K, and at the same time, it possesses much low selectivity (Rabbani and El-Kaderi 2012). Similarly, PECONF shows the highest selectivity towards CO_2 but unfortunately, it shows the same inverse relationship among adsorption and selection (Mohanty, Kull, and Landskron 2011). But there is some special framework of POPs which do not follow this trend. For example, the network of PPN-6 bounded with amino groups is a potential candidate for high uptake capacity, selectivity and stability towards CO_2 from flue gas as compared to other gases and particulate matters (Lu et al. 2012).

The adsorbents may also be sensitive towards other gases from flue gas such as sulphur dioxide, nitrogen dioxide and water. They act as co-adsorbent towards the adsorbate. These competing adsorbents may hold the sorption sites and, as a result, affect the efficiency of the sorbent by blocking the CO_2 selective sites and consequently decreasing the CO_2 uptake capacity. When the polar groups are introduced in the POPs to enhance the CO_2 uptake then the hydrophilic POPs such as HCPs show that the molecules of water strive with the adsorbed CO_2 molecules (Wang et al. 2018). On the other hand, the hydrophobic POPs do not get affected by water vapour. To illustrate the correct uptake capacity of CO_2, the POPs materials should not be examined under dry conditions as they may not show the exact potential from real flue gas uptake (Jia et al. 2019).

13.8 CONCLUSION AND PROSPECTS

In conclusion, the physicochemical stabilities, high surface area, modifiable pore size and structure and easily tuned functionality make the POPs the potential and promising candidates for capturing carbon dioxide. They show a phenomenal versatility for CO_2 capture ability by modifying the physical and chemical parameters of POP materials. Their prestige as organic polymers provides them with a wide library of synthetic organic reactions. In order to functionalize these POP materials, the repository of organic reactions provides multitudinous avenues. Many factors are there to improve the affinity between sorbent and sorbate such as surface area, pore size and functionalized framework. The higher BET surface area of porous polymer network is crucial for CO_2 uptake at high pressure but at low pressure, their efficiencies get reduced. Chemically modified POPs show better performance even at pressure.

Among the other porous materials, POPs are distinguished as the prominent sorbents because of their relatively low cost. Although some POPs utilize expensive noble metals, new POPs can be synthesized at a cheap cost. Furthermore, noticeable progress has been observed in high-yielding reactions.

In the coming decades, the technologies to reduce CO_2 from the environment will be required to overcome the overall level of CO_2 emissions. POPs, these days, are considered as the standard materials for this problem. It has been seen that POPs can be functionalized easily to achieve remarkable CO_2 uptake values.

Although POPs possess good characteristics to be the adsorbent of CO_2 but still limited towards their facile synthesis to be utilized for an industrial scale, high stability is also an issue at chemical conditions; the surface pore size should be less than 100 nm to improve the efficiency; high BET surface area for CO_2 uptake and existence of copious CO_2-philic sites. These issues need to be addressed for the improvement in POPs materials and their application towards CO_2 uptake in order to clean the atmosphere.

REFERENCES

Abdelnaby, Mahmoud M, Kyle E Cordova, Ismail Abdulazeez, Ahmed M Alloush, Bassem A Al-Maythalony, Youcef Mankour, Khalid Alhooshani, Tawfik A Saleh, and Othman Charles S Al Hamouz. 2020. Novel porous organic polymer for the concurrent and selective removal of hydrogen sulfide and carbon dioxide from natural gas streams. *ACS Applied Materials & Interfaces* 12 (42):47984–47992.

Balibar, Sebastien. 2017. Energy transitions after COP21 and 22. *Comptes Rendus Physique* 18 (7–8):479–487.

Ben, Teng, Yanqiang Li, Liangkui Zhu, Daliang Zhang, Dapeng Cao, Zhonghua Xiang, Xiangdong Yao, and Shilun Qiu. 2012. Selective adsorption of carbon dioxide by carbonized porous aromatic framework (PAF). *Energy & Environmental Science* 5 (8):8370–8376.

Ben, Teng, Hao Ren, Shengqian Ma, Dapeng Cao, Jianhui Lan, Xiaofei Jing, Wenchuan Wang, Jun Xu, Feng Deng, and Jason M Simmons. 2009. Targeted synthesis of a porous aromatic framework with high stability and exceptionally high surface area. *Angewandte Chemie* 121 (50):9621–9624.

Bhanja, Piyali, Arindam Modak, and Asim Bhaumik. 2019. Porous organic polymers for CO_2 storage and conversion reactions. *ChemCatChem* 11 (1):244–257.

Budd, PM, ES Elabas, BS Ghanem, S Makhseed, NB McKeown, KJ Msayib, CE Tattershall, and D. Wang. 2004. *Advanced Materials* 16:456–459.

Choi, Hwa Seob, Hyung Joon Jeon, Jung Hoon Choi, Gyu-Heon Lee, and Jeung Ku Kang. 2015. Tailoring open metal sites for selective capture of CO_2 in isostructural metalloporphyrin porous organic networks. *Nanoscale* 7 (45):18923–18927.

Cote, Adrien P, Annabelle I Benin, Nathan W Ockwig, Michael O'Keeffe, Adam J Matzger, and Omar M Yaghi. 2005. Porous, crystalline, covalent organic frameworks. *Science* 310 (5751):1166–1170.

Das, Sabuj Kanti, Piyali Bhanja, Sudipta K Kundu, Saptarsi Mondal, and Asim Bhaumik. 2018. Role of surface phenolic-OH groups in N-rich porous organic polymers for enhancing the CO_2 uptake and CO_2/N_2 selectivity: Experimental and computational studies. *ACS Applied Materials & Interfaces* 10 (28):23813–23824.

Dawson, Robert, Dave J Adams, and Andrew I Cooper. 2011. Chemical tuning of CO_2 sorption in robust nanoporous organic polymers. *Chemical Science* 2 (6):1173–1177.

Dawson, Robert, Andrew I Cooper, and Dave J Adams. 2013. Chemical functionalization strategies for carbon dioxide capture in microporous organic polymers. *Polymer International* 62 (3):345–352.

Dawson, Robert, Andrea Laybourn, Rob Clowes, Yaroslav Z Khimyak, Dave J Adams, and Andrew I Cooper. 2009. Functionalized conjugated microporous polymers. *Macromolecules* 42 (22):8809–8816.

Dawson, Robert, Thanchanok Ratvijitvech, Matthew Corker, Andrea Laybourn, Yaroslav Z Khimyak, Andrew I Cooper, and Dave J Adams. 2012. Microporous copolymers for increased gas selectivity. *Polymer Chemistry* 3 (8):2034–2038.

Del Regno, Annalaura, Aleksandra Gonciaruk, Laura Leay, Mariolino Carta, Matthew Croad, Richard Malpass-Evans, Neil B McKeown, and Flor R Siperstein. 2013. Polymers of intrinsic microporosity containing tröger base for CO_2 capture. *Industrial & Engineering Chemistry Research* 52 (47):16939–16950.

Deng, Q-W, G-Q Ren, Y-J Li, L Yang, S-L Zhai, T Yu, L Sun, W-Q Deng, A Li, and Y-H Zhou. 2020. Hydrogen and CO_2 storage in high surface area covalent triazine-based frameworks. *Materials Today Energy* 18:100506.

El-Mahdy, Ahmed FM, Cheng-Han Kuo, Abdulmohsen Alshehri, Christine Young, Yusuke Yamauchi, Jeonghun Kim, and Shiao-Wei Kuo. 2018. Strategic design of triphenylamine-and triphenyltriazine-based two-dimensional covalent organic frameworks for CO_2 uptake and energy storage. *Journal of Materials Chemistry A* 6 (40):19532–19541.

Gao, Qiang, Xing Li, Guo-Hong Ning, Hai-Sen Xu, Cuibo Liu, Bingbing Tian, Wei Tang, and Kian Ping Loh. 2018. Covalent organic framework with frustrated bonding network for enhanced carbon dioxide storage. *Chemistry of Materials* 30 (5):1762–1768.

Garibay, Sergio J, Mitchell H Weston, Joseph E Mondloch, Yamil J Colón, Omar K Farha, Joseph T Hupp, and SonBinh T Nguyen. 2013. Accessing functionalized porous aromatic frameworks (PAFs) through a de novo approach. *CrystEngComm* 15 (8):1515–1519.

Ge, Rile, Dandan Hao, Qi Shi, Bin Dong, Wenguang Leng, Chang Wang, and Yanan Gao. 2016. Target synthesis of an azo (N=N) based covalent organic framework with high CO_2-over-N_2 selectivity and benign gas storage capability. *Journal of Chemical & Engineering Data* 61 (5):1904–1909.

Hansen, James, Reto Ruedy, Mki Sato, and Ken Lo. 2010. Global surface temperature change. *Reviews of Geophysics* 48 (4).

Hou, Shuangshuang, and Bien Tan. 2018. Naphthyl substitution-induced fine tuning of porosity and gas uptake capacity in microporous hyper-cross-linked amine polymers. *Macromolecules* 51 (8):2923–2931.

Huang, Ning, Gregory Day, Xinyu Yang, Hannah Drake, and Hong-Cai Zhou. 2017a. Engineering porous organic polymers for carbon dioxide capture. *Science China Chemistry* 60 (8):1007–1014.

Huang, Ning, Rajamani Krishna, and Donglin Jiang. 2015. Tailor-made pore surface engineering in covalent organic frameworks: Systematic functionalization for performance screening. *Journal of the American Chemical Society* 137 (22):7079–7082.

Huang, Ning, Ping Wang, Matthew A Addicoat, Thomas Heine, and Donglin Jiang. 2017b. Ionic covalent organic frameworks: Design of a charged interface aligned on 1D channel walls and its unusual electrostatic functions. *Angewandte Chemie International Edition* 56 (18):4982–4986.

Hug, Stephan, Maria B Mesch, Hyunchul Oh, Nadine Popp, Michael Hirscher, Jürgen Senker, and Bettina V Lotsch. 2014. A fluorene based covalent triazine framework with high CO_2 and H_2 capture and storage capacities. *Journal of Materials Chemistry A* 2 (16):5928–5936.

Jansen, Daniel, Matteo Gazzani, Giampaolo Manzolini, Eric van Dijk, and Michiel Carbo. 2015. Pre-combustion CO_2 capture. *International Journal of Greenhouse Gas Control* 40:167–187.

Jena, Himanshu Sekhar, Chidharth Krishnaraj, Guangbo Wang, Karen Leus, Johannes Schmidt, Nicolas Chaoui, and Pascal Van Der Voort. 2018. Acetylacetone covalent triazine framework: An efficient carbon capture and storage material and a highly stable heterogeneous catalyst. *Chemistry of Materials* 30 (12):4102–4111.

Jia, Jiangtao, Zhijie Chen, Hao Jiang, Youssef Belmabkhout, Georges Mouchaham, Himanshu Aggarwal, Karim Adil, Edy Abou-Hamad, Justyna Czaban-Jóźwiak, and Mohamed Rachid Tchalala. 2019. Extremely hydrophobic POPs to access highly porous storage media and capturing agent for organic vapors. *Chem* 5 (1):180–191.

Kong, Xueying, Shangsiying Li, Maria Strømme, and Chao Xu. 2019. Synthesis of porous organic polymers with tunable amine loadings for CO_2 capture: Balanced physisorption and chemisorption. *Nanomaterials* 9 (7):1020.

Krishna, Rajamani, and Jasper M van Baten. 2012. A comparison of the CO_2 capture characteristics of zeolites and metal–organic frameworks. *Separation and Purification Technology* 87:120–126.

Kuhn, Pierre, Markus Antonietti, and Arne Thomas. 2008. Porous, covalent triazine-based frameworks prepared by ionothermal synthesis. *Angewandte Chemie International Edition* 47 (18):3450–3453.

Kweku, Darkwah Williams, Odum Bismark, Addae Maxwell, Koomson Ato Desmond, Kwakye Benjamin Danso, Ewurabena Asante Oti-Mensah, Asenso Theophilus Quachie, and Buanya Beryl Adormaa. 2017. Greenhouse effect: Greenhouse gases and their impact on global warming. *Journal of Scientific Research and Reports*:1–9.

Li, Buyi, Ruini Gong, Wei Wang, Xin Huang, Wang Zhang, Huanmin Li, Chunxiao Hu, and Bien Tan. 2011. A new strategy to microporous polymers: Knitting rigid aromatic building blocks by external cross-linker. *Macromolecules* 44 (8):2410–2414.

Li, Haiying, Bo Meng, Shannon M Mahurin, Song-Hai Chai, Kimberly M Nelson, David C Baker, Honglai Liu, and Sheng Dai. 2015. Carbohydrate based hyper-crosslinked organic polymers with –OH functional groups for CO_2 separation. *Journal of Materials Chemistry A* 3 (42):20913–20918.

Li, Yankai, Li Yang, Xiang Zhu, Jun Hu, and Honglai Liu. 2017. Post-synthesis modification of porous organic polymers with amine: A task-specific microenvironment for CO_2 capture. *International Journal of Coal Science & Technology* 4 (1):50–59.

Livingston, Jasmine E, Eva Lövbrand, and Johanna Alkan Olsson. 2018. From climates multiple to climate singular: Maintaining policy-relevance in the IPCC synthesis report. *Environmental Science & Policy* 90:83–90.

Lu, Weigang, Julian P Sculley, Daqiang Yuan, Rajamani Krishna, Zhangwen Wei, and Hong-Cai Zhou. 2012. Polyamine-tethered porous polymer networks for carbon dioxide capture from flue gas. *Angewandte Chemie* 124 (30):7598–7602.

Luo, Shuangjiang, Qinnan Zhang, Yizhou Zhang, Kevin P Weaver, William A Phillip, and Ruilan Guo. 2018. Facile synthesis of a pentiptycene-based highly microporous organic polymer for gas storage and water treatment. *ACS Applied Materials & Interfaces* 10 (17):15174–15182.

Luo, Yali, Buyi Li, Wei Wang, Kangbing Wu, and Bien Tan. 2012. Hypercrosslinked aromatic heterocyclic microporous polymers: A new class of highly selective CO_2 capturing materials. *Advanced Materials* 24 (42):5703–5707.

Lyngfelt, Anders. 2020. Chemical looping combustion: Status and development challenges. *Energy & Fuels* 34 (8):9077–9093.

Mohanty, Paritosh, Lilian D Kull, and Kai Landskron. 2011. Porous covalent electron-rich organonitridic frameworks as highly selective sorbents for methane and carbon dioxide. *Nature Communications* 2 (1):1–6.

Muhammad, Hafiz Ali, Gilbong Lee, Junhyun Cho, Umair Hassan Bhatti, Young-Jin Baik, and Beomjoon Lee. 2019. Design and optimization of CO_2 pressurization system integrated with a supercritical CO_2 power cycle for the CO_2 capture and storage system. *Energy Conversion and Management* 195:609–619.

Rabbani, Mohammad Gulam, and Hani M El-Kaderi. 2012. Synthesis and characterization of porous benzimidazole-linked polymers and their performance in small gas storage and selective uptake. *Chemistry of Materials* 24 (8):1511–1517.

Rabbani, Mohammad Gulam, Timur Islamoglu, and Hani M El-Kaderi. 2017. Benzothiazole-and benzoxazole-linked porous polymers for carbon dioxide storage and separation. *Journal of Materials Chemistry A* 5 (1):258–265.

Ren, Shijie, Michael J Bojdys, Robert Dawson, Andrea Laybourn, Yaroslav Z Khimyak, Dave J Adams, and Andrew I Cooper. 2012. Porous, fluorescent, covalent triazine-based frameworks via room-temperature and microwave-assisted synthesis. *Advanced Materials* 24 (17):2357–2361.

Rubin, Edward S. 2008. CO_2 capture and transport. *Elements* 4 (5):311–317.

Saleh, Muhammad, Han Myoung Lee, K Christian Kemp, and Kwang S Kim. 2014. Highly stable CO_2/N_2 and CO_2/CH_4 selectivity in hyper-cross-linked heterocyclic porous polymers. *ACS Applied Materials & Interfaces* 6 (10):7325–7333.

Satar, Hind A, Ahmed A Ahmed, Emad Yousif, Dina S Ahmed, Mohammed F Alotibi, and Gamal A El-Hiti. 2019. Synthesis of novel heteroatom-doped porous-organic polymers as environmentally efficient media for carbon dioxide storage. *Applied Sciences* 9 (20):4314.

Socolow, Robert H. 2005. Can we bury global warming? *Scientific American* 293 (1):49–55.

Stanger, Rohan, Terry Wall, Reinhold Spörl, Manoj Paneru, Simon Grathwohl, Max Weidmann, Günter Scheffknecht, Denny McDonald, Kari Myöhänen, and Jouni Ritvanen. 2015. Oxyfuel combustion for CO_2 capture in power plants. *International Journal of Greenhouse Gas Control* 40:55–125.

Tan, Liangxiao, and Bien Tan. 2017. Hypercrosslinked porous polymer materials: Design, synthesis, and applications. *Chemical Society Reviews* 46 (11):3322–3356.

Vega, Fernando, Mercedes Cano, Sara Camino, Luz M Gallego Fernández, Esmeralda Portillo, and Benito Navarrete. 2018. Solvents for carbon dioxide capture. *Carbon Dioxide Chemistry, Capture and Oil Recovery*:142–163.

Wang, Guangbo, Karen Leus, Himanshu Sekhar Jena, Chidharth Krishnaraj, Shuna Zhao, Hannes Depauw, Norini Tahir, Ying-Ya Liu, and Pascal Van Der Voort. 2018. A fluorine-containing hydrophobic covalent triazine framework with excellent selective CO_2 capture performance. *Journal of Materials Chemistry A* 6 (15):6370–6375.

Wang, Han, Danni Jiang, Danlian Huang, Guangming Zeng, Piao Xu, Cui Lai, Ming Chen, Min Cheng, Chen Zhang, and Ziwei Wang. 2019. Covalent triazine frameworks for carbon dioxide capture. *Journal of Materials Chemistry A* 7 (40):22848–22870.

Wang, Tianxiong, and Baohang Han. 2020. Carbazole-based porous organic polymers for carbon dioxide capture and catalytic conversion. *Chinese Science Bulletin* 65 (31):3389–3400.

Wang, Wenjing, Mi Zhou, and Daqiang Yuan. 2017b. Carbon dioxide capture in amorphous porous organic polymers. *Journal of Materials Chemistry A* 5 (4):1334–1347.

Wang, Xiaoyan, Yang Zhao, Lingling Wei, Chong Zhang, and Jia-Xing Jiang. 2015. Nitrogen-rich conjugated microporous polymers: Impact of building blocks on porosity and gas adsorption. *Journal of Materials Chemistry A* 3 (42):21185–21193.

Wang, Xueyuan, Ting He, Junhua Hu, and Min Liu. 2021. The progress of nanomaterials for carbon dioxide capture via the adsorption process. *Environmental Science: Nano*.

Wang, Yuan, Li Zhao, Alexander Otto, Martin Robinius, and Detlef Stolten. 2017a. A review of post-combustion CO_2 capture technologies from coal-fired power plants. *Energy Procedia* 114:650–665.

Wu, Dawei, Jing Liu, Yingju Yang, and Ying Zheng. 2020. Nitrogen/oxygen co-doped porous carbon derived from biomass for low-pressure CO_2 capture. *Industrial & Engineering Chemistry Research* 59 (31):14055–14063.

Xu, Chao, Guipeng Yu, Jiayin Yuan, Maria Strømme, and Niklas Hedin. 2020. Microporous organic polymers as CO_2 adsorbents: Advances and challenges. *Materials Today Advances* 6:100052.

Xu, Guangjuan, Yiang Zhu, Wei Xie, Shuran Zhang, Chan Yao, and Yanhong Xu. 2019. Porous cationic covalent triazine-based frameworks as platforms for efficient CO_2 and iodine capture. *Chemistry–An Asian Journal* 14 (19):3259–3263.

Yang, Zhenzhen, Hongye Zhang, Bo Yu, Yanfei Zhao, Zhishuang Ma, Guipeng Ji, Buxing Han, and Zhimin Liu. 2015. Azo-functionalized microporous organic polymers: Synthesis and applications in CO_2 capture and conversion. *Chemical Communications* 51 (58):11576–11579.

Yuan, Ye, and Guangshan Zhu. 2019. Porous aromatic frameworks as a platform for multifunctional applications. *ACS Central Science* 5 (3):409–418.

Zeng, Yongfei, Ruqiang Zou, and Yanli Zhao. 2016. Covalent organic frameworks for CO_2 capture. *Advanced Materials* 28 (15):2855–2873.

Zhang, Chong, Xiao Yang, Yang Zhao, Xiaoyan Wang, Miao Yu, and Jia-Xing Jiang. 2015. Bifunctionalized conjugated microporous polymers for carbon dioxide capture. *Polymer* 61:36–41.

Zhang, Weichao, Yuanhui Cheng, Chunshuai Guo, Chengpeng Xie, and Zhonghua Xiang. 2018. Cobalt incorporated porous aromatic framework for CO_2/CH_4 separation. *Industrial & Engineering Chemistry Research* 57 (32):10985–10991.

Zhi, Yongfeng, Pengpeng Shao, Xiao Feng, Hong Xia, Yumin Zhang, Zhan Shi, Ying Mu, and Xiaoming Liu. 2018. Covalent organic frameworks: Efficient, metal-free, heterogeneous organocatalysts for chemical fixation of CO_2 under mild conditions. *Journal of Materials Chemistry A* 6 (2):374–382.

Zhou, Hong-Cai, Jeffrey R Long, and Omar M Yaghi. 2012. Introduction to metal–organic frameworks. *Chemical Reviews* 112 (2):673–674.

Zhu, Xiang, Shunmin Ding, Carter W Abney, Katie L Browning, Robert L Sacci, Gabriel M Veith, Chengcheng Tian, and Sheng Dai. 2017. Superacid-promoted synthesis of highly porous hypercrosslinked polycarbazoles for efficient CO_2 capture. *Chemical Communications* 53 (54):7645–7648.

Zou, Lanfang, Yujia Sun, Sai Che, Xinyu Yang, Xuan Wang, Mathieu Bosch, Qi Wang, Hao Li, Mallory Smith, and Shuai Yuan. 2017. Porous organic polymers for post-combustion carbon capture. *Advanced Materials* 29 (37):1700229.

Zulfiqar, Sonia, Muhammad Ilyas Sarwar, and David Mecerreyes. 2015. Polymeric ionic liquids for CO_2 capture and separation: Potential, progress and challenges. *Polymer Chemistry* 6 (36):6435–6451.

14 Antibacterial Applications of Porous Polymers

Jhonatas Rodrigues Barbosa
Federal University of Para (UFPA), Belém, Brazil

Luiza Helena da Silva Martins
Federal Rural University of the Amazon (UFRA), Belém, Brazil

Johnatt Allan Rocha de Oliveira and Raul Nunes de Carvalho Junior
Federal University of Para (UFPA), Belém, Brazil

CONTENTS

14.1 INTRODUCTION

Porous polymers can be synthesized using various polymeric protocols and materials, including synthetic and natural. Porous polymers with antimicrobial activity are applied in the most varied fields, including in the food and pharmaceutical industry. Although there are a variety of synthesis techniques and protocols, the application of the polymer (food or drugs) should govern the entire synthesis route, including the choice of reagents for the polymerization reaction (Min et al., 2021).

Currently, there is a worldwide demand for new, technological, cheaper, and functional antimicrobial products. In this increasingly competitive market, porous polymers with antimicrobial activity are a high-tech option, with functional and adjustable properties. Depending on the composition of the polymer, the antimicrobial activity can be derived from specific antibiotics and bioactive

DOI: 10.1201/9781003169604-14

trapped in the pores or even from the polymer's own molecular structure, which has biological properties. In addition, pores with dimensions at the nanomolecular level provide porous polymers with extra advantages compared to products currently marketed, including a higher rate of antimicrobial activity and better intracellular absorption by pathogens (Tian et al., 2021).

Porous polymers are applied in the food industry to increase the shelf life of food, reduce rates of microbial contamination, and avoid economic losses. Porous antimicrobial polymers applied in food are produced mainly with natural polymers, including polysaccharides of plant, bacterial and fungal origin, or agro-industrial residues (Sousa et al., 2017). In general, antimicrobial porous polymers cannot be toxic and must be biocompatible with human metabolism. Finally, porous antimicrobial polymers are applied in the form of films, aerogel, foams, and particulate materials. In addition, antimicrobial activity is controlled by the physicochemical and structural properties of the polymer, reducing the negative effects of long-term exposure and increasing the viability of food products (Díez-Pascual & Díez-Vicente, 2014).

Porous polymers are also applied in the pharmaceutical industry and medical procedures, increasing the chances of patients recovering and minimizing side effects. In the synthesis process, several protocols and materials are applied for the polymerization reaction, including organic, inorganic, and complex polymeric materials associated with natural products (W. Zhang et al., 2019). The use of porous antimicrobial polymers in surgery, like dressings, has been extensively studied and several materials have already been produced including mixed polymer structures (Zheng, Pan, Liu, & Ren, 2021). Although it is possible to develop numerous materials based on porous polymers with antimicrobial activity, there are important challenges to be considered, including the constant threat of artificial selection of more resistant microorganisms. Therefore, many studies have sought new synthesis strategies and new structure designs for porous polymers with greater efficiency (Yang et al., 2019).

In this context, the main objective of this book chapter is to make a simplified and objective review of the literature, exploring the antimicrobial potentials of porous polymers and their applications. The approach is direct and starts by evaluating the main synthesis strategies for the development of porous polymers with antimicrobial activity. Then, some porous biodegradable and biocompatible polymers are explored as promising alternatives for antimicrobial applications. Finally, we explore in a simplified and objective way the application of pore polymers with antimicrobial activity in the food and pharmaceutical industry.

14.2 DEVELOPMENT OF POROUS POLYMERS WITH ANTIMICROBIAL POTENTIAL

In recent years, the field of polymer science has made great strides, with an unprecedented expansion of various methodologies aimed at the synthesis of new porous polymers, such as direct modeling techniques, direct synthesis methodologies, and self-assembly of block copolymers. The direct modeling methodology is a simple technique, based on the casting processes, and molding of the material, following a design concept, similar to that applied in the production of plastic packaging, however, adapted for the production of materials in the nanoscale (Wu et al., 2012).

The direct synthesis methodology is used for the production of pores even during the manufacture of the polymer, followed by the removal of solvents from the pores by drying techniques. The technique of direct synthesis is applied in the production of polymers with high surface area, permanent porosity, and for the formation of block systems (Wu et al., 2012). The block copolymer self-assembly methodology is applied for the design of mesoporous or macroporous polymers, including microspheres and mesoscale structures. In addition to these aforementioned methodologies, others frequently applied are high internal phase emulsion polymerizations and interfacial polymerization. In this topic, we will cover only the three main methods initially mentioned, and we will make a simplified approach, summarizing the most relevant aspects.

14.2.1 Direct Modeling Methodology

The application of this technique has led to the development of numerous porous polymers, including spherical porous polymers (example: PMAA spherical nanoparticles, poly (methacrylic acid); PCL, polycaprolactone; PS, polystyrene; MF, acid-decomposing formaldehyde melamine particles; among others), porous tubular polymers (example: PC tubular membranes, polycarbonate; TiO_2; Na_2SO_4 nanofibers, ZnO, V_2O_5, among others). In addition to porous ordered polymers (example: PMAA colloidal crystals, poly (methacrylic acid); PS, polystyrene, among others) there are others such as mesoporous silica and silica particles.

There are several requirements for the synthesis of porous polymers with the direct modeling technique. According to a review article published in Chemical Reviews by Wu et al. (2012), we can summarize the main requirements:

- The surface materials of the molds must be compatible with polymeric materials and their reagents used in the synthesis so that the porous polymer formed is a perfect mold.
- When the surface of the molds is not compatible, it is necessary to modify these surfaces to form target sites for the formation of polymers and the growth of polymeric networks.
- The molds must have well-defined, adjustable, and predetermined structures, which help in the construction of rational polymerization projects and the formation of porous polymers.
- Molds must be easy to remove, for example, silica molds that can be removed by chemical corrosion.
- Polymeric networks, regardless of whether they are linear or branched, must be robust enough to withstand the high-energy rates and strong capillary stresses that can disrupt the pores, leading to the collapse and disorder of the structure.
- The wall of the porous polymers must support specific and selectable modifications, such as techniques for building responsive structures or even adding specific chemical groups.

Wu et al. (2012) categorized the direct modeling procedures into two groups, according to the preparation of the polymeric walls, that is, with prefabricated polymers or with monomers. When prefabricated polymers are used, two techniques are applied, that is, infiltration techniques and layer-by-layer assembly. When monomers are used, the following techniques are applied: conventional polymerization, electrochemical polymerization, and controlled/living polymerization techniques. The purpose of this topic is not to detail all the synthesis techniques, but the reader must understand that the listed techniques have subdivisions with their own methodologies.

14.2.2 Direct Synthesis Methodologies

The pores in a polymer are formed by removing molecular components called sacrificial components, which help to design precisely sized pores, including mesopores and macropores. Although there are some disadvantages associated with the use of sacrificial components as a starting material for the development of pores, the direct synthesis technique allows the preparation of numerous porous polymers only using specific polymerization reactions. Two steps can be observed in direct synthesis: the formation of polymers by simple polymerization reactions, followed by removal of the solvent from the pores. The direct synthesis technique is efficient for the production of micropores with a high surface area, including hierarchical porous polymers and mesoporous and macroporous polymers (Sun & Deng, 2005).

Porous polymers produced by the direct synthesis technique can be designed from polymerizable monomers or through polymers with a high degree of crosslinking. The choice of the monomer is crucial and must be done carefully to ensure the stability of the polymerizing structure. Generally, when we want to synthesize microporous polymers, the chosen monomers have two functionalities, that is, to have several target locations for multifunctional reactions and structures with rigid properties to form linear polymers. Another relevant issue is the route of chemical synthesis, which must be efficient to connect several building blocks to the monomer, in this regard the polycondensation technique is often applied (Wang et al., 2019).

The direct synthesis procedures for the formation of responsible micropores can be summarized as follows: the porous polymer is derived from the packaging of the polymeric chain spaces without a firm mesh or by the chain networks of a highly cross-linked polymer. In this case, micropores are formed by the collapse of the pores through capillary pressure and high surface energy. Therefore, the use of rigid and twisted chains, or cross-linked structures, can collapse, bend and twist the polymer chains, resulting in micropores (Fu et al., 2017).

Now for the formation of mesopores and macropores, the process must involve phase separation, induced by reactions, which form three-dimensional polymeric gels. For the formation of macropores, the stiffness of the polymeric network must be reduced, thus the micropores collapse, leading to the packaging of the space forming a gel structure, depending on the phase separation behavior. The use of this technique leads to the formation of mesopores and macropores, due to the compact aggregation of the polymeric network, which can vary from a scale of micrometers to nanometers. In addition, the two techniques can be applied together to form hierarchical porous polymers, combining a structure with micropores, mesopores, and macropores (Xu et al., 2011).

14.2.3 Block Copolymer Self-assembly Methodologies

Block self-assembly is a sophisticated strategy, idealized from the observation that nature uses self-assembly to build complex and functional biological structures. In nature, self-assembly is built from the interactions between different polymers by connections of secondary forces, such as weak interaction forces. Currently, it is considered that the block copolymer strategy developed for self-assembly is one of the most sophisticated, elegant, and ideal strategies for the construction of porous polymers with well-defined mesopores and macropores. In this synthesis strategy, two chemically immiscible polymers are used, covalently linked to each other. The self-assembly process starts with the principle of thermodynamic incompatibility between polymers, which induces phase separation due to the difference in contact energy (Wan, Shi, & Zhao, 2007).

The function of the block copolymers depends on the researcher's interest; therefore, two strategies can be used. At first, the block copolymers are used as a mold for the formation of the pore, that is, they serve as a self-maintained base with prepolymers (or polymeric precursors), and then the block copolymer is removed, forming pores. The other form is more complex, with several steps for the formation of pores. Here, block copolymers serve as a basic polymeric structure for the formation of pores. Then, a series of methods such as selective corrosion, removal of additional components, physical reconstruction of the morphology with solvents, and selective crosslinking are applied for the development of hollow pores or structures. This technique has been extensively studied for the formation of complex structures such as porous polymeric thin films and structured monoliths (Wan, Shi, & Zhao, 2008).

Self-assembly with block copolymers such as the pore model has been used mainly in the production of periodic mesoporous inorganic materials, including silica and carbons. The procedure consists of several steps, including the construction of supramolecular blocks using copolymers in amphiphilic blocks to interact with specific prepolymers through covalent bonds. In addition to the supramolecular structure, prepolymers can be crosslinked to form a stable structure. The pore morphologies can be ordered or disordered, depending on how the procedure is done, but the pores have well-defined sizes, are small, in the range of 3–10 nm, and form large surface areas (Septani et al., 2020).

There are numerous self-assembly models with block copolymers, each with specificity for controlled pore development, although it is not possible to speak in detail with everyone, readers can consult the article Wu et al. (2012), where numerous synthesis models are presented in detail. However, it is important to highlight that the development of polymers with ordered mesopores is possible through self-assembly techniques with block polymers, including various types of resins, such as phenolic resins, melanin resins, urea-phenol-formaldehyde resin, phenol-formaldehyde resin, and resorcinol-formaldehyde resin (Sun, Huang, Lin, & Cheng, 2017).

14.3 SOME POROUS BIODEGRADABLE AND BIOCOMPATIBLE POLYMERS FOR ANTIMICROBIAL APPLICATIONS

The wall material and the synthesis techniques used in the development of porous polymers with antimicrobial properties depend a lot, on where this material will be applied. Thus, the application of porous polymers in the medical and food fields depends on some essential characteristics such as biodegradability, biocompatibility, and low toxicity. We will discuss below the development of some porous polymers that can be functionalized with antimicrobial activity, these are: polymeric aerogels, bioaerogels, and polymeric foams.

14.3.1 Polymeric Aerogels, Bioaerogels, and Polymeric Foams

The use of oil and its derivatives in recent years has contributed to the increase in environmental pollution, as pointed out by recent research. It is suggested that this problem be faced with a new scientific philosophy on the part of the academic and industrial community, considering the increase in productivity associated with the reduction of its negative impacts through new technologies and biodegradable materials. Therefore, the development and use of biodegradable polymers have been strongly encouraged to reduce the waste of strategic resources, minimize environmental impacts, and make individuals aware of the use of new practices, considered more beneficial to the sustainable development of civilization (Freitas et al., 2021).

Biodegradability is the term used to define the biological degradation capacity of material through the metabolism of living organisms to its basic components such as water, oxygen, carbon dioxide, and other basic chemical elements. The development of porous biodegradable polymers, with antimicrobial potential, has been explored and some results with natural polymers have attracted attention. These porous polymers include bioaerogels, polymeric aerogels, and porous foams (Barbosa, da Silva Martins, & de Carvalho Junior, 2021).

Bioaerogels are a class of porous polymers derived from natural polysaccharides and obtained from residues, algae, fungi, bacteria, and plants. Among numerous polymers of carbohydrate obtained, we have chitosan, alginate, glucans, starch, agar, and others such as cellulose and its derivatives (Barbosa & de Carvalho Junior, 2020). Bioaerogels are biodegradable materials, with promising physicochemical properties such as low density, thermal resistance, high porosity, high surface area, ultralight material, low dielectric constant, and a promising candidate as a drug delivery material, including antibiotics and other medicines (Barbosa, Pinto, da Silva Martins, & de Carvalho Junior, 2021).

Polymeric aerogels and polymeric foams can be obtained through synthetic techniques, as previously reported, and are biodegradable materials, formed by modified natural polysaccharides such as carboxymethyl cellulose-coated polypropylene (da Silva Martins et al., 2021) or with other materials like poly (lactic acid) (Jeong, Park, & Kim, 2020; Milovanovic et al., 2019; Standau et al., 2019) polycaprolactone (Campardelli, Franco, Reverchon, & De Marco, 2019; K. Zhang et al., 2019), poly (vinyl alcohol) (Yin et al., 2019), poly (butylene succinate) (Chen, Hu, Ju, & Kuang, 2019), poly (ethylene oxide) (W. Zhang et al., 2019), poly (lactic-co-glycolic acid) (Milovanovic et al., 2019), poly (ethylene glycol) (Asikainen et al., 2019) and poly (propylene carbonate) (Liu et al., 2019).

Polymeric aerogels in conjunction with polymeric foams are highly useful materials, especially considering their applications in the medical, food, aerospace, electronic circuits, and others. Although these porous materials have unique properties and are biodegradable, some structural problems of polymeric aerogels and single-component polymeric foams limit their application (Barbosa, da Silva Martins & de Carvalho Junior, 2021). Therefore, the development of new polymer-based hybrid materials is seen as a new approach to improve the properties of these porous polymers. Thus, these porous polymers could be designed and applied more efficiently, especially considering the recent demands for new materials to load antimicrobial drugs or the demand for the construction of new components for space exploration, including new drugs designed to withstand the conditions of low gravity (Pinto et al., 2020).

14.3.2 Polymeric Aerogels, Bioaerogels, and Bio-based Polymeric Foams are Biocompatible and non-Toxic

Polymeric aerogels, bioaerogels, and bio-based polymeric foams are mostly considered biocompatible and low toxicity materials, mainly because bio-based polymers are used, such as natural polysaccharides (Barbosa, Pinto, da Silva Martins, & de Carvalho Junior, 2021). The fundamental issue for biocompatibility is the application of polymers that are recognized by biological systems and that do not cause damage or deformation to the macromolecules that make up the cells. Thus, natural carbohydrate polymers have associated evolutionary advantages, as the species evolution process has consolidated a wide base of carbohydrate polymers that are recognized by the cellular system. Carbohydrate polymers obtained from natural sources still have other advantages, such as related biological activities such as antioxidant activity (Barbosa et al., 2020; Barbosa, dos Santos Freitas, da Silva Martins, & de Carvalho Junior, 2020), anti-inflammatory (Song et al., 2020), antimicrobial (Jridi et al., 2014) and immunoprotective (Barbosa & de Carvalho Junior, 2021).

Wang et al. (2021) developed a hybrid chitosan/mesoporous silica aerogel with bactericidal properties and presented biocompatibility for applications in hemorrhage media after surgical operations. The highly porous, low-density aerogel was obtained by freeze-drying. After chlorination by the addition of N-halamine precursor monomer, (3-acrylamidopropyl) trimethylammonium chloride (APTMAC) bound to mesoporous silica, the hybrid aerogel showed strong bactericidal activity against *Staphylococcus aureus* and *Escherichia coli* bacteria. The biocompatibility of aerogels is demonstrated by the excellent adhesion of erythrocytes and platelets to the surface of aerogels. Y. Zhang et al. (2019) based on functional nanocrystalline cellulose nanocomposites also produced similar aerogels. The authors also demonstrate that chlorinated aerogels have excellent bactericidal properties against *S. aureus* and *E. coli* bacteria. This type of aerogel can be applied to remove polluting compounds and to reduce the microbial load of water supply systems.

14.4 APPLICATIONS OF POROUS ANTIMICROBIAL POLYMERS IN THE FOOD INDUSTRY

The use of polymers with antimicrobial activity has become an increasingly relevant strategy for food preservation, through the use of these materials for the manufacture of packaging with properties to reduce the processes of food spoilage. As a result, the number of polymer-based packaging developed with desired properties is increasing.

The development of polymers with the incorporation of antimicrobial agents is one of the most viable strategies to control food contamination or target-specific food pathogens to improve and guarantee food safety and quality. Polymers directed to the control of microbiota in food may have been developed in a way that is more and more adjusted to these purposes, especially when we are talking about food packaging, in which polymers can be applied on food contact surfaces or as coatings edible to increase the shelf life of fresh food.

Food-grade polymers can be custom-designed to improve food safety, either as new food packaging materials, which confer active antimicrobial properties, applied to food contact surfaces to prevent cross-contamination, or as edible coatings to increase life fresh produce. The incorporation of antimicrobial agents in food-grade polymers can be used to control the contaminating microbiota and even to direct the microbiological safety of food, with improvements in food quality.

According to Díez-Pascual (2020), these functional composites use the combination of reinforcement phases with antimicrobial agents in a polymer matrix. To obtain the desired function, the active agents chosen as metallic nanoparticles, essential oils, or natural extracts must be defined before the construction of the polymer, according to the desired function. The packaging frequently developed with these polymers can be described as active packaging and among the main examples of interaction with the products include carbon dioxide emitters or absorbers, moisture absorbers, oxygen scavengers, ethylene absorbers, ethanol emitters, flavor release/absorption, and films containing antimicrobials and for smart packaging include time-temperature indicators, leakage indicators, and freshness indicators (Ozdemir & Floros, 2004).

Another way to incorporate the antimicrobial agent into polymers is by melting or by solvent composition. In addition to extrusion, another method of incorporating the agent is injection molding, used for thermally stable antimicrobials, such as metals (Appendini & Hotchkiss, 2002). In the case of thermolabile antimicrobial agents, which could not withstand high polymer processing temperatures, the agents can simply coat the materials, which requires the functionalization of the surface to improve the adhesion of the coatings (Radheshkumar & Münstedt, 2006).

Another technology that has been drawing attention is the application of nanoparticles or inorganic microparticles, which have been widely used with antimicrobial function, due to their high thermal stability, longevity, and a broad spectrum of antifungal and antibacterial activities (Mousavi Khaneghah et al., 2018). According to Imran et al. (2012), the combination of technologies such as nanoencapsulation and immobilization of biopolymer makes it possible to manufacture biodegradable antimicrobial films. Among the inorganic nanoparticles most widely incorporated into packaging, systems include silver, titanium dioxide (TiO_2), zinc oxide (ZnO), copper and gold nanoparticles, nanoclay, and carbon nanotubes, most of which have significant antimicrobial properties (Mousavi Khaneghah et al., 2018).

One of the problems found in antimicrobial films is that the active agent inserted in the constituent polymer may lose part or all of its antimicrobial activity when the film is incorporated. Thus, it is necessary that the polymer used undergo treatment or that the active agents undergo derivatization, before its addition to the polymer, to improve the compatibility between the active agent and the polymer. The immobilization of peptides with antimicrobial activity on the surface of polymeric films is possible with the modification of their surface characteristics with chemical methods (Haynie et al., 1995). This process prevented the transfer or migration of antimicrobial substances from the polymer to the food. According to Cohen et al. (1995) modify the composition of the polymer surface by electron irradiation, so that the surface contains amine groups, exhibited antimicrobial activity about the microorganisms in contact. Antimicrobial agents incorporated into polymers interact with the surface to prevent microbial growth from food sources, in addition to being able to control the diffusion rate of antimicrobial agents (Gharsallaoui et al., 2016).

An important step in the constitution of a polymer is the choice of agents that will make it up, which can improve the properties of the polymer, such as, for example, increasing its moisture barrier. According to Sanchez-Garcia et al. (2008) the crystalline fibers when inserted in the polymers, increase the tortuosity of the material in which they are inserted and lead to slower diffusion processes, with decreased permeability. In packaging films, some agents, such as TiO_2, can reduce the action of UV radiation on food, this aspect is extremely important, as this type and radiation can initiate a series of damaging transformations in food, such as oxidative rancidification.

Another study found that orange juice stored at 4 °C, in films incorporated with P105 (mixture of TiO_2 and silver nanoparticles (10 nm)) showed a significant reduction in the development of *Lactobacillus Plantarum*. Other nanoparticles such as MgO and ZnO, when inserted in the polymer, also had a positive effect in reducing the growth of microorganisms such as *L. Plantarum* and *Campylobacter jejuni* respectively, while ZnO showed bactericidal and not only bacteriostatic activity (Xie et al., 2011; Tang et al., 2012).

Polymeric films for packaging are easy to produce and have excellent performance, which is why they are commonly used (Silvestre et al., 2011). Polymeric food packaging can protect food from deterioration and contamination caused by microorganisms, and still demonstrate good technological properties such as flexibility, resistance, stiffness, and a barrier to oxygen and moisture. This type of antibacterial packaging can be classified as natural antimicrobial packaging and synthetic organic antimicrobial packaging, according to the source of antimicrobial agents used in the polymers.

According to Huang et al. (2019), the development of biodegradable materials has been promising with the improvement of biodegradable antimicrobial polymers such as polylactic acid (PLA), cellulose, starch, and chitosan. In the case of chitosan, the incorporation of essential oil of thyme in films made with both components showed antioxidant and antibacterial activity against four bacterial strains, which can be very interesting for food and even for the medical industry. Another type of material intended for food is cellulose acetate films and poly (vinyl alcohol) nanocrystalline membranes – cellulose, which, when impregnated in supercritical carbon dioxide with thymol, were able to prevent the formation of biofilms, enabling the development of active food packaging (Alvarado et al., 2018).

The impregnation of cellulose nanofibril structures with a natural active molecule, thymol, and the use of supercritical carbon dioxide as a green alternative to the common solvent has proved to be an excellent ecologically correct alternative for the development of polymers with antimicrobial activity. With high potential for application in food, this same technology would allow the impregnation of the polymer with other phenolic compounds, such as gallic acid, eugenol, or linalool (Darpentigny et al., 2020).

The antimicrobial abilities of food packaging depend on antimicrobial agents, which can be classified as natural antibacterial agents and synthetic antibacterial agents based on their sources. Natural antibacterial agents are the agents that will confer the developed packaging, its preservative, and protective properties for food. The use of natural antimicrobials has been strongly recommended because they are considered safer and easier to obtain than synthetic antibacterial agents. Natural agents can be obtained from nature, through sophisticated extraction processes, sometimes from rare and scarce raw materials (Huang et al., 2019).

Thus, with the growth of the population and the growing need for packaging, natural antimicrobial materials have become insufficient. Thus, organic or inorganic synthetic antimicrobial agents have been developed and grown on the market. It is still verified today that the main antibacterial components in food packaging are synthetic organic materials, which mainly contain ethylenediaminetetraacetic acid (EDTA), fungicides, parabens, and other chemicals (Suhr & Nielsen, 2003).

14.5 APPLICATIONS OF POROUS ANTIMICROBIAL POLYMERS IN THE PHARMACEUTICAL INDUSTRY

Many are porous polymers that can be applied to the pharmacy industry, in this topic we will summarize some of the most used ones focusing on some diseases and wounds. We will also talk about its properties, pros, and cons.

PLA – it is generally used to heal wounds, this material has excellent therapeutic properties, it is biocompatible, it has good stability against temperatures, in addition to having an environmental appeal, as it is biodegradable and is environmentally friendly. The limitations of PLA would be its low toughness, low degradation rate, and it is a very hydrophobic compound. However, such limitations can be solved with the use of other polymers in the mixture such as gelatins, poly (L-lactide), or PLLA, which can control water loss and promote fluid drainage, and are also considered biocompatible elements (Phaechamud & Mahadlek, 2015).

Chitosan – comes from the partial deacetylation of chitin, which is a natural polysaccharide responsible for the shell structure of crustaceans. Such polymer has several advantages such as being non-toxic, biodegradable, biocompatible, bioactive, non-antigenic, antimicrobial, and it can be applied in medicine acting mainly as a skin dressing. The use of biocompatible membranes for guided tissue regeneration (GTR) has been used for the treatment of periodontitis; such a structure when coupled with chitosan has excellent biomedical properties, being one of its potential uses. However, this Chitonas-GRT coupling still has limitations such as low bioactivity (which helps in the regeneration of bone tissue), there is also the possibility of the formation of bacterial biofilms, but to overcome this obstacle, calcium and phosphate mineral structures can be coupled to this structure. Silver nanoparticles that in addition to having antimicrobial properties also help in increasing bone tissue regeneration (Hurt, Kotha, Trivedi, & Coleman, 2015).

Porous surface with a recently introduced slippery liquid infusion (SLIPS) – SLIPS was studied by Li et al. (2013) presenting high potential against the opportunistic microorganism *Pseudomonas aeruginosa*, allowing the prevention of biofilm formation due to its low surface energy, in addition to its self-cleaning properties. In the study by Li et al. (2013) the poly (butyl methacrylate-co-ethylene dimethacrylate) infused in the hydrophobic liquid (slippery BMA-EDMA) was used with an antimicrobial character in the mineral BM2 and high long-term stability in aqueous environments. Thus, they were tested in different strains of the opportunistic pathogen *P. aeruginosa* for seven days in a low-nutrient medium.

Antimicrobial peptides (AMPs) – these polymers mimic natural peptides and can act as protectors of the body's defense components against various opportunistic microorganisms, acting mainly in the processes of immunomodulation and inflammation. These polymers act against several microorganisms through mechanisms such as cell lysis, interfering with the metabolism and targeting of cytoplasm components. The structure of these compounds is based on hydrophilic and hydrophobic groups allowing the best solubilization of the molecule in aqueous and lipid media. Its limitation would be the question of the high cost of its production, in addition, it is easily susceptible to proteolysis, and because it is an overly complex molecule, it is a little complicated to understand its goodness. AMPs are promising molecules to act as antimicrobial agents, as they present a low risk of microbial resistance, a broader spectrum of action, are more specific to the target, have high efficacy and good synergism with antibiotics (Li et al., 2021).

Pyridinium polyvinylpyrrolidones – are synthetic polymers made with different counter-anions, this polymer was tested in the pharmacy and biomedicine field as an antiviral against the influenza virus using the plaque assay technique. The polycations in this study showed antiviral activity of up to 95.6% in virucidal efficiency. The test against bacteria on the same material was performed using *E. coli*. It was found that polyvinylpyrrolidones of the pyridinium type play against viruses and bacteria in a wide spectral range of such microorganisms (Xue & Xiao, 2015).

Starch biopolymer – starch is a polysaccharide present in vegetables, it is abundant and widely used as platforms in many industries due to the presence of its hydroxyls (OH), due to its natural origin, it has advantages such as being anti-allergic, non-toxic, and biodegradable, and low cost. When compared to chitin it has a greater anti-allergenic characteristic. Chemically, starch is composed of anhydroglycosis units, with linear (amylose) and branched (amylopectin) structures; the variation between amylose and amylopectin will vary from plant to plant (Venkataraman et al., 2019).

Starch has a high gel-forming capacity and thus can be used to produce biopolymers with applications in the pharmacy when they are coupled with antimicrobial substances that can be essential oils (OEs) among other compounds. As a disadvantage, working with starches still needs improvements regarding the physical properties of the polymeric material as well as its hygroscopicity (Syafiq et al., 2020).

Polymeric micelles – micelles are nanostructures in size from 50 to 200 nm that help in the formation of nanostructures that can be used as encapsulating agents for poorly soluble drugs. Micelle-based polymers have numerous applications such as easy preparation, kinetics-controlled liberation, and good drug delivery agent. Some drugs used to fight cancer, such as paclitaxel, doxorubicin, 5-fluorouracil, 9-nitrocamptothecin, cisplatin, triptorelin, dexamethasone, and xanthone, have already been successfully encapsulated in poly (D, LL-lactate-co-glycolide) (PGLA). The use of multifunctional nanospheres coupled to micelles such as gold (Au) or bio-functionalized CdSe can also be used with good potential for use in the pharmacy area (Moreno-Vega et al., 2012).

Ceramic polymers – They have been widely used in medicine in recent years and their use has been growing more and more when the ceramics are in nanoscales. Ceramic polymers such as hydroxyapatite (HA), zirconia (ZrO_2), silica (SiO_2), titanium oxide (TiO_2), and alumina (Al_2O_3) can be improved using new methodologies that aim to improve their physicochemical properties, in addition to reducing cytotoxic effects on biological systems. The use of ceramics is more applicable when it comes to the release of drugs and drug delivery systems, logically coupled with nanotechnology. As a disadvantage, the use of ceramics was an adverse effect on the host in the tests that were carried out (Fadeel & Garcia-Bennett, 2010).

Porous materials, especially mesoporous materials, help in the dissolution process of the drug, which is poorly soluble when it involves the surface area or crystallinity, when the pore is larger than the drug molecule, the formation of crystalline material is restricted. The drug is retained in its amorphous form; in this form, the drug presents better dissolution rates when compared to its crystalline form. Porous polymers can also help in the purification of drugs, through adsorption techniques. Mesoporous materials can also be used as low-density carriers, some examples are porous silicon dioxide, polypropylene foam powder, porous calcium silicate, magnesium aluminum silicate, and porous ceramic, such materials are capable of providing large areas and improving the dissolution and bioavailability of drugs that have low solubility such as meloxicam, aspirin, and indomethacin (Syafiq et al., 2020).

14.6 CONCLUSIONS

Porous antimicrobial polymers are a promising technology for dealing with infectious processes or as a way to inhibit possible contamination and infections. Although it is a technology based on polymeric networks, porous polymers are manufactured with a set of synthesis tools that include, depending on the purpose, organic, inorganic materials, natural polysaccharides, secondary compounds, and antibiotics.

The advantages of this technology are associated with a better ability to penetrate through the microbial cell membrane, increased rates of inhibition of microbial reproduction, reduced contamination rate, and increased rates of patient recovery. Alternatively, even an improvement in the shelf life of food, reduction of contamination rates, and economic losses.

The great challenges are associated with increasing the chances of artificial selection of more resistant microorganisms. In this scenario, several strategies are being explored, and the scenario seems to be promising for the development of more sophisticated and efficient technologies, reducing the possibilities of artificial selection to a minimum. It is also important to consider the growing exploitation of natural resources such as polysaccharides for the production of porous materials. These features have an extra advantage, including biocompatibility, low toxicity, and important biological activity, such as antioxidant, immunomodulatory, and antimicrobial activity.

CONFLICT OF INTEREST

The authors declare that there is no conflict of interest.

ACKNOWLEDGMENT

The authors would like to thank the Federal University of Para for the space for the development of the research.

REFERENCES

Alvarado, N., Romero, J., Torres, A., de Dicastillo, C. L., Rojas, A., Galotto, M. J., & Guarda, A. 2018. Supercritical impregnation of thymol in poly (lactic acid) filled with electrospun poly (vinyl alcohol)-cellulose nanocrystals nanofibers: Development an active food packaging material. *Journal of Food Engineering*, 217: 1–10.

Asikainen, S., Paakinaho, K., Kyhkynen, A. K., Hannula, M., Malin, M., Ahola, N., ... Seppälä, J. 2019. Hydrolysis and drug release from poly (ethylene glycol)-modified lactone polymers with open porosity. *European Polymer Journal*, 113: 165–175.

Barbosa, J. R., & de Carvalho Junior, R. N. 2020. Occurrence and possible roles of polysaccharides in fungi and their influence on the development of new technologies. *Carbohydrate Polymers*, 246: 116613.

Barbosa, J. R., & de Carvalho Junior, R. N. 2021. Polysaccharides obtained from natural edible sources and their role in modulating the immune system: Biologically active potential that can be exploited against COVID-19. *Trends in Food Science & Technology*, 108: 223–235.

Barbosa, J. R., da Silva Martins, L. H., & de Carvalho Junior, R. N. 2021. Polymer aerogels: Preparation and potential for biomedical application. In *Aerogels II: Preparation, Properties and Applications*, ed. I. R. Mobin, M. I. Ahamed and T. Altalhi, 98, 1–22. Materials Research Foundations.

Barbosa, J. R., dos Santos Freitas, M. M., da Silva Martins, L. H., & de Carvalho Junior, R. N. (2020). Polysaccharides of mushroom Pleurotus spp.: New extraction techniques, biological activities and development of new technologies. *Carbohydrate Polymers*, 229: 115550.

Barbosa, J. R., Freitas, M. M. S., Oliveira, L. C., Martins, L. H. S., Almada-Vilhena, A. O., Oliveira, R. M., ... Junior, R. N. C. 2020. Obtaining extracts rich in antioxidant polysaccharides from the edible mushroom *Pleurotus ostreatus* using binary system with hot water and supercritical CO_2. *Food Chemistry*, 330: 127173.

Barbosa, J. R., Pinto, R. H. H., da Silva Martins, L. H., & de Carvalho Junior, R. N. 2021. Bioaerogels: Synthesis approaches, biomedical applications and cell uptake. In *Aerogels II: Preparation, Properties and Applications*, ed. I. R. Mobin, M. I. Ahamed and T. Altalhi, 98, 43–56. Materials Research Foundations.

Campardelli, R., Franco, P., Reverchon, E., & De Marco, I. 2019. Polycaprolactone/nimesulide patches obtained by a one-step supercritical foaming+ impregnation process. *The Journal of Supercritical Fluids*, 146: 47–54.

Chen, Z., Hu, J., Ju, J., & Kuang, T. 2019. Fabrication of poly (butylene succinate)/carbon black nanocomposite foams with good electrical conductivity and high strength by a supercritical CO_2 foaming process. *Polymers*, 11(11): 1852.

Cohen, J. D., Erkenbrecher Jr, C. W., Haynie, S., Kelley, M. J., Kobsa, H., Roe, A. N., & Scholla, M. H. 1995. Process for preparing antimicrobial polymeric materials using irradiation. U.S. Patent No. 5,428,078. Washington, DC: U.S. Patent and Trademark Office. https://patents.google.com/patent/US5428078A/en

da Silva Martins, L. H., Neto, J. M., Barbosa, J. R., Junior, G. C. A. C., de Oliveira, J. A. R., Rai, M., & Lopes, A. S. 2021. Carboxymethyl cellulose-coated polypropylene films containing essential oil for food preservation. In *Biopolymer-Based Nano Films*, ed. M. Rai and C. Alves dos Santos, 133–147. Elsevier.

Darpentigny, C., Marcoux, P. R., Menneteau, M., Michel, B., Ricoul, F., Jean, B., ... Nonglaton, G. 2020. Antimicrobial cellulose nanofibril porous materials obtained by supercritical impregnation of thymol. *ACS Applied Bio Materials*, 3(5): 2965–2975.

Díez-Pascual, A. M. 2020. Antimicrobial polymer-based materials for food packaging applications. *Polymers*, 12(4): 731.

Díez-Pascual, A. M., & Díez-Vicente, A. L. 2014. High-performance aminated poly (phenylene sulfide)/ZnO nanocomposites for medical applications. *ACS Applied Materials & Interfaces*, 6(13): 10132–10145.

Emamifar, A., Kadivar, M., Shahedi, M., & Soleimanian-Zad, S. 2011. Effect of nanocomposite packaging containing Ag and ZnO on inactivation of *Lactobacillus plantarum* in orange juice. *Food Control*, 22(3–4): 408–413.

Fadeel, B., & Garcia-Bennett, A. E. 2010. Better safe than sorry: Understanding the toxicological properties of inorganic nanoparticles manufactured for biomedical applications. *Advanced Drug Delivery Reviews*, 62(3): 362–374.

Freitas, L. C., Barbosa, J. R., da Costa, A. L. C., Bezerra, F. W. F., Pinto, R. H. H., & de Carvalho Junior, R. N. 2021. From waste to sustainable industry: How can agro-industrial wastes help in the development of new products? *Resources, Conservation and Recycling*, 169: 105466.

Fu, Y. N., Li, Y., Li, G., Yang, L., Yuan, Q., Tao, L., & Wang, X. 2017. Adaptive chitosan hollow microspheres as efficient drug carrier. *Biomacromolecules*, 18(7): 2195–2204.

Gharsallaoui, A., Joly, C., Oulahal, N., & Degraeve, P. 2016. Nisin as a food preservative: Part 2: Antimicrobial polymer materials containing nisin. *Critical Reviews in Food Science and Nutrition*, 56(8): 1275–1289.

Haynie, S. L., Crum, G. A., & Doele, B. A. 1995. Antimicrobial activities of amphiphilic peptides covalently bonded to a water-insoluble resin. *Antimicrobial Agents and Chemotherapy*, 39(2): 301–307.

Huang, T., Qian, Y., Wei, J., & Zhou, C. 2019. Polymeric antimicrobial food packaging and its applications. *Polymers*, 11(3): 560.

Hurt, A. P., Kotha, A. K., Trivedi, V., & Coleman, N. J. 2015. Bioactivity, biocompatibility and antimicrobial properties of a chitosan-mineral composite for periodontal tissue regeneration. *Polimeros*, 25(3): 311–316.

Imran, M., Revol-Junelles, A. M., René, N., Jamshidian, M., Akhtar, M. J., Arab-Tehrany, E., … Desobry, S. 2012. Microstructure and physico-chemical evaluation of nano-emulsion-based antimicrobial peptides embedded in bioactive packaging films. *Food Hydrocolloids*, 29(2): 407–419.

Jeong, E. J., Park, C. K., & Kim, S. H. 2020. Fabrication of microcellular polylactide/modified silica nanocomposite foams. *Journal of Applied Polymer Science*, 137(17): 48616.

Jridi, M., Hajji, S., Ayed, H. B., Lassoued, I., Mbarek, A., Kammoun, M., … Nasri, M. 2014. Physical, structural, antioxidant and antimicrobial properties of gelatin–chitosan composite edible films. *International Journal of Biological Macromolecules*, 67: 373–379.

Li, J., Kleintschek, T., Rieder, A., Cheng, Y., Baumbach, T., Obst, U., … Levkin, P. A. 2013. Hydrophobic liquid-infused porous polymer surfaces for antibacterial applications. *ACS Applied Materials & Interfaces*, 5(14): 6704–6711.

Li, S., Wang, Y., Xue, Z., Jia, Y., Li, R., He, C., & Chen, H. 2021. The structure-mechanism relationship and mode of actions of antimicrobial peptides: A review. *Trends in Food Science & Technology*, 109: 103–115.

Liu, Z., Hu, J., Gao, F., Cao, H., Zhou, Q., & Wang, X. 2019. Biodegradable and resilient poly (propylene carbonate) based foam from high pressure CO_2 foaming. *Polymer Degradation and Stability*, 165: 12–19.

Milovanovic, S., Markovic, D., Mrakovic, A., Kuska, R., Zizovic, I., Frerich, S., & Ivanovic, J. 2019. Supercritical CO_2-assisted production of PLA and PLGA foams for controlled thymol release. *Materials Science and Engineering: C*, 99: 394–404.

Min, T., Sun, X., Yuan, Z., Zhou, L., Jiao, X., Zha, J., … Wen, Y. 2021. Novel antimicrobial packaging film based on porous poly (lactic acid) nanofiber and polymeric coating for humidity-controlled release of thyme essential oil. *LWT*, 135: 110034.

Moreno-Vega, A. I., Gomez-Quintero, T., Nunez-Anita, R. E., Acosta-Torres, L. S., & Castaño, V. 2012. Polymeric and ceramic nanoparticles in biomedical applications. *Journal of Nanotechnology*, 2012: 1–10.

Mousavi Khaneghah, A. M. I. N., Hashemi, S. M. B., Eş, I., Fracassetti, D., & Limbo, S. 2018. Efficacy of antimicrobial agents for food contact applications: Biological activity, incorporation into packaging, and assessment methods: A review. *Journal of Food Protection*, 81(7): 1142–1156.

Ozdemir, M., & Floros, J. D. 2004. Active food packaging technologies. *Critical Reviews in Food Science and Nutrition*, 44(3): 185–193.

Phaechamud, T., & Mahadlek, J. 2015. Solvent exchange-induced in situ forming gel comprising ethyl cellulose-antimicrobial drugs. *International Journal of Pharmaceutics*, 494(1): 381–392.

Pinto, R. H. H., Barbosa, J. R., Lucas Cantão Freitas, I., de Aguiar, Q. D. S., Pires, F. C. S., e Silva, A. P. D. S., & de Carvalho Junior, R. N. 2020. Aerogels envisioning future applications. In *Aerogels I: Preparation, Properties and Applications*, ed. I. R. Mobin, T. A. Rangreez, M. I. Ahamed and R. Boddula, 84, 214–229. Materials Research Foundations.

Radheshkumar, C., & Münstedt, H. 2006. Antimicrobial polymers from polypropylene/silver composites—Ag+ release measured by anode stripping voltammetry. *Reactive and Functional Polymers*, 66(7): 780–788.

Sanchez-Garcia, M. D., Gimenez, E., & Lagaron, J. M. 2008. Morphology and barrier properties of solvent cast composites of thermoplastic biopolymers and purified cellulose fibers. *Carbohydrate Polymers*, 71(2): 235–244.

Septani, C. M., Wang, C. A., Jeng, U. S., Su, Y. C., Ko, B. T., & Sun, Y. S. 2020. Hierarchically porous carbon materials from self-assembled block copolymer/dopamine mixtures. *Langmuir*, 36(40): 11754–11764.

Silvestre, C., Duraccio, D., & Cimmino, S. 2011. Food packaging based on polymer nanomaterials. *Progress in Polymer Science*, 36(12): 1766–1782.

Song, X., Ren, Z., Wang, X., Jia, L., & Zhang, C. 2020. Antioxidant, anti-inflammatory and renoprotective effects of acidic-hydrolytic polysaccharides by spent mushroom compost (*Lentinula edodes*) on LPS-induced kidney injury. *International Journal of Biological Macromolecules*, 151: 1267–1276.

Sousa, A. F., Ferreira, S., Lopez, A., Borges, I., Pinto, R. J., Silvestre, A. J., & Freire, C. S. 2017. Thermosetting AESO-bacterial cellulose nanocomposite foams with tailored mechanical properties obtained by Pickering emulsion templating. *Polymer*, 118: 127–134.
Standau, T., Zhao, C., Murillo Castellón, S., Bonten, C., & Altstädt, V. 2019. Chemical modification and foam processing of polylactide (PLA). *Polymers*, 11(2): 306.
Suhr, K. I., & Nielsen, P. V. 2003. Antifungal activity of essential oils evaluated by two different application techniques against rye bread spoilage fungi. *Journal of Applied Microbiology*, 94(4): 665–674.
Sun, Q., & Deng, Y. 2005. In situ synthesis of temperature-sensitive hollow microspheres via interfacial polymerization. *Journal of the American Chemical Society*, 127(23): 8274–8275.
Sun, Y. S., Huang, W. H., Lin, C. F., & Cheng, S. L. 2017. Tailoring carbon nanostructure with diverse and tunable morphology by the pyrolysis of self-assembled lamellar nanodomains of a block copolymer. *Langmuir*, 33(8): 2003–2010.
Syafiq, R., Sapuan, S. M., Zuhri, M. Y. M., Ilyas, R. A., Nazrin, A., Sherwani, S. F. K., & Khalina, A. 2020. Antimicrobial activities of starch-based biopolymers and biocomposites incorporated with plant essential oils: A review. *Polymers*, 12(10): 2403.
Tang, Z. X., Fang, X. J., Zhang, Z. L., Zhou, T., Zhang, X. Y., & Shi, L. E. 2012. Nanosize MgO as antibacterial agent: Preparation and characteristics. *Brazilian Journal of Chemical Engineering*, 29(4): 775–781.
Tian, X., Ge, X., Guo, M., Ma, J., Meng, Z., & Lu, P. 2021. An antimicrobial bio-based polymer foam from ZnO-stabilised pickering emulsion templated polymerisation. *Journal of Materials Science*, 56(2): 1643–1657.
Venkataraman, S., Lee, A. L., Tan, J. P., Ng, Y. C., Lin, A. L. Y., Yong, J. Y., ... Yang, Y. Y. 2019. Functional cationic derivatives of starch as antimicrobial agents. *Polymer Chemistry*, 10(3): 412–423.
Wan, Y., Shi, Y., & Zhao, D. 2007. Designed synthesis of mesoporous solids via nonionic-surfactant-templating approach. *Chemical Communications*, (9): 897–926.
Wan, Y., Shi, Y., & Zhao, D. 2008. Ordered mesoporous polymers and carbon molecular sieves. *Chemistry of Materials*, 20: 932–945.
Wang, J., Le-The, H., Wang, Z., Li, H., Jin, M., van den Berg, A., ... Eijkel, J. C. 2019. Microfluidics assisted fabrication of three-tier hierarchical microparticles for constructing bioinspired surfaces. *ACS Nano*, 13(3): 3638–3648.
Wang, Y., Yin, M., Zheng, X., Li, W., & Ren, X. 2021. Chitosan/mesoporous silica hybrid aerogel with bactericidal properties as hemostatic material. *European Polymer Journal*, 142: 110132.
Wu, D., Xu, F., Sun, B., Fu, R., He, H., & Matyjaszewski, K. 2012. Design and preparation of porous polymers. *Chemical Reviews*, 112(7): 3959–4015.
Xie, Y., He, Y., Irwin, P. L., Jin, T., & Shi, X. 2011. Antibacterial activity and mechanism of action of zinc oxide nanoparticles against *Campylobacter jejuni*. *Applied and Environmental Microbiology*, 77(7): 2325–2331.
Xu, J., Chen, G., Yan, R., Wang, D., Zhang, M., Zhang, W., & Sun, P. 2011. One-stage synthesis of cagelike porous polymeric microspheres and application as catalyst scaffold of Pd nanoparticles. *Macromolecules*, 44(10): 3730–3738.
Xue, Y., & Xiao, H. 2015. Antibacterial/antiviral property and mechanism of dual-functional quaternized pyridinium-type copolymer. *Polymers*, 7(11): 2290–2303.
Yang, T., Oliver, S., Chen, Y., Boyer, C., & Chandrawati, R. 2019. Tuning crystallization and morphology of zinc oxide with polyvinylpyrrolidone: Formation mechanisms and antimicrobial activity. *Journal of Colloid and Interface Science*, 546: 43–52.
Yin, D., Xiang, A., Li, Y., Qi, H., Tian, H., & Fan, G. 2019. Effect of plasticizer on the morphology and foaming properties of poly(vinyl alcohol) foams by supercritical CO_2 foaming agents. *Journal of Polymers and the Environment*, 27(12): 2878–2885.
Zhang, K., Wang, Y., Jiang, J., Wang, X., Hou, J., Sun, S., & Li, Q. 2019. Fabrication of highly interconnected porous poly (ε-caprolactone) scaffolds with supercritical CO_2 foaming and polymer leaching. *Journal of Materials Science*, 54(6): 5112–5126.
Zhang, W., Hu, J., Zhou, Y., Chen, Y., Yu, F., Hong, C., ... Wang, X. 2019. Latex and a ZnO-based multifunctional material for cardiac implant-related inflammation. *Biomaterials Science*, 7(10): 4186–4194.
Zhang, Y., Yin, M., Lin, X., Ren, X., Huang, T. S., & Kim, I. S. 2019. Functional nanocomposite aerogels based on nanocrystalline cellulose for selective oil/water separation and antibacterial applications. *Chemical Engineering Journal*, 371: 306–313.
Zheng, Y., Pan, N., Liu, Y., & Ren, X. 2021. Novel porous chitosan/N-halamine structure with efficient antibacterial and hemostatic properties. *Carbohydrate Polymers*, 253: 117205.

Index

Page numbers in *italic* indicate figures, page numbers in **bold** indicate tables

E

F

H

I

L

M

N

O

P